Nanotechnology Interventions in Food Packaging and Shelf Life

Nanotechnology has revolutionized agriculture and food technology, improving the shelf life of foods through interventions of nanomaterials in the packaging. Smart materials, biosensors, nanobiosenors, packaging materials, nanocarbon dots, and nanodevices address aspects of the food industry, such as food safety, food security, and packaging and shelf life.

Nanotechnology Interventions in Food Packaging and Shelf Life shows how nanotechnology has the potential to transform food packaging materials in the future. Nanotechnology applied to food packaging can increase the shelf life of foods, minimize spoilage, ensure food safety, and repair damaged packaging.

Key Features

- Sheds light on benefits of nanotechnology in the food packaging industry
- Contains information on utilization of nanocellulose and nanofibrils in food packaging
- Provides an overview of nanosensor applications for shelf-life extension of different food materials

This book presents a comprehensive review of new innovations in nanotechnology, packaging, preservation, and processing of food and food products. It serves as a useful tool for food engineers and technologists in the food packaging industry.

Nanotechnology Interventions in Food Packaging and Shelf Life

Edited by

Aamir Hussain Dar and Gulzar Ahmad Nayik

CRC Press is an imprint of the
Taylor & Francis Group, an **informa** business

First edition published 2023
by CRC Press
6000 Broken Sound Parkway NW, Suite 300, Boca Raton, FL 33487-2742

and by CRC Press
4 Park Square, Milton Park, Abingdon, Oxon, OX14 4RN

CRC Press is an imprint of Taylor & Francis Group, LLC

ISBN: 978-1-032-06274-7 (hbk)
ISBN: 978-1-032-07555-6 (pbk)
ISBN: 978-1-003-20764-1 (ebk)

DOI: 10.1201/9781003207641

Typeset in Times
by KnowledgeWorks Global Ltd.

Dr. Gulzar Nayik dedicates this book to his better half (Mrs. Raihana Bashir).

Contents

Preface

The evolution of nanotechnology and nanomaterials has made great strides in recent years and different nanotechnology-based techniques have played a profound role in food processing. The main approach of the nanotechnology research has been to resolve economic issues in the food processing industries. The utilization of nano-sized materials came into the existence after the development of nano-based technology in the food technology sector. Different problems involved in the safety, packaging, preservation, and economics of foods can be solved by utilizing smart, nano-based materials, such as packaging materials consisting of biosensors, nano-biosensors, and nanocarbon dots and devices. In the field of food packaging, nanotechnology plays an innovative role in enhancing the packaging and preservation properties by creating barriers that detect the presence of pathogens as well as smart biosensors providing required, relevant information from inside the package. Active packaging involves the use of aluminum as an inner coating of snack food packages. Nano-based packaging provides better barrier and mechanical properties, allowing quality maintenance.

Nanostructures have continuously advanced and have been used in a wide variety of applications, and there is limitless potential in this area. The widely divergent opinions between specialists and the general public concerning the practical applicability and usefulness of nanotechnology have slowed the progress.

People are mainly concerned with the potential risks that nano-engineered materials and devices might pose to their health and the environment. Thus, strict and new approaches are considered in terms of health issues and ecology maintenance while designing the nanomaterials used in the processing and packaging of foods. *Nanotechnology Interventions in Food Processing* presents a comprehensive review of new innovations in nanotechnology, packaging, preservation, and processing of food and food products. Nanotechnology has revolutionized the field of agriculture and food technology by improving the shelf-life properties of foods through the incorporation of nanomaterials in packaging. Utilization of nano- or micro-designed packaging materials ensures the safety of food products by allowing the detection of pathogens and contaminants as well as other food-spoiling factors. There is a growing receptiveness by consumers for nano-based food packaging materials, which helps in the improvement of packaging material. Today's international market emphasizes packaging that facilitates not only real-time supply and preservation of food products, but packaging that simplifies the product's end-usage suitability and communication with consumers at different levels. Nano-based packaging materials not only protect the stored foods but also provide the delivery of bioactive compounds. The global food crisis can be also reduced through the concept of nano agri-food industrialization, which involves the improvement in and balancing of water, soil, ecological, and health problems.

This book serves as a useful tool for researchers, engineers, and technologists of food packaging. It will also be an essential and comprehensive reference to postgraduate students and researchers in universities and research institutions who are interested in the area of nanotechnology and its application in food packaging.

Conclusively, this book assists researchers, food scientists, nanotechnologist, and other members working in various food packaging industries. It could be used by university libraries and institutions all around the world as a handbook or ancillary reading for students pursuing bachelor's and master's degrees in food technology, nanoscience, and nanotechnology. All the contributing authors are acknowledged for their constant dedication and contribution to the editorial guidelines and timeline. We are fortunate to have opportunity to collaborate with experts of international repute from Germany, Mexico, Turkey, Pakistan, China, Peru, Greece, and Hungry. We would like to thank colleagues from the production team of Taylor & Francis Group for their constant help during the editing and production process. Finally, we as editors acknowledge to our readers that this book may contain minor errors or gaps. Suggestions, criticism, and comments are always welcome, so please do not hesitate to contact us with any relevant issue.

Dr. Aamir Hussain Dar,
Dr. Gulzar Ahmad Nayik

Contributors

Atka Afzal
Department of Food Sciences
Government College University
Faisalabad, Pakistan

Muhammad Afzaal
Department of Food Sciences
Government College University
Faisalabad, Pakistan

Amir Ali
Nanoscience and Nanotechnology Program
Cinvestav
Mexico City, Mexico

Kashif Ameer
Institute of Food Science and Nutrition
University of Sargodha
Sargodha, Pakistan

Torres-Gomez Andres P
Sustainability of Natural Resources and Energy Programs
Cinvestav-Saltillo
Coahuila, C.P., Mexico

Mohammad Javed Ansari
Department of Botany, Hindu College Moradabad
(Mahatma Jyotiba Phule Rohilkhand University)
Bareilly, UP, India

Nazmin Ansari
Department of Food Technology and Nutrition
Lovely Professional University
Punjab, India

Franklin Ore Areche
Professional School of Agroindustrial Engineering
National University of Huancavelica
Huancavelica, Peru

Qudsiya Ayaz
Division of Food Science and Technology
Sher-e-Kashmir University of Agricultural Sciences and Technology
Shalimar Srinagar, J&K, India

Ahmet Aygun
Finike Vocational School
Akdeniz University
Antalya, Turkey

Rosy Bansal
Department of Food Processing & Engineering
GSSDGS Khalsa College Patiala
Punjab, India

Iqra Bashir
Division of Food Science and Technology
Sher-e-Kashmir University of Agricultural Sciences and Technology
Shalimar Srinagar, J&K, India

Mansoor Ahmad Bhat
Department of Environmental Engineering, Faculty of Engineering
Eskisehir Technical University
Eskisehir, Turkey

Amandeep Kaur Braich
Department of Food Science and Technology
Punjab Agricultural University
Ludhiana, India

R. G. Campos-Montiel
Instituto de Ciencias Agropecuarias
Universidad Autónoma del Estado de Hidalgo
Tulancingo, Hidalgo, México

A. J. Cenobio-Galindo
Instituto de Ciencias Agropecuarias
Universidad Autónoma del Estado de Hidalgo
Tulancingo, Hidalgo, México

Kajal Dorge
Department of Food Technology and Nutrition
Lovely Professional University
Punjab, India

Fernández-Luqueño Fabián
Sustainability of Natural Resources and Energy Programs
Cinvestav-Saltillo
Coahuila, C.P., Mexico

Denis Dante Corilla Flores
Professional School of Agroindustrial Engineering
National University of Huancavelica
Huancavelica, Peru

Eftade O. Gaga
Department of Environmental Engineering, Faculty of Engineering
Eskisehir Technical University
Eskisehir, Turkey

Kadir Gedik
Department of Environmental Engineering, Faculty of Engineering
Eskisehir Technical University
Eskisehir, Turkey

L. González-Montiel
Instituto de Ciencias Agropecuarias
Universidad Autónoma del Estado de Hidalgo
Tulancingo, Hidalgo, México

Monika Hans
Department of Food Science & Technology
Padma Shri Padma Sachdev Govt. PG College for Women Gandhi Nagar
J&K, India

I. Hernández-Soto
Instituto de Ciencias Agropecuarias
Universidad Autónoma del Estado de Hidalgo
Tulancingo, Hidalgo, México

Jovencio Ticsihua Huaman
National University of Huancavelica Dos de Mayo
Acobamba, Peru

Bushra Ishfaq
Food Technology Section
Post-Harvest Research Centre Ayub Agricultural Research Institute
Faisalabad, Pakistan

Yash D. Jagdale
MIT School of Food Technology
MIT ADT University
Pune, Maharashtra, India

Kiran Jeet
Electron Microscopy and Nanoscience Laboratory
Punjab Agricultural University
Ludhiana, Punjab, India

Guihun Jiang
School of Public Health
Jilin Medical University
Jilin, China

Ioannis K. Karabagias
Department of Food Science & Technology, School of Agricultural Sciences
University of Patras
Agrinio, Greece

Gurkirat Kaur
Electron Microscopy and Nanoscience Lab
Punjab Agricultural University
Punjab, India

Manpreet Kaur
Department of Chemistry
Punjab Agricultural University
Ludhiana, Punjab, India

Ramandeep Kaur
Department of Mathematics, Statistics and Physics
Punjab Agricultural University
Ludhiana, Punjab, India

Muhammad Younas Khan
Department Pharmacognosy
Islamia University of Bahawalpur
Punjab, Pakistan

Sipper Khan
University of Hohenheim
Institute of Agricultural Engineering
Tropics and Subtropics Group
Stuttgart, Germany

Muhammad Nadeem
Institute of Food Science and Nutrition
University of Sargodha
Sargodha, Pakistan

Vikas Nanda
Department of Food Engineering and Technology
Sant Longowal Institute of Engineering & Technology
Longowal, Sangrur, Punjab, India

Gulzar Ahmad Nayik
Department of Food Science & Technology
Govt. Degree College Shopian
J&K, India

Bushra Niaz
Department of Food Sciences
Government College University
Faisalabad, Pakistan

Shahid Mahmood
Institute of Food Science and Nutrition
University of Sargodha
Sargodha, Pakistan

Ghulam Mueen-ud-Din
Institute of Food Science and Nutrition
University of Sargodha
Sargodha, Pakistan

Mian Anjum Murtaza
Institute of Food Science and Nutrition
University of Sargodha
Sargodha, Pakistan

Pérez-Moreno Andrea
Sustainability of Natural Resources and Energy Programs
Cinvestav-Saltillo
Coahuila, C.P., Mexico

Pérez-Hernández Hermes
El Colegio de la Frontera Sur (CONACYT)
Agroecología, Unidad Campeche
Campeche, Mexico

E. Pérez-Soto
Instituto de Ciencias Agropecuarias
Universidad Autónoma del Estado de Hidalgo
Tulancingo, Hidalgo, México

Suleyman Polat
Food Engineering Department
Cukurova University
Adana, Turkey

Miguel Angel Quispe Solano
Faculty of Food Industry Engineering
National University of the Center of Peru
Huancayo, Peru

Shafiya Rafiq
Amity Institute of Biotechnology
Amity University Rajasthan
Jaipur, India

Alfonso Ruiz Rodriguez
Professional School of Agroindustrial Engineering
National University of Huancavelica
Huancavelica, Peru

Farhan Saeed
Department of Food Sciences
Government College University
Faisalabad, Pakistan

Amna Sahar
Department of Food Engineering
University of Agriculture
Faisalabad, Pakistan

Aysha Sameen
National Institute of Food Science and Technology
University of Agriculture
Faisalabad, Pakistan

Sarabia-Castillo Cesar R
Sustainability of Natural Resources and Energy Programs
Cinvestav-Saltillo
Coahuila, C.P., Mexico

Yasir Abbas Shah
Department of Food Sciences
Government College University
Faisalabad, Pakistan

Ayaz Mukarram Shaikh
Institute of Food Science
University of Debrecen
Debrecen, Böszörményi, Hungary

Haamiyah Sidiq
Division of Food Science and Technology
Sher-e-Kashmir University of Agricultural Sciences and Technology
Shalimar Srinagar, J&K, India

Dilpreet Singh
Department of Food Engineering and Technology
Sant Longowal Institute of Engineering & Technology
Longowal, Sangrur, Punjab, India

Priyanka Suthar
Department of Food Science and Technology
Dr. Y. S. Parmar University of Horticulture and Forestry
Nauni, Himachal Pradesh, India

Tayyaba Tariq
National Institute of Food Science and Technology
University of Agriculture
Faisalabad, Pakistan

Sajad Mohd Wani
Division of Food Science and Technology
Sher-e- Kashmir University of Agricultural Sciences and Technology
Shalimar Srinagar, J&K, India

Syeda Saniya Zahra
Department Pharmacognosy
Shifa Tameer-e-Millat University
Islamabad, Pakistan

Aiman Zehra
Division of Food Science and Technology
Sher-e-Kashmir University of Agricultural Sciences and Technology
Shalimar Srinagar, J&K, India

Chang-Cheng Zhao
School of Life Science
Zhengzhou University
Henan, China

About the Editors

Dr. Aamir Hussain Dar is currently working as Assistant Professor in the Department of Food Technology, Islamic University of Science & Technology, Awantipora, J&K, India. He completed his Master's Degree in Agricultural Processing & Food Engineering from Sam Higginbottom University of Agriculture, Technology and Sciences, formerly Allahabad Agricultural Institute, Allahabad, India and a PhD from Sant Longowal Institute of Engineering & Technology, Longowal, Sangrur, Punjab, India. Dr. Aamir has published many peer-reviewed research and review papers in reputed journals. He has also published many book chapters and delivered a number of presentations in national and international conferences. Dr. Aamir also serves as editorial board member and reviewer of several journals related to food science and technology. Dr. Aamir has supervised many students at bachelor's and master's levels.

Dr. Gulzar Ahmad Nayik completed his Master's Degree in Food Technology from Islamic University of Science & Technology, Awantipora, Jammu and Kashmir, India and a PhD from Sant Longowal Institute of Engineering & Technology, Longowal, Sangrur, Punjab, India. He has published sixty peer-reviewed research and review papers, authored or co-authored thirty book chapters, and edited eight books with publishers including Springer, Elsevier, and Taylor & Francis. Dr. Nayik has also published a textbook on food chemistry and nutrition, and has delivered numerous presentations at various national and international conferences, seminars, workshops, and webinars. Dr. Nayik was short-listed twice for the prestigious Inspire-Faculty Award in 2017 and 2018 from the Indian National Science Academy, New Delhi, India. He was nominated for India's prestigious National Award (Indian National Science Academy Medal for Young Scientists, 2019–20). Dr. Nayik also fills the roles of editor, associate editor, assistant editor, and reviewer for many food science and technology journals. He has received many awards, appreciations, and recognitions and holds membership in various international societies and organizations. Dr. Nayik is currently editing several book projects with Elsevier, Taylor & Francis, Springer Nature, and the Royal Society of Chemistry.

Part I

Role of Nanotechnology in Food Packaging

1 Application of Nanotechnology in Food Packaging and Food Quality

Dilpreet Singh and Vikas Nanda
Department of Food Engineering and Technology,
Sant Longowal Institute of Engineering &
Technology, Longowal, Sangrur,
Punjab, India

CONTENTS

1.1 INTRODUCTION

Nanotechnology is a novel and fast-evolving technology that enables us to alter and manufacture structures of materials at the molecular scale, sometimes atom by atom, until structure on the nanometre scale is attained. Nanotechnology is a sustainable

DOI: 10.1201/9781003207641-2

method for the development and production of different types of materials or devices used for testing purposes, such as quality analysis of water, food, and etcetera (Asadi and Mousavi, 2006). Nanotechnology is described in terms of modification and rearrangement of matter on the nanoscale scale (i.e., fewer than 100 nanometres) to develop objects having fundamentally new characteristics and functionalities (Sanchez and Sobolev, 2010). Drexler et al. (1991) give two classifications of nanotechnology. One is the 'top-down' approach in which the big molecular size particles are broken down to nanometre size, retaining their original characteristics without any control on an atomic level (for an instance, electronics item with miniaturising components). Another is the 'bottom-up' approach in which various processes and assemblies are used for engineering new materials with atoms and small molecular compounds in the range of a nanometre. The nanoparticles have a great impact on the material's physical properties and final substance; properties such as strength, decrease in weight of the material, and increase in thermochemical properties of the material (Nasrollahzadeh et al., 2019).

Consumers are showing significant interest in health benefits and quality of food. This has been encouraging scientists to work on improving both the quality and nutritive values of food. The use of materials based on nanoparticles has surged in the food industry, as they are non-toxic and contain several essential nutrients (Roselli et al., 2003). Gupta et al. (2015) conducted a study related to acceptance of nanotechnology by consumers and concluded that people showed a positive attitude towards new approaches. In his study, various applications of nanotechnology such as in water filtration, as surface cleaner and even in food nutrition all are associated with general public well-being, and they have many benefits related to society, health, and the environment. Moreover, the field of Food Science and Technology has been experiencing a paradigm shift with the application of nanotechnology, as various domains of food science, such as food processing, food packaging, new product development, food microbiology, as well as safety and quality control of the food, are incorporating nanoparticles (Singh et al., 2017). The uses of nanotechnology in the food business may be divided into two categories: food nanostructured ingredients and food nanosensing. Food nanostructured substances include a wide range of applications from food processing to food packaging. These nanoparticles can be used as food additives, carriers for smart nutrient delivery, anti-caking agents, antimicrobial agents, fillers to improve mechanical strength and durability of packaging materials, and so on. Food nanosensing can be used to improve food quality and safety evaluation (Ezhilarasi et al., 2013).

Food packaging is directly related to food quality. A food package has key functions in the safety of the packaged material during storage, transportation, sales, and end use. Development in packaging has taken place with the advancement in food processing. There are packages with specific functions used to provide long shelf stability based on the physicochemical properties of various foods, such as meat, fruits and vegetables, cereal products, and dairy products. Active packaging, modified atmosphere packaging (MAP), and edible films/coatings are a few functional packaging systems (Han, 2005). The nano-based packaging is both 'smart' and 'active.' These types of packaging provide various benefits over traditional packaging, such having high mechanical strength or being a barrier to gasses and liquids, and more advanced packaging can detect pathogens in a packaged atmosphere and even have antimicrobial action. The application of nanoparticles to packaging materials also improves thermal resistance to extreme high and low temperatures (Singh et al., 2017).

Packaging based on nanotechnology falls under three categories. Under the first, 'improved packaging,' physical properties, such as elasticity, gas barrier quality, thermal stability, and moisture retention capacity, are improved by incorporation of nanoparticles (NPs) in a polymer matrix of film or some other kind of packaging. 'Intelligent packaging' has a greater function in communicating to consumers. It indicates the quality of packaged food by giving visual feedback or a signal to the consumer relating to external and internal changes or impurities. 'Active packaging' can help in maintaining the quality of food. Incorporating active components of nanotechnology in food packaging systems, the NPs interact with internal and external factors of the food system, leading to antimicrobial action, thereby increasing the shelf life of the food and improving the quality and safety of food. In the case of intelligent packaging systems, sensors based on nanotechnology are incorporated into the food packaging and they sense the change in the composition of food, such as a variation in gas composition in headspace, or any other undesirable biological and chemical changes in the packed food (Sharma et al., 2017).

The application of emerging nanotechnology in food packaging has ensured the higher quality of the packaged food in relation to human health. Packaging based on nanotechnology has a greater impact on the overall quality of the food that we consume. Food commodities, such as fruits and vegetables, cereals and pulses, meat and poultry, and dairy products, have different physical and chemical properties, and therefore, different kinds of packages are developed to provide effective packaging systems.

To store a food material for longer time intervals, packaging must have various functions. It has to protect the food from extrinsic matter: invading of insects and rodents, light, oxygen, microbial attach, and moisture gain or loss. The packaging must be inert, non-toxic, cheap, lightweight, easy to dispose of or recycle, able to withstand harsh conditions during processing, and able to face the extremes of the supply chain. It must detect changes in temperature during storage and transportation, any microbial growth, and the freshness of the food. All these functions are possible with the application of NPs-based packaging systems.

In this chapter, various type of nanomaterials used for development of different kind of packaging system will be examined.

1.2 DIFFERENT TYPES OF NANOMATERIALS USED FOR DEVELOPING NANO-BASED FOOD PACKAGING

1.2.1 Polymer Nanocomposites

Polymer nanocomposites (PNCs) are emerging materials in the development of new packaging materials to solve problems faced with traditional types of packaging. Clay and silicate nanoplatelets (see infrared), silica (SiO_2) nanoparticles, carbon nanotubes, graphene, starch nanocrystals, cellulose-based nanofibers or nano-whiskers, chitin or chitosan nanoparticles, and other inorganics can all be used as filler materials. Though the most obvious application of PNCs in the food business is to improve polymer barrier qualities, PNCs are also tougher, more flame resistant, and have better thermal properties (e.g., melting points, degradation, and glass transition temperatures) than nanoscale filler containing control polymers. Changes in surface wettability and hydrophobicity have also been reported. Some of these physical property improvements might be rather spectacular. A layer-by-layer assembly technique, for example, was used to create a PNC material composed of clay nanoplatelets dispersed within cross-linked polyvinyl acetate (PVA) with a modulus (stiffness) of 106 ± 11 GPa—nearly two orders of magnitude greater than 'virgin' PVA and comparable to the stiffness of some grades of Kevlar. When employed as a coating, a similar manufacturing approach was used to build clay/poly(ethyleneimine) PNCs that conserved the weave structure of cotton textiles throughout extended burning periods (Duncan, 2011). Because PNCs use fewer polymer to provide packaging materials with comparable or even superior mechanical properties, they should provide the food packaging sector with greater down-gauging options as well as with cost savings and waste reductions. When a nanofiller is mixed together with the biocompatible polymer polylactic acid (PLA), the PLA bio-nanocomposite has a faster rate of biodegradation than PLA without such additives (Sinha Ray et al., 2002). Hence, the food quality would be retained by packing the food in a moisture barrier, gas barrier, and more tear resistant material. Additionally, the environmental impact of plastic-based package materials will be reduced due to a higher degree of biodegradability.

1.2.2 Nano-Clay

Nano-clay is mostly recovered from volcanic ash or rocks. It has various applications in food packaging or in the development of a variety of smart and active packaging systems. There is a hydrated magnesium aluminium silicate layered clay named montmorillonite which belongs to the family of 2:1 layered silicates. Its structure has a crystal shape with share edge which share octahedral sheets of magnesium and aluminium-hydroxide packed in two tetrahedral layers of silica. The clay platelets have a net negative surface charge due to an isomorphous substitute in the layer of nanoclay. The platelet system is used for balancing the imbalance charge by exchangeable cations (usually alkali or alkaline earth cations). The weak electrostatic force is used to link parallel layers with each other. The introduction of montmorillonite within a matrix of polymer leads to an inorganic crystalline structure that is non-permeable in nature. This causes a decline in gas permeability of a film layer because of an increase in obstacles and enlonging the pathway of gases (Enescu et al., 2015). These functional properties of the packaging system have application in packaging of various types of food, such as beverages; dairy products, like cheese; meat; cereals; fruit juices; etc. Bentonite clay is used as a plastic material for food packaging, and within limits of 5% w/w it is used as colour carrier additive (Enescu et al., 2019).

1.2.3 Nanoscaled Cellulose

Nanoscaled cellulose is also known as either micro-fibrils or cellulose nano-whiskers. It has alternating amorphous and crystalline strings. In the crystalline region, stability of cellulose, a polymer, is attained by an internal network of hydrogen bonds. The complex hydrogen bond provides extreme stiffness in the cellulose chain fibre. Due to the high aspect ratio, little quantity of nanocellulose can enhance the significant rigidity of a polymer. Hydrolysis of delignified coconut husk is performed to produce cellulose nano-whiskers (Rosa et al., 2010); however, nano-whiskers produced from bamboo fibre showed significantly improved results in Young's modulus and fracture stress (Kim and Lim, 2009). Additionally, nanocellulose is a good alternative to petroleum-based materials because of its greater biodegradability (De Azeredo, 2009). Packaging material formulated with nanoscaled cellulose can protect the food material from external impacts during transportation and storage, save the food from physical damage, and help in maintaining the desirable quality.

1.2.4 Carbon Nanotubes

Carbon nanotubes (CNTs) are used in antimicrobial and intelligent packaging; polymerisation with (PVOH), polypropylene (PP), nylon, polylactic acid (PLA), etc. Single atom thick nanotubes and several concentric nanotubes are two basic types of CNTs. They provide both high-tensile strength and elastic modulus. Abdelhalim et al. (2013) described it as a process in which CNT-based gas sensors are sprayed on packets to make a transparent, thin film that is surrounded by wireless chips. When there is any element or a microbe population deposition exceeding a defined limit in a food product, or the food is about to spoil, then the package gives a signal through an identifying mark that can be read by the consumer or a salesperson. This makes this type of intelligent packaging more convenient and easier for maintaining the quality of food.

1.2.5 Nanostarch

Nanostarch is a trustworthy filler for the production of packaging materials with its increased flexibility. It can improve mechanical strength as well as provide barrier properties because of its high vulnerability towards hydration. Nanostarch is a biodegradable, renewable, non-poisonous, and natural polysaccharide. It has various applications in food and the food industry, such as for making economical and eco-friendly packaging materials. Nanostarch is obtained by separating crystalline and an amorphous phase through acid or enzyme hydrolysis, precipitation, and mechanical micro-fluidiser treatment (Kim and Lim, 2009; Liu et al., 2009; Pérez-Pacheco et al., 2016; Putaux et al., 2003). In acid hydrolysis, mild acid treatment is given to the slurry below gelatinisation temperature, as mild acid can act on the amorphous region more easily as compared to crystalline. This treatment can provide us with a nanostarch filler. Moreover, by modification of starch structures, many physico-chemical properties of the material can be changed. Cationic starch molecules are used for the development of starch-based nano-biopolymer binder to replace synthetic binders in packaging materials (Chaudhary et al., 2020). Packaging made from this nanomaterial is low cost and can be used for packaging cheese, cookies, and different hygroscopic products because of high barrier properties that retain the quality of the product.

1.2.6 Nano-Silver

Nano-silver particles are used in developing active packaging materials; antimicrobial activity of the material helps in reducing microbial growth in the packed food. Growth of *Escherichia Coli*, *Enterococcus faecalis*, *Staphylococcus aureus* and *epidermidis*, *Vibrio cholera*, *Pseudomonas aeruginosa* and *putida and fluorescens and oleovorans*, *Shigella flexneri*, *Bacillus anthracis*, *Proteus mirabilis*, *Salmonella enterica*, *Micrococcus luteus*, *Listeria monocytogenes*, and *Klebsiella pneumonia* can be controlled by the application of a nano-silver coating in the packaging, through the release of silver particles to the food system from the container (Enescu et al., 2019). The particular action of these NPs helps in maintaining the quality and microbial stability of food.

1.2.7 Titanium Nitride

Titanium nitride (TiN) in the range of 20 mg/kg is permitted in polyethylene terephthalate (PET) bottles to increase the thermal stability. This is declared to be not toxic by the European Food Safety Authority (EFSA, 2008). Titanium dioxide (TiO_2) is among the top five nanomaterials used in the food industry. TiO_2 is used for pigmentation of glass bottles and plastic containers because it provides high brightness and smoothness to the surface of the package. It offers protection from UV rays, so it is fit to use for packaging of light-sensitive food (Nešić et al., 2018). In addition to protection from light, TiO_2 also reduces a significant number of *E. coli* from the food surface (Chawengkijwanich and Hayata, 2008).

1.2.8 Magnesium Oxide Nanoparticles

The nanoparticle MgO (2% wt) is used in the production of PLA biopolymer films, and it showed barriers to O_2 and greater tensile strength. Moreover, within 24 hours of treatment with 2% wt MgO-based PLA biopolymer film, there was a significant reduction in *E. coli* population and a 25% reduction in moisture retention (Swaroop and Shukla, 2018). To increase the shelf life of a product, nano-sized selenium particles are incorporated into multilayer PET bottles. The selenium particles in the range of 50–60 nm are incorporated in a multilayer system through a water-based carrier, which showed potential antioxidant activity by the package through the scavenging various free radicals in the system. This multilayer system acts as active packaging, as it can help in the prevention of oxidation of food even without being in contact with it (Vera et al., 2016).

1.2.9 Chitosan Nanoparticles

The heteropolysaccharide nature of chitosan gives it excellent biodegradability and biocompatibility with metal complexation. In addition, the antimicrobial activity of chitosan is due to its polycationic nature (Arora and Padua, 2016). Ionic gelation is used for the development of chitosan nanoparticles, where amino groups carrying positive charge in chitosan get polyanions engagement through electrostatic interactions and form cross-linkers (Ahmed and Aljaeid, 2016). The mechanical and barrier properties of the packaging material can be improved by using chitosan NP additives in HPMC (hydroxypropyl methylcellulose) (De Moura et al., 2009). Edible films consisting of chitosan NPs have a significant enhancement in physicochemical and microbiological quality and can be used for fresh-cut fruits and vegetables (Ramos et al., 2012). Moreover, another antimicrobial film consisting of chitosan NPs and polyvinyl alcohol has antimicrobial activity against *E. coli*, *S. aureus*, and *B. subtilis*, and it is used for increasing the shelf life of tomatoes (Tripathi et al., 2009). Growth of *E. coli*, *S. aureus*, *P. aeruginosa*, *Aspergillus niger*, and *Candida albicans* in food can be prevented by the application of films based on cinnamaldehyde/chitosan, chitosan/gold, and chitosan/silver (Youssef et al., 2014). There are many active package-based antibacterial bag developed from chitosan/polyethylene, and they cause significant inhibition of aerobic mesophilic bacteria, yeast, and mould. Additionally, they can help in preserving the colour, texture, and pH of chicken drumsticks (Soysal et al., 2015).

1.2.10 O_2 Sensors

For packaging of oxygen-sensitive food products, a modified atmosphere packaging system is developed by eliminating the O_2 from the container, limiting the concentration of O_2 in headspace or by replacing the system with flushing of N_2. Therefore, detection of O_2 in the package container is made possible with the application of photosensitive or UV-based colourimetric O_2 sensors that use TiO_2 NPs encapsulated in a medium reduced by methylene blue, which induce photosensitisation of triethanolamine (Lee et al., 2005). Mills and Hazafy (2009) developed TiO_2 nanoparticle-based oxygen indicator MBRd that is methylene blue in reduced state, and MBox is methylene blue in an oxidising state that detects a reduction in critical temperature of stored packaged food. Mihindukulasuriya and Lim (2014) developed a membrane by an electrospinning method, which is an UV-based active O_2 indicator. TiO_2, methylene blue, and glycerol are used for encapsulation in electrospun polyfibre, which increased the sensitivity of the membrane to O_2.

1.2.11 CO_2

In the case of fruits and vegetables, it is very important to monitor the physicochemical changes, such as the production of CO_2 and ethylene during respiration. To maintain the natural quality of the packaged fruits and vegetables, CO_2 indicators are used for monitoring of freshness of the food. To detect CO_2, gas permeable sachets are filled with an indicator based on chitosan NPs having an additive of single strength indicator (2-amino-2-methyl-1-propanol (AMP)) (Lim et al., 2013).

1.2.12 Time, Temperature, and Humidity Indicators

Temperature and humidity have a great role in the quality of food. Increased or decreased storage temperature of packages throughout the supply chain can harm the overall quality of packaged commodities. To track the temperature history of packed commodities, time-temperature indicators (TTIs) of colourimetric indicators based on triangular Ag nanoplates for detection of temperature and time history are used. The thermodynamic chemistry of Ag nanoplates makes it possible to detect a temperature change. Ag nanoplate has sharp corners, strong in-plane dipole resonance mode in the visible spectrum of radiation. With increased storage time, the corners of Ag nanoplates start rounding, leading to a slow change in blue colour from cyan, depicting the peal position of resonance, a reaction that is temperature dependent (Zeng et al., 2010). Temperature changes in packaged food can also be detected by colour-changing strips based on Au NPs. When a package temperature crosses into a freezing temperature, then the colour of the test strip becomes red, which is a qualitative indication of temperature. Additionally, this reaction is reversible for frozen packaged food (Ghaani et al., 2016).

1.2.13 Freshness Indicator

To have correct information about a food product's quality during a particular storage period, freshness indicators are used. For detection of meat and meat product spoilage, plastic film coated with transition metals (Cu and Ag) and with layer thickness, of 1–10 nm is used. These compounds are sensitive to volatile substances, like the sulphide and amine produced during the spoilage of meat by microbes, or to other chemical spoilage. The coat started turning when volatile sulphides are released from spoiling meat (Fuertes et al., 2016; Smolander et al., 2004). Bioelectronic nose based on peptide receptors is used for the detection of trimethylamine to analyse the freshness of seafood (Lim et al., 2013). A carbon-based nanotube is used to detect CO_2 and ethylene in fruits and vegetables and is used to indicate freshness (Esser et al., 2012).

1.2.14 Other Nano-Based Food Packaging with Antimicrobial Activity

Any antimicrobial activity of a compound must have specific targets, similar in cell physiology to a microbial cell. They can directly interfere with cell wall development, DNA replication, and transcription. The special properties that are exerted by nanoparticles are physical, chemical, biological, dielectric, electrical, thermal, mechanical, and optical. Moreover, antimicrobial activity is enhanced by oxides of nanoparticles (Suski et al., 2012). The enzymatic activity and DNA synthesis of microbial cells and other cell organelles are disrupted by modified nanomaterials, such as silver nanoparticles. Even though the microbe is not able to survive because of damage to the cell wall due to the transmission of positively charged zeta potential of nanoparticle through the cell membrane, this charge also interfaces with replication in the cell (Hahn et al., 2012). Another mode of action of nanomaterials is to produce an antimicrobial effect by producing H_2O_2 and to produce efux-infux in a cell. The negative charge of the bacterial cell and positive ion on the surface of the nanoparticle showed a significant impact on the respiratory enzymes, an action that causes ionic efux-infux within a cell and leads to its death (Chiang et al., 2012). Additionally, when microbial cells encounter nanoparticles, the production of reactive oxygen species and oxidation of respiratory enzymes take place, which affects the physiological function of cells and also damages the DNA. With an interruption in respiratory function in mitochondria, molecular O_2 levels start decreasing through a transmission cycle of electron-proton, which leads to the formation of hydroxyl radicals, superoxide anionic radicals, and hydrogen peroxide (H_2O_2), which is deadly for microbes (Stadtman and Berlett, 1997).

Cellular constituents face oxidative stress. There is generation of protein ridicule in a cell, as well as lipid peroxidation, breakdown of cell DNA-strand, modification of nucleic acid, damage to communication channels and cellular constituents, and activation of redox-sensitive transcription factors leading to gene expression modulation. All this can cause cytotoxicity and genotoxicity effects because of generation of ROS (reactive oxygen species). It has been concluded in many types of research that D-alanine metabolism in S. mutants can be altered by the application of NPs of TiO_2 in package films (Kumar et al., 2014). Similar effects of CuO-NPs were seen in inhibition of *Paracoccus denitrifcans* by affecting enzyme activity of microbe. And, electron transfer for denitrification is inhibited by regulating proteins that are related to nitrogen metabolism and electrons. NADH dehydrogenase and cytochrome, which are major electron transport proteins, are blocked by CuO-NPs, which leads to lysis of the microbe in food (Barabaszová et al., 2020).

1.3 SAFETY CONCERNS AND ETHICAL ISSUES REGARDING NPS-BASED FOOD PACKAGING

The implementation of NP-based packaging systems has led to concerns regarding the safety of the food material packed in it. The benefits of nano-based food have been well established, but the health implications are still being studied for feasibility and verification, and their use in packaging materials has yet to be recognised due to safety issues that may relate to the transfer of nanoparticles from the container to the food product. Migration of NPs from the metal surface of the container into food, or the direct contact of the container with the food, may cause toxicological impact in mammalian and other animal systems. These effects are irritation in the skin due to inflammation or digestive malfunctioning. This effect is also called toxic potential (Tervonen et al., 2009).

Although people are willing to accept new technology in food areas, cost-effectiveness is a major concern for both consumers and the producer. Such is the case with the application of nanoscience. The cost of product per package will increase and profit would decline, which might have a huge impact on the sales, as well as on acceptance and implementation of new technology. For a product to be cost-effective and profitable to both producer and consumer, it is estimated that the package cost should be only 10% of the product (Dainelli et al., 2008).

NPs of ZnO showed genotoxicity in the epidermal tissue of the human body. Consuming food in which NPs of metal oxide migrated from the container meant they were observed in the oral cavity (Aschberger et al., 2011; Sharma et al., 2009). According to the study, toxicity of smaller NPs is much greater than larger ones because of more surface area and greater interaction with biological molecules (Chithrani et al., 2006; Varela et al., 2012).

Under the European Regulation Act, it is mandatory to analyse the overall migration of NPs in food packaging based on nano science, so that consumers can be protected from any harmful effects of NPs (Kruijf et al., 2002). In addition, Food and Drug Administration (FDA) regulations and the European Act have mandated the maximum permissible migration of NPs in packaged food so that harm human health will be avoided in any case (Chowdhury, 2008; De Jong et al., 2005).

1.4 CONCLUSION

The food industry is experiencing a paradigm shift. Consumer demands are changing along with the lifestyle changes, and food quality is becoming a greater concern for both producers and consumers. Trends and needs for packed food have been increasing, which creates a need for innovation in the food packaging system. Nano-based food packaging systems have gained popularity among the population, and the application of nanoscience has improved the overall properties of food packaging, such as physicochemical properties, consumer appeal, eco-friendly nature, quality control, etc. When anything has a positive side there are will be a few cons that will raise many questions. Nanoparticles showed toxicity and many adverse effects on human health. Therefore, food regulation bodies are focusing on outlining strict regulations before any kind of new technology in food system is implemented. Even in nano-based packaging, the FDA has several regulations related to nanoparticle migration in food from the package. Perhaps, people will be more likely to adopt the improved technology of nano-based food packaging as it assures them of superior food quality and increased biodegradability.

REFERENCES

Abdelhalim, A., Abdellah, A., Scarpa, G., and Lugli, P. 2013. Fabrication of carbon nanotube thin films on flexible substrates by spray deposition and transfer printing. *Carbon* 61: 72–79.

Ahmed, T. A., and Aljaeid, B. M. 2016. Preparation, characterization, and potential application of chitosan, chitosan derivatives, and chitosan metal nanoparticles in pharmaceutical drug delivery. *Drug Design, Development and Therapy* 10: 483–507.

Asadi, G., and Mousavi, S. M. 2006. Application of nanotechnology in food packaging. In 13th World Congress of Food Science & Technology 2006: 739–739.

Arora, A., and Padua, G. W. 2010. Nanocomposites in food packaging. *Journal of Food Science* 75(1): R43–R49.

Aschberger, K., Micheletti, C., Sokull-Klüttgen, B., and Christensen, F. M. 2011. Analysis of currently available data for characterising the risk of engineered nanomaterials to the environment and human health—Lessons learned from four case studies. *Environment International* 37(6): 1143–1156.

Barabaszová, K., Holešová, S., Bílý, M., and Hundáková, M. 2020. CuO and CuO/vermiculite based nanoparticles in antibacterial PVAc nanocomposites. *Journal of Inorganic and Organometallic Polymers and Materials* 30(10): 4218–4227.

Chaudhary, P., Fatima, F., and Kumar, A. 2020. Relevance of nanomaterials in food packaging and its advanced future prospects. *Journal of Inorganic and Organometallic Polymers and Materials* 30(12): 5180–5192.

Chawengkijwanich, C., and Hayata, Y. 2008. Development of TiO_2 powder-coated food packaging film and its ability to inactivate Escherichia coli in vitro and in actual tests. *International Journal of Food Microbiology* 123(3): 288–292.

Chiang, H. M., Xia, Q., Zou, X., Wang, C., Wang, S., Miller, B. J., Howard, P. C., Yin, J. J., Beland, F. A., Yu, H., and Fu, P. P. 2012. Nanoscale ZnO induces cytotoxicity and DNA damage in human cell lines and rat primary neuronal cells. *Journal of Nanoscience and Nanotechnology* 12(3): 2126–2135.

Chithrani, B. D., Ghazani, A. A., and Chan, W. C. 2006. Determining the size and shape dependence of gold nanoparticle uptake into mammalian cells. *Nano Letters* 6: 662–668.

Chowdhury, S. R. 2008. Some important aspects in designing high molecular weight poly (l-lactic acid)–clay nanocomposites with desired properties. *Polymer International* 57(12): 1326–1332.

Dainelli, D., Gontard, N., Spyropoulos, D., Zondervan-van den Beuken, E., and Tobback, P. 2008. Active and intelligent food packaging: Legal aspects and safety concerns. *Trends in Food Science and Technology* 19: 103–112.

De Azeredo, H. M. 2009. Nanocomposites for food packaging applications. *Food Research International* 42(9): 1240–1253.

De Moura, M. R., Aouada, F. A., Avena-Bustillos, R. J., McHugh, T. H., Krochta, J. M., and Mattoso, L. H. 2009. Improved barrier and mechanical properties of novel hydroxypropyl methylcellulose edible films with chitosan/tripolyphosphate nanoparticles. *Journal of Food Engineering* 92(4): 448–453.

Drexler, K. E., Peterson, C., and Pergamit, G. 1991. *Unbounding the Future: the Nanotechnology Revolution*. New York: William Morrow.

Duncan, T. V. 2011. Applications of nanotechnology in food packaging and food safety: Barrier materials, antimicrobials and sensors. *Journal of Colloid and Interface Science* 363(1): 1–24.

EFSA. 2008. 21st list of substances for food contact materials. Scientific opinion of the panel on food contact materials, enzymes, flavourings and processing aids. *The EFSA Journal* 888–890: 1–14. http://www.efsa.europa.eu/en/efsajournal/doc/888.pdf. EFSA, 2009.

Enescu, D., Cerqueira, M. A., Fucinos, P., and Pastrana, L. M. 2019. Recent advances and challenges on applications of nanotechnology in food packaging. A literature review. *Food and Chemical Toxicology* 134: 110814.

Enescu, D., Frache, A., and Geobaldo, F. 2015. Formation and oxygen diffusion barrier properties of fish gelatin/natural sodium montmorillonite clay self-assembled multilayers onto the biopolyester surface. *RSC Advances* 5(75): 61465–61480.

Esser, B., Schnorr, J. M., and Swager, T. M. 2012. Selective detection of ethylene gas using carbon nanotube-based devices: Utility in determination of fruit ripeness. *Angewandte Chemie International Edition* 51(23): 5752–5756.

Ezhilarasi, P. N., Karthik, P., Chhanwal, N., and Anandharamakrishnan, C. 2013. Nanoencapsulation techniques for food bioactive components: A review. *Food Bioprocess Technology* 6: 628–647.

Fuertes, G., Soto, I., Carrasco, R., Vargas, M., Sabattin, J., and Lagos, C. 2016. Intelligent packaging systems: Sensors and nanosensors to monitor food quality and safety. *Journal of Sensors* 2016. DOI: https://doi.org/10.1155/2016/4046061

Ghaani, M., Cozzolino, C. A., Castelli, G., and Farris, S. 2016. An overview of the intelligent packaging technologies in the food sector. *Trends in Food Science & Technology* 51: 1–11.

Gupta, N., Fischer, A. R. H., and Frewer, L. J. 2015. Ethics, risk and benefits associated with different applications of nanotechnology: A comparison of expert and consumer perceptions of drivers of societal acceptance. *NanoEthics* 9(2): 93–108.

Hahn, A., Fuhlrott, J., Loos, A., and Barcikowski, S. 2012. Cytotoxicity and ion release of alloy nanoparticles. *Journal of Nanoparticle Research* 14(1): 1–10.

Han, J. H. 2005. New technologies in food packaging: Overview. In *Innovations in Food Packaging* (pp. 3–11). Academic Press, California, USA.

Huang, X., and Netravali, A. 2009. Biodegradable green composites made using bamboo micro/nano-fibrils and chemically modified soy protein resin. *Composites Science and Technology* 69(7–8): 1009–1015.

de Jong, W. H., Roszek, B., and Geertsma, R. E. 2005. Nanotechnology in medical applications: Possible risks for human health. *RIVM Publications Repository, (RIVM report 265001002)*: 2–46 http://hdl.handle.net/10029/7266

Kim, J. Y., and Lim, S. T. 2009. Preparation of nano-sized starch particles by complex formation with n-butanol. *Carbohydrate Polymers* 76(1): 110–116.

Kumar, P. S. M., Francis, A. P., and Devasena, T. 2014. Biosynthesized and chemically synthesized titania nanoparticles: Comparative analysis of antibacterial activity. *Journal of Environmental Nanotechnology* 3(3): 73–81.

Kruijf, N. D., Beest, M. V., Rijk, R., Sipiläinen-Malm, T., Losada, P. P., and Meulenaer, B. D. 2002. Active and intelligent packaging: Applications and regulatory aspects. *Food Additives & Contaminants* 19(S1): 144–162.

Lee, S. K., Sheridan, M., and Mills, A. 2005. Novel UV-activated colorimetric oxygen indicator. *Chemistry of Materials* 17(10): 2744–2751.

Lim, J. H., Park, J., Ahn, J. H., Jin, H. J., Hong, S., and Park, T. H. 2013. A peptide receptor-based bioelectronic nose for the real-time determination of seafood quality. *Biosensors and Bioelectronics* 39(1): 244–249.

Liu, D., Wu, Q., Chen, H., and Chang, P. R. 2009. Transitional properties of starch colloid with particle size reduction from micro-to nanometer. *Journal of Colloid and Interface Science* 339(1): 117–124.

Mihindukulasuriya, S. D. F., and Lim, L. T. 2014. Nanotechnology development in food packaging: A review. *Trends in Food Science and Technology* 40(2): 149–167.

Mills, A., and Hazafy, D. 2009. Nanocrystalline SnO2-based, UVB-activated, colourimetric oxygen indicator. *Sensors and Actuators B: Chemical* 136(2): 344–349.

Nasrollahzadeh, M., Sajadi, S. M., Sajjadi, M., and Issaabadi, Z. 2019. An Introduction to Nanotechnology. *Interface Science and Technology* 28: 1–27

Nešić, A., Gordić, M., Davidović, S., Radovanović, Ž, Nedeljković, J., Smirnova, I., and Gurikov, P. 2018. Pectin-based nanocomposite aerogels for potential insulated food packaging application. *Carbohydrate Polymers* 195: 128–135.

Pérez-Pacheco, E., Canto-Pinto, J. C., Moo-Huchin, V. M., Estrada-Mota, I. A., Estrada-León, R. J., and Chel-Guerrero, L. 2016. Thermoplastic starch (TPS)-cellulosic fibers composites: Mechanical properties and water vapor barrier: A review. *Composites from Renewable and Sustainable Materials* 85: 89–107.

Putaux, J. L., Molina-Boisseau, S., Momaur, T., and Dufresne, A. 2003. Platelet nanocrystals resulting from the disruption of waxy maize starch granules by acid hydrolysis. *Biomacromolecules* 4(5): 1198–1202.

Ramos, O. L., Fernandes, J. C., Silva, S. I., Pintado, M. E., and Malcata, F. X. 2012. Edible films and coatings from whey proteins: A review on formulation, and on mechanical and bioactive properties. *Critical Reviews in Food Science and Nutrition* 52(6): 533–552.

Rosa, M. F., Medeiros, E. S., Malmonge, J. A., Gregorski, K. S., Wood, D. F., Mattoso, L. H. C., and Imam, S. H. 2010. Cellulose nanowhiskers from coconut husk fibers: Effect of preparation conditions on their thermal and morphological behavior. *Carbohydrate Polymers* 81(1): 83–92.

Roselli, M., Finamore, A., Garaguso, I., Britti, M. S., and Mengheri, E. 2003. Zinc oxide protects cultured enterocytes from the damage induced by *Escherichia coli*. *The Journal of Nutrition* 133(12): 4077–40.

Sharma, C., Dhiman, R., Rokana, N., and Panwar, H. 2017. Nanotechnology: An untapped Resource for food packaging. *Frontiers in Microbiology* 8: 1735. DOI: https://doi.org/10.3389/fmicb.2017.01735

Sharma, V., Shukla, R. K., Saxena, N., Parmar, D., Das, M., and Dhawan, A. 2009. DNA damaging potential of zinc oxide nanoparticles in human epidermal cells. *Toxicology Letters* 185(3): 211–218.
Singh, Trepti, Shukla, Shruti, Kumar, Pradeep, Wahla, Verinder, Bajpai, Vivek K., and Rather, Irfan A. 2017. Application of nanotechnology in food science: Perception and overview. *Frontiers in Microbiology* 8: 1501.
Sinha Ray, S., Yamada, K., Okamoto, M., and Ueda, K. 2002. Polylactide-layered silicate nanocomposite: A novel biodegradable material. *Nano Letters* 2(10): 1093–1096.
Smolander, M., Hurme, E., Koivisto, M., and Kivinen, S. 2004. Indicator. In (Vol. International Patent WO2004/102185 A1).
Soysal, C., Bozkurt, H., Dirican, E., Güçlü, M., Bozhüyük, E. D., and Uslu, A. E., et al. 2015. Effect of antimicrobial packaging on physicochemical and microbial quality of chicken drumsticks. *Food Control* 54: 294–299.
Stadtman, E. R., and Berlett, B. S. 1997. Reactive oxygen-mediated protein oxidation in aging and disease. *Chemical Research in Toxicology* 10(5): 485–494.
Suski, J. M., Lebiedzinska, M., Bonora, M., Pinton, P., Duszynski, J., and Wieckowski, M. R. 2012. Relation between mitochondrial membrane potential and ROS formation. In *Mitochondrial Bioenergetics* (pp. 183–205). Humana Press, New York.
Sanchez, F., and Sobolev, K. 2010. Nanotechnology in concrete–A review. *Construction and Building Materials* 24(11): 2060–2071.
Swaroop, C., and Shukla, M. 2018. Nano-magnesium oxide reinforced polylactic acid biofilms for food packaging applications. *International Journal of Biological Macromolecules* 113: 729–736.
Tervonen, T., Linkov, I., Figueira, J. R., Steevens, J., Chappell, M., and Merad, M. 2009. Risk-based classification system of nanomaterials. *Journal of Nanoparticle Research* 11(4): 757–766.
Tripathi, S., Mehrotra, G. K., and Dutta, P. K. 2009. Physicochemical and bioactivity of cross-linked chitosan–PVA film for food packaging applications. *International Journal of Biological Macromolecules* 45(4): 372–376.
Varela, J. A., Bexiga, M. G., Aberg, C., Simpson, J. C., and Dawson, K. A. 2012. Quantifying size-dependent interactions between fluorescently labeled polystyrene nanoparticles and mammalian cells. *Journal of Nanobiotechnology* 10(1): 1–6.
Vera, P., Echegoyen, Y., Canellas, E., Nerín, C., Palomo, M., Madrid, Y., and Cámara, C. 2016. Nano selenium as antioxidant agent in a multilayer food packaging material. *Analytical and Bioanalytical Chemistry* 408: 6659–6670.
Youssef, A. M., Abdel-Aziz, M. S., and El-Sayed, S. M. 2014. Chitosan nanocomposite films based on Ag-NP and Au-NP biosynthesis by *Bacillus Subtilis* as packaging materials. *International Journal of Biological Macromolecules* 69: 185–191.
Zeng, J., Roberts, S., and Xia, Y. 2010. Nanocrystal-based time–temperature indicators. *Chemistry–A European Journal* 16(42): 12559–12563.

2 Nanotubes as Packaging Tool

Sarabia-Castillo Cesar R.[1], *Amir Ali*[2], *Pérez-Hernández Hermes*[3], *Pérez-Moreno Andrea*[1], *Torres-Gómez Andrés P.*[1], *Ayaz Mukarram Shaikh*[4], *Yash D. Jagdale*[5], *and Fernández-Luqueño Fabián*[1]
[1]Sustainability of Natural Resources and Energy Programs, Cinvestav-Saltillo, Coahuila, C.P., Mexico
[2]Nanoscience and Nanotechnology Program, Cinvestav, Mexico City, Mexico
[3]El Colegio de la Frontera Sur (CONACYT), Agroecología, Unidad Campeche, Campeche, Mexico
[4]Institute of Food Science, University of Debrecen, Böszörményi, Hungary
[5]MIT School of Food Technology, MIT ADT University, Pune, Maharashtra, India

CONTENTS

2.1 INTRODUCTION

Carbon nanotubes (CNTs) are cylinder-shaped molecules of single-layer carbon atoms folded up into a tube (graphene). These nanotubes can be single walled (SWCNT) with a diameter of less than one nanometer (nm) or multi walled (MWCNT) with a diameter ranging from less than one nanometer to more than 100 nm. They can be a few micrometers long to several millimeters long, or even longer.

DOI: 10.1201/9781003207641-3

CNTs exhibit outstanding electrical, thermal, and mechanical capabilities despite their small size. They are lighter in weight than most other materials, but they are also stronger. They may be manufactured as both either very conductive or semiconductive, depending on the use. They may be manufactured from nano-sized tiny catalyst particles, and can be as long as tens of millimeters in length when correctly assembled. Since CNTs became a popular research subject in the early 1990s, a great deal of effort has been devoted to investigating the different features of this novel material. CNTs have been offered as alternative materials with exceptional promise in various applications, including electronics, chemical sensors, composite materials, and mechanical sensors/actuators, amongst other things.

Because of their attributes, biodegradability, and non-toxic form, biopolymers are employed in food packaging to combat environmental risks, such as pesticides. In addition to these advantages, they have several disadvantages, including poor mechanical qualities and limited water resistance. Since their discovery in the 1970s, nanomaterials (NMs) have piqued the public's interest due to their extraordinary properties, which have made them the standard for food packaging applications. Nanomaterials improve the mechanical, thermal, and gas barrier properties of food packaging without compromising their ability to be non-toxic and biodegradable. The most common nanomaterials used in food packaging are montmorillonite, zinc oxide covered silicate, kaolinite, silver nanoparticles (NPs), and titanium dioxide NPs. These NMs-coated films act against oxygen, CO_2, and various other contaminants. In addition, they possess the capacity to scavenge oxygen, have antibacterial action, and are temperature tolerant. The most challenging parts of creating these nanocomposites are ensuring that they are evenly distributed throughout the polymer matrix and that they are compatible with one another. Therefore, there is a growing need for improvements in the performance of nano-packaging materials, including mechanical stability, degradability, and the effectiveness of their antibacterial properties, among other factors (Chaudhary et al., 2020).

By incorporating functional nanoparticles into polymer matrices, it is feasible to improve food packaging materials' mechanical and barrier properties. By enhancing the flexibility, durability, temperature, humidity resistance, and flame resistance of packaging materials, the addition and modification of different nanomaterials may boost the shelf life and quality of food commodities. To protect food items and provide a mass-transfer barrier, a variety of natural and inorganic-organic nanofillers, such as cellulose nanocrystals, NPs, and cellulose-NPs, as well as inorganic nanomaterials, such as clay and layered silicates, mesoporous silica nanoparticles, metal and metal oxide NPs, and edible coatings or edible films in the form of films or thin layers, are used. Edible coatings are put directly on foods, whereas non-edible coatings serve as protective containers. These coatings are used in cheese, agricultural and bakery products, and meat processing to provide color, enzymes, tastes, and antioxidants to food (Nile et al., 2020).

According to Rezic et al. (2017), packaging materials can be classified into four subclasses: a) passive packaging—refers to the use of covering materials in conventional packaging, b) active packaging—materials that actively respond to internal and external biological changes (for example, materials that may absorb oxygen and extend the shelf life of the packaging material), c) intelligent packaging—materials with cutting-edge design that are convenient for consumers, and d) smart packaging—these involve the use of chemicals or technology (electronic, mechanical, or other) in conjunction with one another to create items that are functional and beneficial, as well as products that add innovative functionalities to the overall product. Besides this, the blending of polymers and polymer nanocomposites are two of the most common methodologies utilized to generate hybrid materials for food packaging. Nanocomposites are composites or hybrid materials made up of dispersed nanometer-sized particles in a polymer matrix, and they are used in a variety of applications. CNTs are the nanomaterials that are most often employed for this purpose.

2.2 SINGLE-, DOUBLE-, OR MULTI-WALLED CARBON NANOTUBES TO PRESERVE FOOD

The short shelf life of fresh food negatively impacts the economy, health, and environment. The food industry depends on the devolvement of technologies to preserve food for consumers by changing the package conditions to include preventing the food degradation and growth of the pathogen microorganisms. The Food and Agriculture Organization (FAO) proposes active packaging that depends on the integration of bioactive components. The nanoscale materials are used in films. Tensile strength (TS) and elongation at the breaker (EB) are the most critical parameters for utilization of material for making films. CNTs in polymer nanocomposites with zinc oxide and halloysite nanotubes are examples (Li et al., 2021).

Good food packaging plays an important role in extending the storage life of food for better and safer quality. The food industry consumes about 40% of plastic produced and uses 50% of it for packaging; this represents an environmental issue. There are compound materials for packaging, with two or more phases; one continues, and the other disperses. The dispersed phase commonly has nanomaterials sized below 100 nm. The films made with CNTs take advantage of the allotropic behavior of carbon and have several advantages, such as thermoplastic and thermostable versatility, better TS, unidimensional structures, and compatibility with other chemical compounds (Landa-Salgado et al., 2017).

Consumer interest in the positive effect on the food contributes to developments in the packaging of the food and its environmental impact. The principal target in packaging is the preservation of food until it reaches the final consumer, where it must be attractive and easily managed. Antibacterial and antioxidant effects of fresh chicken packaging were found to be beneficial in studies performed by Ojeda et al. (2019).

The packaging of food materials is a crucial component in the food processing segment. The importance of easy transportation and handling, consumer-friendly products, protection against microbial contaminants, and real-time information drive development of new packaging methods. The demand for active packaging is advancing research into the use of environmentally friendly materials, such as CNTs. CNTs combined with allyl isothiocyanate will impede *Salmonella choleraesuis* growth for over 40 days and with polyethylene film would prevent fungal growth for over 90 days. Many fruits and vegetables are sensitive to ethylene, so it needs to be eliminated by producing ethylene scavengers in packages for extended storage life (Gaikwad et al., 2020; Sharma et al., 2017).

Carbon nanotubes have graphite layers with diameters of 0.4 nm to 10 nm; single-walled (SWCNT) are 0.4 and 2 nm in diameter, while hexagonal, packed and bound, multi-walled (MWCNT)/(DWCNT), two or several layers, are 1 to 3 nm in diameter. There are three different syntheses: the chemical vapor deposition method, the discharge method, and the laser ablation method. Carbon nanotubes have excellent chemical and physical properties, like high tensile strength (TS), ultra-light weight, and thermal stability. In addition, the antibacterial and antifungal properties make carbon nanotubes excellent for making films to preserve food safety (Anzar et al., 2020; Khanna and Islam, 2018).

CNTs are pure carbon forms, approximately 1 nm in diameter and 1 to 100 μm in length. Graphene sheets are rolled into cylinders. The tensile strength of carbon nanotubes is approximately 100 times greater than that of steel of the same diameter (Deshpande and Mahendru, 2018). Moreover, CNTs possess excellent mechanical properties, making them ideal as reinforcement for polymer matrices. A hierarchical composite material is both a composite material and a polymeric nanocomposite. This material has better properties than single ones; the properties of hybrid materials mean that they can be better used to enhance the quality of the package (Vázquez-Moreno et al., 2019; Zhang et al., 2021).

SWCNTs are hollow cylinders. In the case of some SWCNTs, growth chiral depends on the temperature of synthesis or catalyst. Therefore, the tube has to control the chirality when using SWCNTs property to obtain excellent quality for making several materials suitable for packaging and that do not affect human health (Magnin et al., 2018).

The use of biological methods, like enzymatic methods, to obtain chitosan is an eco-friendly option. Using glycerin at 6% to intersect is the most optimal dose, with a MWCNT of 1.5%. The biomaterials with chitosan and MWCNTs are helpful for films, in order to permeate water. They can be used as films for food packaging. Also, the color of these films is black, and this material can absorb energy (Cedillo et al., 2019; Portillo et al., 2020).

Polystyrene (PS)/MWCNT of 6 to 7 nm diameter combined with dodecylamine (DDA) showed high permeability. A double-walled carbon nanotube (DWCNT) of 1.6 nm diameters with a Si_3N_4 film showed better properties. Studies showed that films with CNTs combined with other materials or elements yield better results for making better films (Armenta-Armenta and Espinoza-Gómez, 2020).

MWCNTs with other materials, like L-polylactic acid (PLA) and polypropylene (PP), have been used as suitable biodegradable food packaging because, in combination, their elements can increase their properties; that is, TS (Azizi et al., 2019). Composite materials (CM) with MWCNTs and PLA were used in food packaging and showed better rigidity. That is good, because it increases the mechanical stress range. Add MWCNTs to increase TS, EB, and flexural modulus. Under low temperature, odor, and sour maintenance, the PLA provides rigidity, transparency, and thermal seal. The MWCNTs degradation temperature is 621°C. It is essential to mix PLA at a temperature of 180°C (Murillo-Vargas et al., 2021).

Several nanoparticles that are lighter and stronger are used in active packaging. Single-walled nanotubes have antibacterial properties and have been used to detect the existence of pesticides on the surfaces of vegetables and fruits. However, studies report that SWCNT used in packaging results in toxic effects on human skin and lungs. Potentially toxic nanomaterials need to include another barrier layer to avoid coming into contact with food (Lugani et al., 2021; Primožič et al., 2021).

Double-walled carbon nanotubes (DWCNTs) have single-walled and multi-walled properties, depending on the type of cylinder. They have high stability under aggressive chemical, mechanical, and thermal conditions. The friction between layers in DWCNTs depends on the diameter of the outer shell. The diameter varies from 2.4 to 17 nm in the outer and inner shells, respectively. The space between them is about 0.35 nm. This is important because the stability of the material is dependent on that. It is essential to control the temperature synthesis because the material needs to be stable for food packaging in order to prevent any issue with human health (Wu et al., 2017).

Polymer matrices with CNTs help with environmental issues because they are biodegradable packaging materials. In the future, they will be used for biosensors. In addition, CNT introduced into biopolymers can increase their properties.

Nanotechnology can be used in all aspects of the food process, but one of the most important is packaging, where it can be used to control microbial contaminants, reduce economic losses from storage, and improve food quality. Health safety necessitates studies about the effects on human health and how to eliminate nanomaterials that are toxic for the body. The effect on the human body depends directly on the properties of the nanomaterials and their stability, so it is essential to analyze how they work as food packaging materials and their interaction with food. The consumer needs to have accurate information about the materials' behavior.

2.3 FILMS, FIBERS, OR BIOCOMPOSITES BASED ON NANOTUBES EXTEND THE FOOD SHELF LIFE

To satisfy the demand for food and at the same time preserve food safety, technologies have been developed that allow food to be packaged and wrapped for its preservation and the reduction of its deterioration, increasing its shelf life while preserving its nutritional quality. The physicochemical, mechanical, and antimicrobial properties of carbon nanotubes, whether single-walled carbon nanotubes or multi-walled carbon nanotubes, have been studied for their implementation in the production of packaging and containers by incorporating them into synthetic polymers or biopolymers. This will give new properties to these materials, such as flexibility, thermal stability, mechanical resistance, gas and moisture barrier, all of which allow for the preservation of food while protecting it from physical, chemical, and biological deterioration as well as facilitating its transport and storage (Almeida et al., 2015; Asgari et al., 2014; Hosseini and Jafari, 2020).

In particular, the incorporation of carbon nanotubes can significantly improve packaging materials due to their excellent mechanical properties—increased tensile strength, resistance to UV radiation, and thermal stability (Asgari et al., 2014; Cui et al., 2020)—properties that allow improvements in the preservation, quality, and shelf life of food. An example of this is reported by Cui et al. (2020), who conclude that the implementation of a biocomposite synthesized from polylactic acid, poly(ε-caprolactone), carbon nanotubes, and cinnamaldehyde for the development of packaging films could be used to reduce microbial contamination and improve the quality of perishable foods. Furthermore, their research found that this film showed properties that can decrease food spoilage by improving UV resistance, controlling the release of antimicrobial agents, such as cinnamaldehyde, and increasing the tensile strength, elastic modulus, and elongation at the break of the film.

In the same vein, Dias et al. (2013) developed a cellulose-based film incorporating allyl isethionate and carbon nanotubes, which was used to store cooked chicken meat, and 40 days after storage, microbial contamination, meat oxidation, and color changes decreased. This was due to the diffusion of the allyl isethionate into the meat and its retention in the film thanks to the carbon nanotubes. In addition, the modifications made by incorporating these compounds improved the film's mechanical properties, making the film effective for storing this type of food for up to 40 days.

On the other hand, Pattanshetti et al. (2020) developed gelatin-based films reinforced with multi-walled carbon nanotubes (derived from recycled polymers) and coated with garlic microparticles, which presented lower water and oil absorption. This decreased the swelling capacity of gelatin and improved the resistance to these compounds, due to the hydrophobic properties of MWCNTs, in addition to increasing the antibacterial activity attributed to the incorporation of the coating made with garlic microparticles, which in turn also decreases the unintentional migration of nanotubes to food.

Similarly, Kavoosi et al. (2014) synthesized films based on gelatin and multiwalled carbon nanotubes and found similar results related to water resistance, where they observed a decrease in solubility and lower water absorption capacity and significant increases in the mechanical properties of the gelatin films. In addition, these gelatin/nanotube films showed antimicrobial activity, so the development of this type of films could potentially be used as an alternative to traditional materials used in food packaging.

Concerning the development of materials for these purposes, synthetic polymeric films in which carbon nanotubes are incorporated have also been investigated to develop materials capable of preserving food for a longer period and reducing waste. However, there are still some drawbacks related to changes in the sensory properties of some foods, as is the case reported by Asgari et al. (2014), who studied low-density polyethylene films with different concentrations of CNTs. These served as packaging for dates due to their antimicrobial properties, demonstrating their effectiveness in extending the shelf life of the. However, the packaging considerably modified the flavor quality of the dates.

On the other hand, Wen et al. (2021) prepared films based on polyvinyl alcohol modified with MWCNTs and doped with ZnO, concluding that this type of film could be used in food preservation because, in their tests, they observed less water loss in vegetables when stored in these films. In addition, they showed that the films could slow the growth of bacteria in chicken meat and that there was no migration of Zn into the meat, suggesting that they may have potential application in the field of food packaging.

However, despite the improvement in the physicomechanical and structural properties of some polymers due to their modification with CNTs for their application in the food industry, research has emerged to evaluate the degree of toxicity of these types of materials modified with CNTs and whether they are suitable for food packaging applications, such as that reported by Shahbazi et al. (2017). They modified starch films with MWCNTs and hydroxylated MWCNTs, resulting in films more resistant to tensile strength, more water-resistant, and exhibiting improved water barrier properties. However, when performing cytotoxicity tests, they found that both modified films were toxic when used directly on food. Due to the improved properties of the films, however, they could be used in secondary food packaging without interacting directly with food.

In general terms, it could be said that for the application of nanotechnology to indirect food packaging, doubts regarding its safety and possible contamination of food have yet to be clarified. However, the development achieved so far should not be discarded or set aside since what has been researched could be mainly applied to the elaboration of secondary or tertiary packaging thanks to the physical-mechanical properties acquired by incorporating carbon nanotubes into their matrices.

2.3.1 Titanium Dioxide Nanotubes (TNT)

The oxidation of fatty acids in meat products is known to cause discoloration and rancidity. Oxygen scavengers are typically metal, enzyme, or acid based. While metallic powders are an economically viable alternative, they rarely require high humidity and high temperatures to activate. Current studies have evaluated titanium dioxide nanotubes for use as oxygen scavengers at room temperature, testing their absorption rates in various ranges of relative humidity (RH) (0, 35, 70, 100% RH), observing that in five minutes percentages greater than 50% of the capacity to absorb oxygen from the established environment. However, it tends to decrease due to condensation or water vapor (Tulsyan et al., 2017).

On the other hand, films based on naturally grown ingredients, like proteins and polysaccharides, tend to inhibit the oxidation of lipids. Furthermore, their hydrophilic nature reduces their efficiency under humid conditions. To improve this deficiency, Feng et al. (2019) used the hydrophobic property of the edible whey protein coating in an assembly of semi-flexible nanofibers (WPNF), whose characteristics were associated with the high antibacterial activity presented by TiO_2 nanotubes (TNT). Thus, they obtained a potential WPNF/TNT nanocomposite film to preserve raw and chilled meat. Likewise, the surface area of TNT is greater than that of TiO_2 nanoparticles. However, due to its high forbidden band (~3.0–3.2 eV), the photocatalytic activity of TNTs under UV irradiation (<380 nm wavelength) is usually shorter. Riahi et al. (2021) used the addition of copper oxides to reduce the bandgap and improve the photocatalytic activity of TNTs. They prepared two nanocomposites, TNT-CuO and TNT-Cu_2O, whose antimicrobial properties were evaluated alongside TiO_2 nanoparticles to inhibit gram-positive (*L. monocytogenes*) and gram-negative (*E. coli*) bacteria. In the results obtained, it was observed that the compounds decorated with copper oxides obtained the most significant antimicrobial activity; TNT-Cu2O, the one with the best results against both strains, suggests an excellent candidate to enhance the shelf life of bananas. Díaz-Visurraga et al. (2010) prepared composite films by depositing titanium nanotubes (TNTs) in a chitosan matrix (CS), producing film-forming solutions of 0.05 and 0.10% [w/v]. When evaluating the cellular reaction of the microorganisms, it was observed that the films of HMW CS (molecular weights of 400.00 Da) and 0.1% TNTs show evidence of the most effective sterile activity against *S. aureus*.

2.3.2 Halloysite Nanotubes (HNTs)

In food packaging, antimicrobial films can significantly increase the shelf life of food. As mentioned above, metals and metal oxides rather than organic materials, such as enzymes, are considered antimicrobial agents (Lomate et al., 2018). However, the application of nanotechnology as carriers of organic agents in polymer film packaging has offered better mechanical, thermal, and biological properties in bio-nano composites, as described in Table 2.1.

The high volatility of essential oils has been detected as one of the main limitations for their use in active packaging. Encapsulated halloysite nanotubes (HNTs) present an optimal alternative. In this sense, halloysite nanotubes (HNTs) have been utilized as nanofiller for bio-nano composites. They have high biocompatibility, and

TABLE 2.1
Halloysite Nanotubes in Bionanocomposites Films to Improve Mechanical Properties

Bionanocomposite	Method	Concentrations	Mechanical Properties	References
Polylactic acid (PLA)/halloysite nanotubes (HNTs)	Casting	0, 1.5, 3.0, 4.5, and 6.0 wt.% HNTs	The optimum properties of the films were supposed to be 3.0% by weight, improving their thermal, mechanical, and barrier properties. Obtaining the potential to extend the shelf life of packaged cherry type tomatoes.	Risyon et al. (2020)
Polyvinyl alcohol (PVA)/starch (ST)/glycerol (GL)/halloysite nanotube (HNT)	Casting	0.25, 0.5, 1, 3 and 5 wt. % HNTs	Water vapor permeability increased linearly when the temperature and relative humidity gradient increased from 25 to 55° C and 10% –70%, respectively.	Abdullah et al. (2019)
Potassium permanganate-impregnated HNTs (P-HNTs)/ polyethylene (LDPE)	Mixed	1:99, 3:97 y 5:95 (p / p) P-HNTS / HNTs	The loading of 1 g/100 g of P-HNT increased the thermal stability, enhancing the freshness of the cherry tomatoes that were wrapped with the nanocomposite film of 1% P-HNT/LDPE (P-HLNF).	Joung et al. (2021)
Chitosan (CS)/ halloysite nanotubes (HNTs)/ tea polyphenol (TP)	Casting	The SEM showed the nanocomposite film with a CS/HNTs ratio of 6:4 and a TP content of 10% (C6H4-TP10)	The water vapor barrier property of the nanocomposite film was extended because of the tortuous channels created by the HNTs.	Wang et al. (2021)

(Continued)

TABLE 2.1 *(Continued)*
Halloysite Nanotubes in Bionanocomposites Films to Improve Mechanical Properties

Bionanocomposite	Method	Concentrations	Mechanical Properties	References
Alkaline halloysite nanotubes (ALK-HNTs)/ polyethylene (LDPE)	Mixed	1, 3, and 5 wt% of ALK-HNTs	The oxygen barrier property was enhanced and ethylene gas adsorption was induced by 220%	Boonsiriwit et al. (2020)
Carboxymethyl cellulose (CMC)/ acid-treated halloysite nanotube (Hal-A)	Mixed	3 g of CMC and 0.06 g (2 wt% based on CMC) of Hal	The functionalized Hal-A exhibited vigorous antimicrobial activity against *L. monocytogenes*, and *E. coli*. The CMC-based film showed a significant increase in water vapor barrier and mechanical and thermal stability characteristics after forming a composite with Hal.	Wang and Rhim (2017)

their hydroxyl groups on the surface make them relatively hydrophobic. They have a cylindrical structure of clay minerals. Hence, they are versatile carriers or vehicles for active compounds in evaporation, especially in dry packaging films (Suppiah et al., 2019). For example, Lee and Park (2015) have evaluated HNTs in the encapsulation and release of thyme oil (TO), extracted from *Thymus vulgaris* L., generating a TO/HNT nanocomposite coated with eudragit polymer (EPO). The release studies were carried out in an oven at 25°C for 96 hours, and a rapid release of TO into the air was observed; that is, at 12 hours, almost 79.73% was released. After encapsulating the TO in HNT coated with EPO, the TO's release decreased so that 45.27% of the amount was released after 24 hours. Kim et al. (2019) used HNTs to encapsulate clove oil (CO) in an effort to generate films resistant to *Plodia interpunctella* (Hübner) by applying polypropylene and polyethylene films. The CO was attached with HNT and polyethyleneimine (PEI) utilizing the vacuum extraction process, generating a film of HNTs/CO/layer by layer (LBL) that maintained the repellent properties for up to 60 days; later, the properties were reduced. However, the literature recognizes that the minimum CO released must be 70% to report insecticidal properties. These investigation results reported a residual amount of CO of 76.6% at 60 days.

2.3.3 Carbon Nanotubes and Multi-Walled Carbon Nanotubes

The surface of CNTs is usually centralized to improve its dispersion and interface with the polymeric matrix. Due to their hydrophobic adsorption toward CNTs and hydrophilic toward aqueous media, surfactants provide excellent stability to CNTs, allowing them to improve their dispersion in aqueous conditions. Using different ionic surfactants, such as sodium dodecyl sulfate (SDS), cetyltrimethylammonium bromide (CTAB), and sodium cholate (SC), to enhance the dispersion of MWCNT within a starch matrix, Alves et al. (2021) observed an increase of 75% in tensile strength and 60% in Young's modulus in the films treated with SC, compared to the rest of the surfactants. Furthermore, the addition of CTAB provided the highest antioxidant activity values, reaching 30% in 1.5 h. Similarly, biofilms designed with poly (3-hydroxybutyrate-co-3-hydroxyvalerate) (PHBV) bio-nano compounds reinforced with MWCNT have reported better thermal, barrier, and migration stability of PHBV. Compared to pure PHBV, the tensile strength and Young's modulus of the nanocomposite film were improved by 88% and 172%, respectively, and the maximum decomposition temperature of the nanocomposite film was 22.3% higher than that of pure PHBV (Yu et al., 2014).

A homogeneous dispersion of charges is required to incorporate carbon-based nanomaterials into biopolymeric matrices. Electrospinning is a versatile technique for generating fibers with submicron diameters. It allows a very high surface area to volume ratio and makes them quite attractive for designing fabrics with biological agents. Liu et al. (2019) incorporated carbon nanotubes (CNT) into the chitosan (CS)/polylactic acid (PLA) compound by electrospinning, thereby obtaining PLA/CNTs/CS smooth surface nanofibers. This nanocomposite was tested to conserve strawberries, determining better mechanical properties, solubility, and swelling ratio, reaching maximum values when using CS at 7% by weight.

2.3.4 Graphene Nanomaterials

The graphene nanosheets (GNs) show compelling immersion in the visible regions. As a result, graphene oxide has gained prominence as a platform for broad-spectrum biocompatible antimicrobial agents (Han et al., 2021). In the case of hydrogels, these polymeric materials with cross-linked 3D networks can provide unique advantages, such as biocompatibility and hydrophilicity, opening up the field of tissue engineering (Fan et al., 2021). For example, Konwar et al. (2016) produced iron oxide-coated graphene oxide nanomaterials (GIO) to manufacture a network system in chitosan hydrogel. They managed to create films with adjustable surface properties; by varying the degrees of percentage of charge of the GIO nanomaterials on the hydrogel matrix, manipulating hydrophobicity. The results report stated that chitosan-iron oxide coated graphene oxide nanocomposite hydrogel films showed significant antimicrobial activity against bacterial strains of *S. aureus*, *E. coli*, and *C. albicans*. Similarly, graphene nanosheets have been tested with clove essential oil (COs) in polylactide (PLA), highlighting the influence of graphene oxide (GO) in avoiding the surface porosity of the PLA/CO plasticized film. This compound revealed excellent antibacterial activity against *S. aureus* with 30% CO content (Arfat et al., 2018).

However, for all this innovation to be accepted by the food industry and by consumers themselves, the new packaging materials that contain nanomaterials must not only show high efficiency in the preservation of food, but they must also ensure that the high reactivity and the potential for toxicity, due to their high surface-volume ratio, is minimal (Omerović et al., 2021). The migration of nanoparticles from packaging to food is an important aspect that involves a complex diffusion process influenced by the concentration gradient of the additive as well as its solubility, temperature, and contact time. Velichkova et al. (2017) used 2% by weight carbon nanofillers to compare the release of graphene nanoplates (GNP) and multi-walled carbon nanotubes (MWCNT) in polymeric films composed of polylactic acid (PLA) and polypropylene (PP), under different time and temperature (4h at 90°C) conditions, and evaluated the total migration in food simulants of ethanol and acetic acid using gravimetry. The hydrolytic dissolution of the PLA polymer in the food simulants was reported to cause the migration of the GNPs (> 100 nm) from the PLA/GNP/MWCNT films to the simulating solvents, otherwise with the MWCNTs, which remained in the matrix when bound. On the other hand, the polymer PP results reported a slight case of swelling in the ethanol solvent, allowing the release of some MWCNT.

2.4 BIODEGRADABLE AND ECO-FRIENDLY CARBON NANOTUBES FOR PACKAGING FOOD

Packaging polyethylene (PE) material is the most commonly used petroleum-dependent polymer and one of the principal sources of inorganic waste pollution (Zhong et al., 2020). Nevertheless, advances in nanoscience and nanotechnology have enabled scientists to understand and manipulate metallic and non-metallic materials at the nanoscale, thus obtaining the properties desired for food packaging. In this sense, it is known that CNTs have unique characteristics, such as high electrical conductivity, thermal stability, mechanical strength, aspect ratio, and surface area. Among the carbon-based materials are MWCNT, SWCNT, carbon nanofibers (CNF), graphene nanoplatelets (GNP), graphite nano stacks (GR), graphene oxide (GO), and reduced graphene oxide (rGO), among others (Carvalho and Conte Junior, 2020).

According to a search carried out on the Web of Science, it is surprising to find that, to date, there is much research related to CNT and food, with a total of 4,230 articles found (Figure 2.1). In addition, it was found that from 2005 to 2021, there has been an

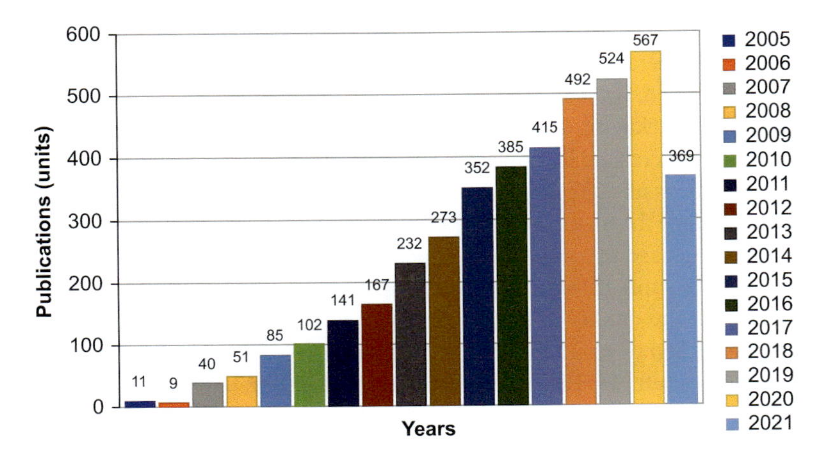

FIGURE 2.1 Publications that relationship of CNT NPs with food packaging and reservation, food quality, and food safety.

increase in research on the relationship of this nanomaterial (CNT) as food packaging and preservation, food quality, and food safety. When we searched the field of topics related to the words "nano," "carbon," and "food packaging," it returned information from 526 articles (2002–2021).

We must clarify that, although the development of biocomposites to extend the shelf life of food products has been mentioned in the previous section, we focus here on the effects of carbon-based nanomaterials with biodegradable, or environmentally friendly, characteristics. In this sense, despite the fascinating characteristics of CNTs, the literature suggests that the smooth and inert surfaces of CNTs promise poor interfacial interactions between CNTs and polymeric matrices. For this reason, scientists have adopted methodologies that allow for modifying and improving the compatibility between CNTs and polymeric matrices (Xu, 2018). Natural biodegradable biopolymers and polysaccharides derived from starch, cellulose, chitosan, and proteins of plants and animals, as well as those obtained through microbial fermentation (e.g., poly(hydroxy alkanoates [PHA]), poly(hydroxybutyrate [PHB]) are included in the ecological methodology (as raw material, biomass) (Zhong et al., 2020). Therefore, to obtain an ecological and active packaging that avoids the deterioration of food due to microbial growth, researchers propose that the packaging consist of a matrix of the film (biodegradable polymers). An active or reinforcing agent such as CNTs, among others (Kumar and Kumar, 2019), can also be effective against the COVID-19 virus (present in food products) (Carvalho and Conte Junior, 2020).

In recent years, CNTs and GO combined with polymers have been evaluated to develop packaging, or smart assets, capable of improving and enhancing the quality of food that requires storage. However, the release of GO and CNTs into polymeric matrices is alarming for scientists and technologists. Some authors have highlighted that, during their degradation phase, GO and CNTs can represent a risk for the environment and human health, since ultraviolet radiation, or burning, causes nanocomposites to penetrate the soil or stay in the air or reach groundwater (Kotsilkov et al., 2018). Indeed, when the authors evaluated the release of GNP and MWCNTs from biodegradable poly (lactic acid) nanocomposite films, they observed that the combustion at 500 and 650°C generated ash containing CNTs. Nevertheless, at 850°C the CNTs were completely degraded, but with GNP residues. The authors suggest that due to poor management of food packaging waste (burning in the open air and fire accidents), nanoparticles will be produced and harm human health (Kotsilkov et al., 2018). A recent review by Makgabutlane et al. (2021) discusses the ecological possibilities and methodological processes to reduce the impact of CNTs on the environment and to decrease production costs while being used in the industry, principally for packaging food.

The incorporation of antimicrobial films as reinforcements in food packaging to maintain the quality and safety of fresh food has been evaluated. Studies by Li et al. (2019) demonstrate that the synergistic effect of poly (3-hydroxybutyrate-co-3-hydroxyvalerate [PHBV])/cellulose nanocrystal-graphene (CNCs and GO), obtained by chemical grafting (covalent bond, 1% by weight), exhibited excellent antibacterial activity (100%) compared to PHBV nanocomposites and CNC with GO (non-covalent bond 1:0.5% by weight). In addition, the authors demonstrated that the ternary nanocomposite presented the lowest level of migration in both food simulants. The authors suggest that this type of ternary nanocomposite produced from bacterial fermentation of maize plant stems has antibacterial potential when used as food packaging (Li et al., 2019).

On the other hand, recent findings have shown that the combination of materials with non-toxic, biodegradable, and non-polluting characteristics, combined with CNTs, appears to be one of the best films for packaging applications (Wen et al., 2021). Indeed, poly (vinyl alcohol) (PVA) films with ZnO-doped MWCNTs showed an attractive strength of 116%. Furthermore, the compound was better in terms of hydrophobicity and antibacterial activity than pure PVA film. On the other hand, the compound retained more water in the vegetables. Therefore, the authors indicated that MWCNTs-ZnO/PVA nanocomposite films are potentially promising food packaging (Wen et al., 2021). Also, other biocomposites were manufactured with an ecological methodology consisting of the PHBV as a matrix and CNT as nanofillers. This experiment showed that the dielectric and electromagnetic interference shielding properties of PHBV/CNT nanocomposite is much higher than pure PHBV. Furthermore, the results suggest that the biodegradability of the nanocomposite is an alternative to replacing petroleum-based plastics (Luo et al., 2021).

Some authors find that, due to the lack of studies, many knowledge gaps remain regarding the creation of protein-derived SWCNT bio-nano compounds and their use in food packaging. As far as it is known, the main problem is attributed to the scattering factor of the protein matrix. Therefore, new research is required to propose a novel and environmentally friendly methodology for creating biodegradable CNT films for food packaging (Zubair and Ullah, 2020).

Regarding GO, considered a thick carbon leaf, authors have shown that the green synthesis of rGO from green tea—which contains high levels of polyphenols such as flavonoids, gallic acid, and tannic acid—is a novel technique that is low cost and friendly to the environment (Vatandost et al., 2020). When tea extract was used as a reducing agent and stabilizer for rGO synthesis, it was shown that the rGO modified carbon paste electrode (rGO/CPE) was adequate for the electrochemical oxidation of sunset yellow (SY). SY is an additive in pharmaceutical, cosmetic, and food products, among others, and when consumed in excess it is considered a pathogen. Therefore, the studies revealed that due to the large surface area of the rGO/CPE, it could be considered a potential sensor to determine the amount of SY in food products (Vatandost et al., 2020).

Although in this section we showed the latest advances of environmentally friendly carbon-based NMs that can be used in food packaging, it is necessary to note that there are nanocomposites that contain CNT and are promising for lengthening the shelf life of fresh produce (Liu et al., 2019). For instance, the nanocomposite fiber polylactic acid (PLA)/CS/MWCNT turned out to be effective when the content of CS was 7% by weight, since it showed higher antimicrobial activity against *E. coli, S. aureus, Botrytis cinerea, and Rhizopus.* In addition, the tests showed that the nanocomposite was able to preserve strawberry fruit for a maximum of 8 days (Liu et al., 2019). On the other hand, the PLA nanocomposite loaded with bald essential oil (CLO) and GO showed an antimicrobial effect against *S. aureus* and *E. coli.* The authors suggest that the integration of plasticized PLA into the nanocomposite places it as one of the promising biopolymers to be used in packaging and for protecting food against pathogens. However, more studies are required to determine the release kinetics of clove oils during storage, and, consequently, to determine the safety and shelf life of the product (Arfat et al., 2018).

On the other hand, the nanocomposites containing CNTs and other biopolymers are excellent antivirals; in fact, against the COVID-19 virus, as mentioned above. For instance, studies by Sanmartín-Santos et al. (2021) demonstrate that calcium alginate with CNF (0.1% w/w) elicited an antiviral action against a double-stranded DNA virus model. Also, the nanocomposite exerted antibacterial properties against *S. aureus*. Therefore, the authors suggest that because alginate is effective against RNA viruses similar to SARS-CoV-2, the calcium alginate (Ca-SA)/CNF nanocomposite can be a promising bionanomaterial for fighting the COVID-19 virus in the areas of biomedical and, without a doubt, in food packaging.

Chitosan (CS) is obtained from chitin and is the second most abundant polymerized carbon in the environment (Ababneh and Hameed, 2021). The use of CS as a film for food packaging and pathogen control is widely reported in the literature (Salgado-Cruz et al., 2021). Their combination with other biological or non-toxic nanomaterials is currently being tested for their impact on the environment and humans but show promise for improving the quality of food and lengthening its shelf life of food (Sarojini et al., 2019). The composite of biodegradable films made from Mahua oil-based polyurethane (PU) and CS, incorporated with 5% zinc oxide NPs, improved the nano's antibacterial properties, barrier properties, and hydrophobicity biofilm. In addition, they extended the shelf life of carrot pieces by up to nine days. The authors suggest that incorporating Zn oxide acts as an antibacterial agent, which shows that the nanocomposite containing PU based on bio polyols is adequate for food packaging and has a low environmental impact (Sarojini et al., 2019).

Similarly, the Pattanshetti et al. (2020) study demonstrated that gelatin (collagen peptide, a soluble protein obtained by partial hydrolysis of collagen) loaded with MWCNT, both covered with garlic microparticles, improve antibacterial activity (*Staphylococcus aureus* and *Escherichia coli*). The use of garlic particles solves the toxicity problem caused by MWCNT due to migration in use of the nanocomposite within food packaging. Other examples of ecological synthesis have been reported with CS and GO materials and are to be evaluated as food packaging. For instance, the researchers observed that GO promotes mechanical reinforcement in the CS matrix. This generated an increase in the resistance to attraction and low water solubility. These properties are sought in carbon-based nanomaterials for food packaging (Barra et al., 2019).

This section has commented on materials and nanocomposites made of carbon and with biological materials. Recently, the carbon points (CD) synthesized (hydrothermal method) from rosemary leaves (*Rosmarinus officinalis* L.) and in combination with PVA were evaluated as coatings for the storage of bananas and to extend the shelf life of the fruit. The results showed that the CD/PVA nanocomposite was much more efficient than the simple PVA. In addition, the nanocomposite (1 mL) caused inhibition against gram-positive and gram-negative yeast strains. Although there is little knowledge about the combination of biological materials, new research explorations of CD/PVA are needed for application in the future food industry (Eskalen et al., 2021).

Tree gums belong to the class of carbohydrate polymers that have not been thoroughly explored in the manufacture of food packaging films (Venkateshaiah et al., 2021). Indeed, using gum kondagogu (GK), films were formed that hydrophobically modified (HMGK) by reacting GK and DDSA in an aqueous medium. They prove to be an excellent oxygen barrier in low humidity conditions when packaging products. Furthermore, they exhibited antibacterial properties against *S. aureus* and *E. coli*. The highlight of this research is that the modified films (HMGK) showed an excellent degradation of 98.2 ± 1.7% in 28 days (Venkateshaiah et al., 2021). Other examples can be seen in Table 2.2; mainly the combination of bio-nano materials with CNTs, such as films for food packaging and various purposes.

TABLE 2.2
Promising Examples of Carbon-Based Nanomaterials and Their Applications in Food Packaging and Food Industries

Nanoparticles	Effects	References
Graphene oxide (GO)	Using coffee films with exposure to GO, the NMs caused a considerable decrease in the number of bacteria for both *Pseudomonas aeruginosa* and *Streptococcus mutans*. The authors suggest that carbon-based NM (GO) can be promising as commercial bactericidal films in the food industry.	Mitura et al. (2021)
Polyhydroxybutyrate combined with biodegradable graphene nanoplatelets (PHB/Gr-NPs)	A fourfold increase in the shelf life of foods, such as potato chips and dairy products, sensitive to moisture and oxygen was observed in a simulation study. Furthermore, the authors demonstrated that, with the soil test for 30 days, the nanocomposite is 100% biodegradable.	Manikandan et al. (2020)
Nanocomposite film based on activated carbon (AC) and cellulose nanofiber (CNF) (AgNPs/AC-CNF)	The nanocomposite inhibited the growth of both *Staphylococcus aureus* and *Escherichia coli*, considered food pathogens.	Sobhan et al. (2020)
Incorporation of SiO_2 to enhance PVA/chitosan (CS) biodegradable films	Biodegradable films show promise in acting as an oxygen and water barrier for food packaging, to extend the shelf life of cherries. When tested on the soil, the films showed excellent biodegradability, demonstrating a 60% weight loss within 30 days.	Yu et al. (2018)
Carbon nanomaterials incorporated into biodegradable starch polymers of acrylic acid, methyl methacrylate (MMA), acrylonitrile (AN), 2-Ethylhexyl acrylate (2-EHA), and Ethyl acrylate (EA)	The starch-g-AA-MMA-AN polymer compound experienced a decrease in permeability and solubility to water vapor. In addition, it turned out that this group of polymers with CNC incorporated presents biodegradability and resistance against microbes. Therefore, this carbon nanocomposite can be used as a biodegradable packaging material.	Iqbal et al. (2021)

It can be concluded from this section that nanotechnology applied to food packaging has no limits. New trends in nanotechnology applied to food packaging and food security have the task of providing new technological alternatives that avoid damage to human health and the environment. Several investigations continue to search for the best methods of ecological synthesis and the best organic and other bio-nano materials that, in combination with CNTs, can lengthen the life of food after packaging.

2.5 DRAWBACKS OF NANOTUBES AS PACKAGING SYSTEMS

One of the primary disadvantages of carbon nanotubes is their inability to dissolve in water. To address this issue, the surface of CNTs has been modified through fictionalization with various hydrophilic molecules and chemistries, which have improved the solubility and biocompatibility of CNTs in water. In addition to improving mechanical and barrier qualities, graphene and carbon nanotube nanoparticles can enhance the functioning of biodegradable polymers for use in packaging applications. On the other hand, the growth in the production and use of nanomaterials raises the likelihood of human and environmental exposure to such nanoparticles in the future.

Modern research is focusing on the use of different types of carbon nanotubes and graphene in polymer nanocomposites for the development of smart, active, and intelligent packaging that can improve the quality and safety of food, solve the problem of food storage, and provide information to the consumer about the quality of packaged food. Nanoscale materials such as graphene and MWCNTs in biodegradable polymers are among the most persistent and potentially valuable nanoscale materials to have emerged in recent years. They are increasingly being investigated for their potential to improve the thermal, mechanical, barrier, and functionality of food packaging materials. Therefore, more nanoscience research efforts are required that concentrate solely on the possible danger posed by nanomaterials such as graphene and CNTs, given the rising exposure of consumers and the environment to nanoparticle-containing packaging. In addition, the specific nanoscale size and shape of graphene and CNTs, with a high aspect ratio and large surface area, will increase the risk of their mobility in the environment. Graphene and carbon nanotubes are airborne, non-soluble in water, and absorptive in soil, increasing the risk of their mobility in the environment (Kotsilkov et al., 2018).

Bioplastic packaging is commonly used these days for wrapping items ranging from food to electronics, to protect these from dust, germs, and water vapor and to extend the shelf life of the products they contain. Polyactic acid (PLA) is one of the many different biopolymers employed. It has the mechanical properties and cost-effectiveness required for biodegradable food packaging. On the other hand, PLA packaging has poor water vapor and oxygen barrier qualities compared to various petroleum-derived alternatives. The inclusion or encapsulation of graphene and CNTs into PLA packaging to concurrently improve both the water vapor and oxygen barrier qualities is a significant problem in PLA packaging.

The incorporation of nanomaterials (NMs) into the formulation of food packaging has provided new and more efficient options for preserving food quality and extending food shelf life, all of which have had a significant positive impact on the sensory perceptions of customers, as well as causing a reduction in food waste and, as a result, an improvement in economic performance (Alfei et al., 2020). Regardless, despite clear demonstration of the benefits that could result from the widespread use of CNTs in the food industry, and because the application of NMs-based food packaging on an industrial scale is constantly increasing, it is necessary to emphasize the importance of conducting short- and long-term toxicity studies, both in terms of their impact on the environment and on humans, to ensure the health of consumers and to shape sustainable development.

It has to be stated that the main disadvantage of the use of CNTs in the food packaging industry is their extensive use and the merchandizing of them without enough evidence regarding the potential toxicity of CNTs to humans or the environment.

2.6 CONCLUSIONS

The use of nanotubes and nanotechnological developments in the food industry can significantly enhance food quality and safety. Excellent new materials with exciting characteristics can completely transform today's food business and food markets. However, the number of times people are exposed to engineered NMs from food packaging materials increases. Unfortunately, this is not accompanied by the appropriate level of awareness and limits prescribed by safety regulations regarding the material's toxicological properties. The findings of several studies indicate that the commercial use of nanotechnologies in food must be treated with more caution and be subjected to a more rigorous evaluation.

The novel food packaging based on nanotechnology targets a tiny global market niche and allows it to claim all rights in the food sector. Although the food nanotechnology business is still in its infancy, the amount of yearly research and investment in the area is continuously increasing, both in academic and industrial institutions. To produce nanoparticles or CNTs for food packaging, there are several technical and functional obstacles to overcome. These include food safety requirements, standardization, and a lack of skilled workers. A significant knowledge gap exists regarding nanomaterial migrations from packaging into food matrices, possible interactions between nanomaterials and biological systems, and the concentrations of NMs in the environment, all of which have implications for human health and the environment. A better understanding of the toxicity characteristics of NMs and their physicochemical properties concerning various modes of exposure to humans and the environment is required to develop appropriate and reliable techniques for their identification, characterization, and quantification in complex food matrices, as well as disposal strategies for them.

On the other hand, innovative food packaging will be the future instrument for revolutionizing global agricultural production and food chain systems. Many elements of innovative food packaging are desired in the food packaging sector, ranging from simple quality control of food goods to biosecurity and safety applications. This kind of packaging is critical in this time of global food insecurity, significant postharvest losses, population growth throughout the world, climate change, and rising competition for water for agricultural and home use, among other factors.

It is essential to participate more actively in the research and future adaptation of innovative packaging technologies that will aid in the reduction of food loss in food supply chains. This is especially true during the supply chain's storage, processing, packaging, transportation, and distribution phases.

ACKNOWLEDGMENT

This work was supported by 'Ciencia Básica SEP-CONACyT' project 287225, 'Fondo FONCYT-COECYT-Convocatoria 2019-C13, Efecto de Nanopartículas de Uso Agrícola Sobre el Desarrollo de la Planta de Maíz (Zea mays L.) y las Propiedades Fisicoquímicas y Biológicas del Suelo (COAH-2019-C13-C006)', 'Fondo FONCYT-COECYT-Convocatoria 2021-C15, Bio- y nano-remediación de suelo y agua, contaminados con metales pesados, del municipio de San Juan de Sabinas, Coahuila de Zaragoza, México, (COAH-2021-C15-C095)', CONACyT-SINANOTOX PN-2017-01-4710, and the Sustainability of Natural Resources and Energy Programs from Cinvestav-Saltillo.

REFERENCES

Ababneh, H., and Hameed, B. H. 2021. Chitosan-derived hydrothermally carbonized materials and its applications: A review of recent literature. *International Journal of Biological Macromolecules* 186: 314–327.

Abdullah, Z. W., Dong, Y., Han, N., and Liu, S. 2019. Water and gas barrier properties of polyvinyl alcohol (PVA)/starch (ST)/glycerol (GL)/halloysite nanotube (HNT) bionanocomposite films: Experimental characterisation and modelling approach. *Composites Part B: Engineering* 174: 107033.

Alfei, S., Marengo, B., and Zuccari, G. 2020. Nanotechnology application in food packaging: A plethora of opportunities versus pending risks assessment and public concerns. *Food Research International* 137: 109664.

Almeida, A. C. S., Franco, E. A. N., Peixoto, F. M., Pessanha, K. L. F., and Melo, N. R. 2015. Application of nanothecnology in food packaging. *Polímeros* 25: 89–97.

Alves, Z., Abreu, B., Ferreira, N. M., Marques, E. F., Nunes, C., and Ferreira, P. 2021. Enhancing the dispersibility of multiwalled carbon nanotubes within starch-based films by the use of ionic surfactants. *Carbohydrate Polymers* 273: 118531.

Anzar, N., Hasan, R., Tyagi, M., Yadav, N., and Narang, J. 2020. Carbon nanotube—A review on synthesis, properties and plethora of applications in the field of biomedical science. *Sensors International* 1: 100003.

Arfat, Y. A., Ahmed, J., Ejaz, M., and Mullah, M. 2018. Polylactide/graphene oxide nanosheets/clove essential oil composite films for potential food packaging applications. *International Journal of Biological Macromolecules* 107: 194–203.

Armenta-Armenta, M. E., and Espinoza-Gomez, J. H. 2020. Membranas compuestas a base de poliestireno con nanotubos de carbono y éteres corona. *Rev. Aristas Cienc. Básica Aplicada* 7: 366–370.

Asgari, P., Moradi, O., and Tajeddin, B. 2014. The effect of nanocomposite packaging carbon nanotube base on organoleptic and fungal growth of Mazafati brand dates. *International Nano Letters* 4(1): 98.

Azizi, S., Azizi, M., and Sabetzadeh, M. 2019. The role of multiwalled carbon nanotubes in the mechanical, thermal, rheological, and electrical properties of PP/PLA/MWCNTs nanocomposites. *Journal of Composites Science* 3(3): 64.

Barra, A., Ferreira, N. M., Martins, M. A., Lazar, O., Pantazi, A., Jderu, A. A., Neumayer, S. M., Rofriguez, B. J., Enăchescu, M., Ferreira, P., and Nunes, C. 2019. Eco-friendly preparation of electrically conductive chitosan—Reduced graphene oxide flexible bionanocomposites for food packaging and biological applications. *Composites Science and Technology* 173: 53–60.

Boonsiriwit, A., Xiao, Y., Joung, J., Kim, M., Singh, S., and Lee, Y. S. 2020. Alkaline halloysite nanotubes/low density polyethylene nanocomposite films with increased ethylene absorption capacity: Applications in cherry tomato packaging. *Food Packaging and Shelf Life* 25: 100533.

Cedillo, J. J., Castañeda, A. O., Solanilla, J. F., Esparza, S. C., Acosta, D. E., and Sáenz, A. 2019. Obtención de nanomateriales a base de quitosano/nanotubos de carbono de pared múltiple. *Revista Iberoamericana De Polímeros* 20(2): 54–60.

Carvalho, A. P. A., and Conte Junior, C. A. 2020. Green strategies for active food packagings: A systematic review on active properties of graphene-based nanomaterials and biodegradable polymers. *Trends in Food Science & Technology* 103: 130–143.

Chaudhary, P., Fatima, F., and Kumar, A. 2020. Relevance of nanomaterials in food packaging and its advanced future prospects. *Journal of Inorganic and Organometallic Polymers and Materials* 30(12): 5180–5192.

Cui, R., Jiang, K., Yuan, M., Cao, J., Li, L., Tang, Z., and Qin, Y. 2020. Antimicrobial film based on polylactic acid and carbon nanotube for controlled cinnamaldehyde release. *Journal of Materials Research and Technology* 9(5): 10130–10138.

Deshpande, P., and Mahendru, A. 2018. A review of single wall carbon nanotube: Structure and preparation. *International Journal of Scientific and Technology Research* 7: 132–135.

Dias, M. V., Soares, N., de, F. F., Borges, S. V., de Sousa, M. M., Nunes, C. A., de Oliveira, I. R. N., and Medeiros, E. A. A. 2013. Use of allyl isothiocyanate and carbon nanotubes in an antimicrobial film to package shredded, cooked chicken meat. *Food Chemistry* 141(3): 3160–3166.

Díaz-Visurraga, J., Meléndrez, M., García, A., Paulraj, M., and Cárdenas, G. 2010. Semitransparent chitosan-TiO2 nanotubes composite film for food package applications. *Journal of Applied Polymer Science* 116(6): 3503–3515.

Eskalen, H., Çeşme, M., Kerli, S., and Özğan, Ş 2021. Green synthesis of water-soluble fluorescent carbon dots from rosemary leaves: Applications in food storage capacity, fingerprint detection, and antibacterial activity. *Journal of Chemical Research* 45(5–6): 428–435.

Fan, Q., Wang, G., Tian, D., Ma, A., Wang, W., Bai, L., Chen, H., Yang, L., Yang, H., Wei, D., and Yang, Z. 2021. Self-healing nanocomposite hydrogels via Janus nanosheets: Multiple effects of metal–coordination and host–guest interactions. *Reactive and Functional Polymers* 165: 104963.

Feng, Z., Li, L., Wang, Q., Wu, G., Liu, C., Jiang, B., and Xu, J. 2019. Effect of antioxidant and antimicrobial coating based on whey protein nanofibrils with TiO2 nanotubes on the quality and shelf life of chilled meat. *International Journal of Molecular Sciences* 20(5): 1184.

Gaikwad, K. K., Singh, S., and Negi, Y. S. 2020. Ethylene scavengers for active packaging of fresh food produce. *Environmental Chemistry Letters* 18(2): 269–284.

Han, J., Feng, Y., Liu, Z., Chen, Q., Shen, Y., Feng, F., Liu, L., Zhong, M., Zhai, Y., Bockstaller, M., and Zhao, Z. 2021. Degradable GO-nanocomposite hydrogels with synergistic photothermal and antibacterial response. *Polymer* 230: 124018.

Hosseini, H., and Jafari, S. M. 2020. Introducing nano/microencapsulated bioactive ingredients for extending the shelf-life of food products. *Advances in Colloid and Interface Science* 282: 102210. https://doi.org/10.1016/j.cis.2020.102210

Iqbal, S., Nadeem, S., Bano, R., Bahadur, A., Ahmad, Z., Javed, M., Al-Anazy, M. M., Qasier, A. A., Laref, A., Shoaib, M., Liu, G., and Qayyum, M. A. 2021. Green synthesis of biodegradable terpolymer modified starch nanocomposite with carbon nanoparticles for food packaging application. *Journal of Applied Polymer Science* 138(25): 50604.

Joung, J., Boonsiriwit, A., Kim, M., and Lee, Y. S. 2021. Application of ethylene scavenging nanocomposite film prepared by loading potassium permanganate-impregnated halloysite nanotubes into low-density polyethylene as active packaging material for fresh produce. *LWT* 145: 111309.

Kavoosi, G., Dadfar, S. M. M., Dadfar, S. M. A., Ahmadi, F., and Niakosari, M. 2014. Investigation of gelatin/multi-walled carbon nanotube nanocomposite films as packaging materials. *Food Science & Nutrition* 2(1): 65–73.

Khanna, S., and Islam, N. 2018. Carbon nanotubes-properties and applications. *Organic and Medicinal Chemistry International Journal* 7: 555705.

Kim, J., Yoon, C. S., Na, J. H., and Han, J. 2019. Prolonged insecticidal activity of clove oil-loaded halloysite nanotubes on *Plodia interpunctella* infestation and application in industrial-scale food packaging. *Journal of Food Science* 84(9): 2520–2527.

Konwar, A., Kalita, S., Kotoky, J., and Chowdhury, D. 2016. Chitosan–Iron oxide coated graphene oxide nanocomposite hydrogel: A robust and soft antimicrobial biofilm. *ACS Applied Materials & Interfaces* 8(32): 20625–20634.

Kotsilkov, S., Ivanov, E., and Vitanov, N. K. 2018. Release of graphene and carbon nanotubes from biodegradable poly(lactic acid) films during degradation and combustion: Risk associated with the end-of-life of nanocomposite food packaging materials. *Materials* 11(12): 2346.

Kumar, K. P., and Kumar, V. R. 2019. Carbon nano tube in polymer nanocomposites. *Materials Today: Proceedings* 18: 4067–4073.

Landa-Salgado, P., Cruz-Monterrosa, R. G., Hernández-Guzmán, F. J., and Reséndiz-Cruz, V. 2017. Nanotecnología en la industria alimentaria: bionanocompuestos en empaques de alimenticios. *Agroproductividad* 10(10): 34–40.

Lee, M. H., and Park, H. J. 2015. Preparation of halloysite nanotubes coated with Eudragit for a controlled release of thyme essential oil. *Journal of Applied Polymer Science* 132(46): 42771.

Li, F., Yu, H.-Y., Wang, Y.-Y., Zhou, Y., Zhang, H., Yao, J.-M., Abdalkarim, S. Y. H., and Tam, K. C. 2019. Natural biodegradable poly(3-hydroxybutyrate-co-3-hydroxyvalerate) nanocomposites with multifunctional cellulose nanocrystals/graphene oxide hybrids for high-performance food packaging. *Journal of Agricultural and Food Chemistry* 67(39): 10954–10967.

Li, Q., Ren, T., Perkins, P., Hu, X., and Wang, X. 2021. Applications of halloysite nanotubes in food packaging for improving film performance and food preservation. *Food Control* 124: 107876.

Liu, Y., Wang, S., Lan, W., and Qin, W. 2019. Fabrication of polylactic acid/carbon nanotubes/chitosan composite fibers by electrospinning for strawberry preservation. *International Journal of Biological Macromolecules* 121: 1329–1336.

Lomate, G. B., Dandi, B., and Mishra, S. 2018. Development of antimicrobial LDPE/Cu nanocomposite food packaging film for extended shelf life of peda. *Food Packaging and Shelf Life* 16: 211–219.

Lugani, Y., Sooch, B. S., Singh, P., and Kumar, S. 2021. 8—Nanobiotechnology applications in food sector and future innovations. In Ray, R. C. (Ed.), *Microbial Biotechnology in Food and Health*, 197–225. Academic Press.

Luo, J., Sun, W., Zhou, H., Zhang, Y., Wen, B., and Xin, C. 2021. Bioderived and biodegradable poly(3-hydroxybutyrate-co-3-hydroxyvalerate) nanocomposites based on carbon nanotubes: Microstructure observation and EMI shielding property improvement. *ACS Sustainable Chemistry & Engineering* 9(32): 10785–10798.

Magnin, Y., Amara, H., Ducastelle, F., Loiseau, A., and Bichara, C. 2018. Entropy-driven stability of chiral single-walled carbon nanotubes. *Science* 362: 212–215.

Makgabutlane, B., Nthunya, L. N., Maubane-Nkadimeng, M. S., and Mhlanga, S. D. 2021. Green synthesis of carbon nanotubes to address the water-energy-food nexus: A critical review. *Journal of Environmental Chemical Engineering* 9(1): 104736.

Manikandan, N. A., Pakshirajan, K., and Pugazhenthi, G. 2020. Preparation and characterization of environmentally safe and highly biodegradable microbial polyhydroxybutyrate (PHB) based graphene nanocomposites for potential food packaging applications. *International Journal of Biological Macromolecules* 154: 866–877.

Mitura, K., Kornacka, J., Kopczyńska, E., Kalisz, J., Czerwińska, E., Affeltowicz, M., Kaczorowski, W., Kolesińska, B., Frączyk, J., Bakalova, T., Svobodová, L., and Louda, P. 2021. Active carbon-based nanomaterials in food packaging. *Coatings* 11(2): 161.

Murillo-Vargas, F., Jiménez-Villalta, G., Esquivel-Alfaro, M., and Vega-Baudrit, J. R. 2021. Ácido l-poliláctico (PLA) y na-notubos de carbono de pared múltiple (NTCPM) con potenciales aplicaciones industriales. *Revista Colombiana De Química* 50(1): 20–39.

Nile, S. H., Baskar, V., Selvaraj, D., Nile, A., Xiao, J., and Kai, G. 2020. Nanotechnologies in food science: Applications, recent trends, and future perspectives. *Nano-Micro Letters* 12(1): 45.

Ojeda, G. A., Arias Gorman, A. M., Sgroppo, S. C., Ojeda, G. A., Arias Gorman, A. M., and Sgroppo, S. C. 2019. Nanotecnología y su aplicación en alimentos. *Mundo Nano. Revista Interdisciplinaria En Nanociencias y Nanotecnología* 12(23): 1e–14e.

Omerović, N., Djisalov, M., Živojević, K., Mladenović, M., Vunduk, J., Milenković, I., Knežević, N. Z., Gadjanski, I., and Vidić, J. 2021. Antimicrobial nanoparticles and biodegradable polymer composites for active food packaging applications. *Comprehensive Reviews in Food Science and Food Safety* 20(3): 2428–2454.

Pattanshetti, A., Pradeep, N., Chaitra, V., and Uma, V. 2020. Synthesis of multi-walled carbon nanotubes (MWCNTs) from plastic waste & analysis of garlic coated gelatin/MWCNTs nanocomposite films as food packaging material. *SN Applied Sciences* 2(4): 730.

Primožič, M., Knez, Ž., and Leitgeb, M. 2021. Nanotechnology in food science—food packaging. *Nanomaterials* 11(2): 292–321.

Portillo, J. J. C., Facio, A. O. C., González, S. C. E., and Galindo, A. S. 2020. Estudio del entrecruzamiento de películas a base de quitosano con glicerina y la integración de nanotubos de carbono de pared múltiple. *Afinidad: Revista De Química Teórica y Aplicada* 77(590): 151–157.

Rezic, I., Haramina, T., and Rezic, T. 2017. 15—Metal nanoparticles and carbon nanotubes—Perfect antimicrobial nano-fillers in polymer-based food packaging materialsa. In Grumezescu, A. M. (Ed.), *Food Packaging*, 497–532. Academic Press.

Riahi, Z., Priyadarshi, R., Rhim, J.-W., Hong, S.-I., Bagheri, R., and Pircheraghi, G. 2021. Titania nanotubes decorated with Cu(I) and Cu(II) oxides: Antibacterial and ethylene scavenging functions to extend the shelf life of bananas. *ACS Sustainable Chemistry & Engineering* 9(19): 6832–6840.

Risyon, N. P., Othman, S. H., Basha, R. K., and Talib, R. A. 2020. Characterization of polylactic acid/halloysite nanotubes bionanocomposite films for food packaging. *Food Packaging and Shelf Life* 23: 100450.

Salgado-Cruz, M., de la, P., Salgado-Cruz, J., García-Hernández, A. B., Calderón-Domínguez, G., Gómez-Viquez, H., Oliver-Espinoza, R., Fernández-Martínez, M. C., and Yáñez-Fernández, J. 2021. Chitosan as a coating for biocontrol in postharvest products: A bibliometric review. *Membranes* 11(6): 421.

Sanmartín-Santos, I., Gandía-Llop, S., Salesa, B., Martí, M., Lillelund Aachmann, F., and Serrano-Aroca, Á 2021. Enhancement of antimicrobial activity of alginate films with a low amount of carbon nanofibers (0.1% w/w). *Applied Sciences* 11(5): 2311.

Sarojini, K. S., Indumathi, M. P., and Rajarajeswari, G. R. 2019. Mahua oil-based polyurethane/chitosan/nano ZnO composite films for biodegradable food packaging applications. *International Journal of Biological Macromolecules* 124: 163–174.

Shahbazi, M., Rajabzadeh, G., and Sotoodeh, S. 2017. Functional characteristics, wettability properties and cytotoxic effect of starch film incorporated with multi-walled and hydroxylated multi-walled carbon nanotubes. *International Journal of Biological Macromolecules* 104: 597–605.

Sharma, C., Dhiman, R., Rokana, N., and Panwar, H. 2017. Nanotechnology: An untapped Resource for food packaging. *Frontiers in Microbiology* 8: 1735.

Sobhan, A., Muthukumarappan, K., Wei, L., Van Den Top, T., and Zhou, R. 2020. Development of an activated carbon-based nanocomposite film with antibacterial property for smart food packaging. *Materials Today Communications* 23: 101124.

Suppiah, K., Leng, T. P., Husseinsyah, S., Rahman, R., Keat, Y. C., and Heng, C. 2019. Thermal properties of carboxymethyl cellulose (CMC) filled halloysite nanotube (HNT) bio-nanocomposite films. *Materials Today: Proceedings* 16: 1611–1616.

Tulsyan, G., Richter, C., and Diaz, C. A. 2017. Oxygen scavengers based on titanium oxide nanotubes for packaging applications. *Packaging Technology and Science* 30(6): 251–256.

Vatandost, E., Ghorbani-HasanSaraei, A., Chekin, F., Naghizadeh Raeisi, S., and Shahidi, S.-A. 2020. Green tea extract assisted green synthesis of reduced graphene oxide: Application for highly sensitive electrochemical detection of sunset yellow in food products. *Food Chemistry: X* 6: 100085.

Vázquez-Moreno, J. M., Sánchez-Hidalgo, R., Sanz-Horcajo, E., Viña, J., Verdejo, R., and López-Manchado, M. A. 2019. Materiales compuestos jerárquicos. *AEMAC* 3: 1–5.

Velichkova, H., Petrova, I., Kotsilkov, S., Ivanov, E., Vitanov, N. K., and Kotsilkova, R. 2017. Influence of polymer swelling and dissolution into food simulants on the release of graphene nanoplates and carbon nanotubes from poly(lactic) acid and polypropylene composite films. *Journal of Applied Polymer Science* 134(44): 45469.

Venkateshaiah, A., Havlíček, K., Timmins, R. L., Röhrl, M., Wacławek, S., Nguyen, N. H. A., Černík, M., Padil, V. V. T., and Agarwal, S. 2021. Alkenyl succinic anhydride modified tree-gum kondagogu: A bio-based material with potential for food packaging. *Carbohydrate Polymers* 266: 118126.

Wang, L.-F., and Rhim, J.-W. 2017. Functionalization of halloysite nanotubes for the preparation of carboxymethyl cellulose-based nanocomposite films. *Applied Clay Science* 150: 138–146.

Wang, Y., Yi, S., Lu, R., Sameen, D. E., Ahmed, S., Dai, J., Qin, W., Li, S., and Liu, Y. 2021. Preparation, characterization, and 3D printing verification of chitosan/halloysite nanotubes/tea polyphenol nanocomposite films. *International Journal of Biological Macromolecules* 166: 32–44.

Wen, Y.-H., Tsou, C.-H., de Guzman, M. R., Huang, D., Yu, Y.-Q., Gao, C., Zhang, X.-M., Du, J., Zheng, Y.-T., Zhu, H., and Wang, Z.-H. 2021. Antibacterial nanocomposite films of poly(vinyl alcohol) modified with zinc oxide-doped multiwalled carbon nanotubes as food packaging. *Polymer Bulletin* 2021: 1–20.

Wu, C.-D., Fang, T.-H., and Tung, F.-Y. 2017. Interface friction of double-walled carbon nanotubes investigated using molecular dynamics. *Micromachines* 8(3): 84.

Xu, D. 2018. Carbon nanotubes (CNTs) composite materials and food packaging. In *Composites Materials for Food Packaging* (pp. 235–249). John Wiley & Sons, Ltd.

Yu, H.-Y., Qin, Z.-Y., Sun, B., Yang, X.-G., and Yao, J.-M. 2014. Reinforcement of transparent poly(3-hydroxybutyrate-co-3-hydroxyvalerate) by incorporation of functionalized carbon nanotubes as a novel bionanocomposite for food packaging. *Composites Science and Technology* 94: 96–104.

Yu, Z., Li, B., Chu, J., and Zhang, P. 2018. Silica in situ enhanced PVA/chitosan biodegradable films for food packages. *Carbohydrate Polymers* 184: 214–220.

Zhang, S., Wang, Q., Li, D., and Ran, F. 2021. Single-walled carbon nanotubes grafted with dextran as additive to improve separation performance of polymer membranes. *Separation and Purification Technology* 254: 117584.

Zhong, Y., Godwin, P., Jin, Y., and Xiao, H. 2020. Biodegradable polymers and green-based antimicrobial packaging materials: A mini-review. *Advanced Industrial and Engineering Polymer Research* 3(1): 27–35.

Zubair, M., and Ullah, A. 2020. Recent advances in protein derived bionanocomposites for food packaging applications. *Critical Reviews in Food Science and Nutrition* 60(3): 406–434.

3 Nanocellulose for Food Packaging Applications

Kiran Jeet[1], *Ramandeep Kaur*[2], *and Manpreet Kaur*[3]
[1]Electron Microscopy and Nanoscience Laboratory, Punjab Agricultural University, Ludhiana, Punjab, India
[2]Department of Mathematics, Statistics and Physics, Punjab Agricultural University, Ludhiana, Punjab, India
[3]Department of Chemistry, Punjab Agricultural University, Ludhiana, Punjab, India

CONTENTS

3.1 INTRODUCTION

Cellulose is quite possibly the most plentiful and widely available regular polymer accessible on this planet. It is a significant source of renewable material, primarily in the form of plant fibre. The average production per year of cellulose, as reported by Habibi (2014), was in excess of 7.5×10^{10} tons. Cellulose has benefited humans in numerous ways for thousands of years, as a source of fibrous materials or composite materials, such as wood, despite the fact that people just recently learned about its chemical composition, structure, and molecular conformation. Because of

DOI: 10.1201/9781003207641-4

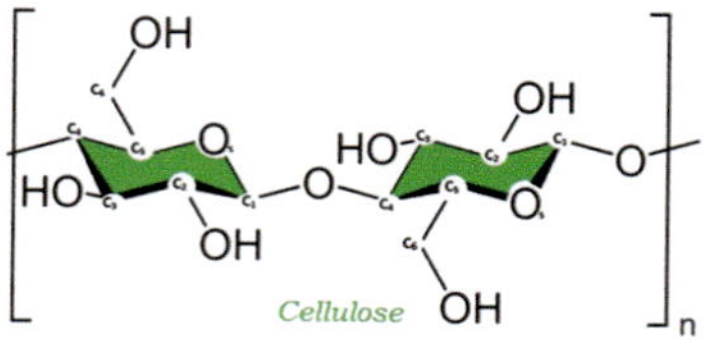

FIGURE 3.1 Schematic illustration of cellulose.

its non-polluting nature, its non-toxicity, biodegradability, biocompatibility, ease of modification, and renewability, cellulose has been utilised for varied applications benefiting our day today life to a significant extent.

A French chemist, Anselme Payen, in 1838 extracted cellulose (a chemical substance related to polysacchrides) from a plant source and determined its chemical composition. Cellulose, being a natural polymer, is made up of repeated segments of D-anhydroglucopyranose linked by 1,4 glycosidic links at C1 and C4 positions forming a cell-wall skeleton. Figure 3.1 gives schematic illustration of cellulose structure. A molecule of cellulose is made up of a reducing end group at C1 position with a free hemiacetal or aldehyde, a non-reducing end group at C4 position with a free hydroxyl group, and internal glucose rings connected at C1 and C4 positions (Giri et al., 2021). Because of the large chain lengths, the internal glucose units predominate. There are three -OH groups in each internal anhydroglucopyranose unit. The -OH at position C6 is a primary alcohol, whereas the -OH present at positions C2 and C3 are secondary alcohols (Kadla and Gilbert, 2000). It is the hydrogen bonding between -OH groups and the presence of O_2 atoms in nearby ring molecules that provide stability to the linkage and help to form linear shape cellulose chain. The position of "-OH" group present in the cellulose structure provides all possible locations from where the cellulose can be chemically modified to successfully alter/modulate its structure. The presence of hydroxyl group at the C6 location is reported to be the most reactive and effective position at which the structure alteration can be performed (Roy et al., 2009). The high specific strength, high modulus of elasticity, and good crystalline properties of cellulose are due to presence of -OH groups in it.

On the basis of raw material utilised for manufacturing of cellulose, the level of polymerisation (no. of D- anhydroglucopyranose units present in a polymer) varies widely. It ranges from 10,000 to 15,000 (John and Thomas, 2008), depending upon the source from which the cellulose is derived. It is reported that approximately 10,000 units of cellulose are obtained from glucose derived from wood (Fang et al., 2020) and 15,000 units are from cotton-derived glucose (Sjostrom, 1993).

Regarding its origin, cellulose is the abundant biopolymer on this planet. It exists in the cell walls of all green plants as their structural component. The cell walls of plants also contain hemicellulose, lignin, and a tiny quantity of extractives in addition to cellulose. Hemicelluloses and lignin are amorphous in form, whereas cellulose has a lot of crystalline areas. During the extraction method, the non-cellulosic components of fibres are removed. Chemicals are easily absorbed and hydrolysed in amorphous regions, but the homogeneity of crystalline regions prevents chemical absorption (Klemm et al., 2006). Despite the fact that wood species remain the primary source of cellulose, non-wood plants are becoming a more popular biomass source for production of cellulose. Because of their inexpensive cost and lower lignin content, they have gathered significant interest within the fabrication of cellulose.

3.2 SOURCES OF CELLULOSE

There are numerous sources from which cellulose can be obtained, such as wood pulp (Ang et al., 2020), cotton fibre (Bharimalla et al., 2017), jute (Rahman et al., 2014), hemp (Manian et al., 2021), corn stalk (Zheng et al., 2012), sisal fibres (Moran et al., 2008), kenaf (Sulaiman et al., 2015), flax (Malyushevskaya et al., 2021), rice straw (Ibrahim et al., 2013), arecanut husk (Perumal et al., 2021), saw dust (Wan BaderulHisan and Mohd Amin, 2017), nypa fruticans trunk (Nang An et al., 2020), maize husk (Gbenga et al., 2013), wheat straw (Liu et al., 2019), linen, coconut husk fibre (Rosa et al., 2010), potato tubers (Dufresne et al., 2000), banana rachis (Velazquez et al., 2020), bamboo (Okahisa and Sakata, 2019), sugarcane bagasse (Oliveira et al., 2016), marine animals (Brinchi et al., 2013), algae (Chen et al., 2016), fungi (Imran et al., 2016), bacteria (Raghunathan, 2013), and invertebrates (Linton, 2020). Figure 3.2 indicates different types of cellulose material classified on the basis of their origin.

3.2.1 Plants and Plant Residues

Plants are the most likely to be used as the raw materials of cellulose since they are widely available and relatively inexpensive. Wood mash and cotton strands are the primary sources of cellulose (Klemm et al., 2005). Cellulose content in cotton fibres is more than 90% (Koh, 2011) while bast fibres, wood, and herbaceous fibres contain only 70–75%, 45–50%, and 30–40% of cellulose, respectively. Cotton fibres have the greatest molecular mass and composition of all plant fibres, which results in production of highly crystalline, orientated, and fibrillar cellulose material (Hsieh, 2007). In the fibre layer underneath the bark of dicotyledenous plants, bast filaments produce packages, or strands, that act as hawsers. (Cook, 2001). Cotton is the most

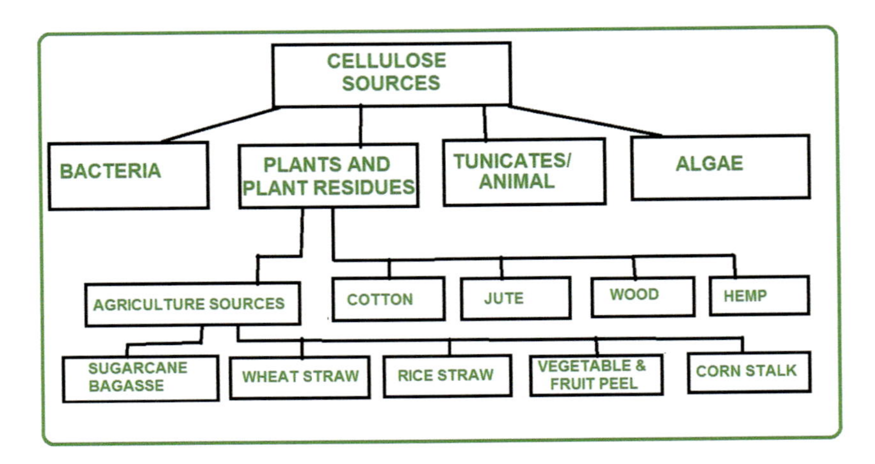

FIGURE 3.2 Classification of different sources for extraction of cellulose.

commonly utilised material for making cellulose; however, a vast assortment of different materials, like jute and ramie, are also notable hotspots for cellulose creation (George and Sabapathi, 2015). Other plants that create cellulose incorporate hydrophytes, grasses, and leaves and stems of other plants. Agricultural waste products like rice and wheat straw, sugarcane bagasse, sawdust, cotton, and other crop waste are also used to make cellulose. Harvesting, processing, and extraction are currently possible using large-scale industrial infrastructure and services.

3.2.2 Bacteria

Certain bacteria, such as *Komagataeibacter xylinus*, found as an adulterant during fermentation of vinegar are well recognised for synthesising cellulose from a wide range of nitrogen and carbon sources (Masaoka et al., 1993). *K. xylinus* generates cellulose microfibrils that float on the surface of the growing media as transparent, horizontal, and broad pellicles (George et al., 2005). These cellulose pellicles are made out of pure cellulose, a considerable amount of water, and other medium elements. The contaminants in the cellulose pellicle can be hydrolysed and removed using dilute alkaline solutions. Acid treatment (alkali) and washings are performed on dried and processed cellulose pellicles to obtain pure cellulose membranes. Microbially produced cellulose is distinct from plant-derived cellulose in terms of purity, distinct structure at nanoscale, higher dimensional stability, better mechanical strength, and enhanced water holding capacity (Brown, 1989). As for the subatomic recipe and polymeric construction, microbial cellulose is like plant cellulose; notwithstanding, the two structures contrast in the course of action of glycosyl units inside the unit cells of the crystallites, resulting in a more noteworthy crystallinity in the microbial cellulose. Microbial cellulose has a higher degree of polymerisation and better characteristics than does plant cellulose.

3.2.3 Tunicates

Being marine invertebrates, tunicates are widely recognised for generating a huge amount of cellulose. These creatures have a thick, rugged mantle that is high in cellulose. The tunicate tissue use cellulose as a skeletal structure (Zhao and Li, 2014). Cellulose is made by these creatures using chemical edifices found in the epidermal layer. There are numerous tunicate species in nature, and the qualities of the cellulose produced vary. Although the construction and attributes of cellulose microfibrils produced by different species are by and large indistinguishable, minor varieties in the cellulose microfibril creation interaction can affect the microfibrils' definitive qualities. *Halocynthia roretzi* (Sugiyama et al., 1991; Yuan et al., 2006;), *Halocynthia papillosa* (Helbert et al., 1998), *and Metandroxarpa uedai* (Moon et al., 2011) are among the most researched species.

3.2.4 Algae

Many algae have cellulose as a key component of their cell walls, and some pigmented algae are also known to produce cellulose. However, of the algae, green algae are more favoured for cellulose extraction. Cladophorales and Siphonocladales are two groups of cellulose-producing algae (Mihranyan, 2011). The cellulose manufactured from Valonia or Cladophora has extremely high degree of crystallinity—up to 95% (George and Sabapathi, 2015). The characteristics of cellulose microfibrils produced may vary depending on the biosynthetic technique used by various species.

3.3 NANOCELLULOSE

Production of cellulose at nanoscale (called nanocellulose) from varied sources has gained remarkable attention due to their use for development of new materials. The cellulose nanoparticles, or nanofibres, can be recovered from fibrils by acid (sulphuric) hydrolysis of delignified hemicellulose-free cellulose. Alkali treatment of raw materials used for fabrication of cellulose makes it hemicellulose free. Then, acid hydrolysis breaks down the cell walls. After removal of hemicellulose, the material is given a bleaching treatment to extract the lignin present in it. The insoluble residue left after these treatments is cellulose. The leftover cellulose is again acid hydrolysed to obtain nanocellulose.

3.3.1 Synthesis of Nanocellulose

There are a number of source materials that can be utilised for the production of nanocellulose (NC). For each source material, there are various approaches that are suitable for synthesising nanocellulose: acid hydrolysis, ionic liquid medium, and enzymatic route are commonly used methods. To eliminate amorphous regions from crystalline regions, acid hydrolysis is done, as crystalline regions are more resistant to hydrolysis than are amorphous regions. Crystalline regions exhibit the properties desired in CNCs. The preparation of nanocellulose from ionic liquid medium involves three steps: a) alkali treatment of raw fibre, b) bleaching treatment, and c) dissolving the cellulose obtained in prepared ionic liquid. Enzymatic hydrolysis has been shown to be a potential alternative to chemical catalysis for obtaining NC, as the process is not only suitable with biological approaches but it also results in production of a nanostructure that exhibits excellent thermal stability and provides the possibility of being put to use easily (De Aguiar et al., 2020). An underlying mechanical (crushing, processing) or physicochemical (autohydrolysis, alkali, bleaching) pretreatment to remove cellulose is typically used in studies describing the use of raw materials that have not yet been processed (disposing of hemicellulose, lignin, and different parts) (Michelin et al., 2020). The percentage yield is calculated as per the equation mentioned below.

$$\%\ \text{Yield} = [\text{Dry weight(nanocellulose)(g)/Dry weight(crude fibre)(g)}] \times 100$$

3.3.2 Types of Nanocellulose

Depending on the material used for the fabrication of cellulose, nanocellulose can be divided into three main categories: cellulose nanocrystals (CNCs), cellulose nanofibrils (CNFs), and micro fibrillated cellulose (MFC). In addition to the three major

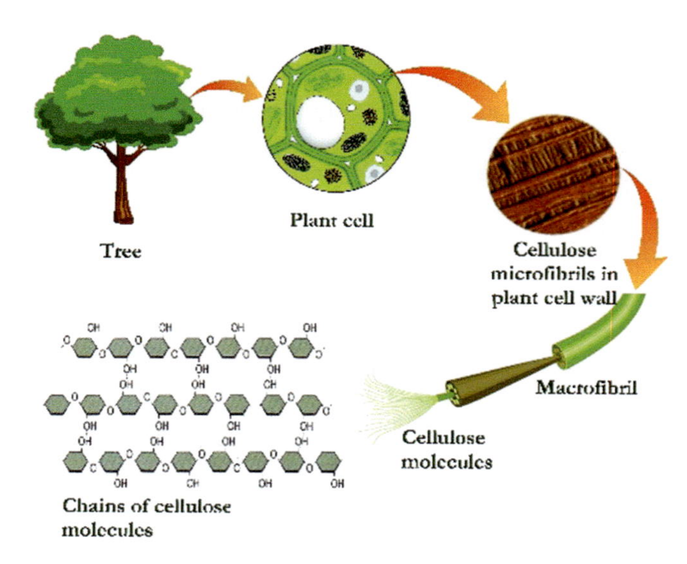

FIGURE 3.3 Schematic illustration of how the cellulose is extracted from the source and how it further gets converted into nanocellulose.

categories, two more types, bacterial nanocellulose (BNC) and spherical nanocellulose (SNC), are recently also being discussed. Figure 3.3 shows the schematic illustration of production of nanocellulose.

3.3.2.1 Cellulose Nanocrystals (CNCs)

Bulk cellulose has a mixture of highly arranged crystalline areas and some disordered (amorphous) parts, in different amounts (Newman and Hemmingson, 1995) depending upon the source from which it is extracted. The exceptionally translucent bits of the cellulose microfibrils can be removed when these microfibrils are added to a suitable combination of mechanical, synthetic, and compound medicines, bringing about the development of cellulose nanocrystals. CNCs are unbending, pole-like particles made from practically immaculate translucent cellulose chain sections. Nebulous areas that are reliably scattered along the microfibrils are for the most part removed using solid corrosive hydrolysis. Strong acids can easily enter low-order amorphous areas and hydrolyse them while leaving the crystalline regions undisturbed (Marchessault et al., 1961). A significant decline in the degree of polymerisation was seen as an outcome of acid hydrolysis, which reached a threshold known as the level-off degree of polymerisation (LODP). CNCs have a critical polydispersity in sub-atomic weight and no sign of the LODP, which may be attributable to the absence of a normal dispersion of the undefined areas (Habibi et al., 2010).

3.3.2.2 Cellulose Nanofibrils (CNFs)

Cellulose nanofibrils are crystalline with amorphous sections that are 3 nm wide and with lengths in the micron range (Usmani et al., 2018). They are made out of cellulose microfibres. Mechanical fibrillation of cellulose biomass produces CNFs (Liu et al., 2016). CNFs were first produced in 1983 (Turbak et al. and Herrick et al.)

using high pressure homogenisation (HPH) to extract NFC from wood. Because the material is made from plant fibres, its manufacturing and disposal have a low environmental effect. It is lightweight, has an elastic modulus comparable to high-strength aramid fibre, thermal expansion comparable to glass, and excellent barrier capabilities against oxygen and other gases, among other qualities. Cellulose nanofibre is a material with potential application in a variety of sectors, including food-based applications, packaging material, filtration-based applications, electronics for device fabrication, pharmaceuticals and health care products, beauty products, and many more, due to its structure and physical characteristics (Dungani et al., 2017). CNFs produced from softwoods and hardwoods have been found to have crystallinity ranging from 40 to 80% in most cases (Nair et al., 2014). Regardless of the way that CNCs have a more noteworthy crystallinity than CNFs, CNCs have better O_2 penetrability than CNF. This is mainly due to expanded ensnarement of CNFs inside film, which builds gas particles dispersion course.

3.3.2.3 Bacterial Nanocellulose (BNC)

Bacterial cellulose is a nanofibrillar extracellular polysaccharide generated by a wide range of bacteria when they grow statically and also when they are immersed in liquid and cultivated by agitation. Bacteria create bacterial cellulose (BNC) on a variety of carbon-rich media; however, the efficiency of BNC synthesis varies significantly among different growth substrates. During the exhausting, energy-consuming route of cellulose synthesis, the substrate provides energy to bacterial metabolism (Ward, 2015). Every carbon atom that the bacterial cell metabolises into glucose can theoretically be utilised to make cellulose. BNC is a nanofibrillar substance possessing a unique set of characteristics, such as extended crystallinity in the range 84–89% (Lee et al., 2021) with high polymerisation degree; a large aspect ratio of fibres, with an average range 20–100 nm; greater flexibility and tensile strength, with average value 200–300 MPa and Young's modulus of 15–30 GPa; and the best of all is the excellent water-holding capacity, with an average estimation of about 100 times of its own weight (Amara et al., 2021). BNC is a non-cytotoxic, non-genotoxic, and extremely biocompatible substance due to its great purity, which includes the lack of lignin and hemicellulose. BNC is produced in bacterial membrane. Bacteria then transport BNC as fibrils comprised of D-glucose units joined by 1,4-glycosidic linkages via the pores of the cell membrane. The chain is expelled from the cell and is linear (Ullah et al., 2019).

3.3.2.4 Microfibrillated Cellulose (MFC)

Wood mash filaments were utilised as the primary material for the manufacture of a fibrillated substance known as microfibrillated cellulose (MFC). MFC can be used as a rheological modifier in a variety of areas, ranging from food commodities to pharmaceuticals, coatings and distempers, cosmetics, and pharmaceuticals (Turbak et al., 1983; Herrick et al., 1983). As a feature of the creation of cellulose–built up nanocomposites, research has been done by various authors (Cash et al., 1999; Cavaille et al., 1997; Gousse et al., 2002; Grunert and Winter, 2002; Kim et al., 2002; Sassi and Chanzy, 1995), grafting with polyethylene glycol (Araki et al., 2001) and surfactant adsorption (Bonini et al., 2002; Heux et al., 2000). Ross Colvin, and Sowden (1985) described a beating-based homogenisation technique for exposing the structure of cellulose fibres.

3.3.2.5 Spherical Nanocellulose (SNC)

To create spherical particles, the material must be freely accessible during the hydrolysis. Acid hydrolysis is used to make spherical nanocellulose (SNC), which is the same process used to make CNC. The diameter of SNC is generally between 10 and 200 nanometres. It's been hypothesised that these spheres are made up of self-assembled short rods because of their great proclivity for forming aggregates. Due of their comparable characteristics, spherical cellulose has been proposed as a possible replacement for CNC in many goods (Zhanga et al., 2007; Lu and Hsieh, 2010).

3.3.3 Characterisation of Nanocellulose

The structure, morphology, and detailed size of the nanocellulose are determined through high resolution imaging (HRI), such as transmission electron microscope (TEM) and scanning electron microscope (SEM). The smaller nanocellulose particles can be better viewed under TEM and atomic force microscopy (AFM). The chemical constitution of cellulose samples can be determined using Fourier transform infrared spectroscopy (FTIR). The major objective of IR spectroscopy is elucidation of the structure of a sample, including identification of functional groups present in the sample. X-ray diffraction analysis is generally performed to identify interplanar distances of crystal structures and characteristic parameters of the unit cell.

Theoretical method is applied to determine the average size of small crystallites based on the Scherrer equation (Scherrer, 1918).

$$D = K\lambda / (B \cos \theta_{hkl})$$

B: width of the peak at the half of maximum height (in radians),
θ_{hkl}: diffraction angle of the peak

Shape factor K is usually taken close to 1.

Nuclear magnetic resonance (NMR) spectroscopy is employed to study various parameters of nanocrystallites, and also used to decipher the amorphous crystalline contents in cellulose and nanocellulose samples. In addition to techniques mentioned, Raman spectroscopy (RS) is used to study the crystallinity of the sample. The index of crystallinity is calculated from the ratio of the RS-peaks height at 380 cm^{-1} and 1096 cm^{-1} (Agarwal et al., 2010). Figure 3.4 gives a summary of the characterisation technique used to decipher the structure of nanocellulose.

The surface charge, stability, and particle size of nanocellulose samples are identified using DLS (dynamic light scattering) and ZP (zeta potential) measurements. DLS and ZP measurements are essential features when the nanocellulose material is used for particular applications. In case of packaging-related applications, this characterisation becomes necessary while synthesising the composite of nanocellulose materials with other nanomaterial fillers. Three thermoanalytical procedures—thermogravimetric analysis, differential scanning calorimetry, and differential thermal analysis—are used in cellulose studies to identify physical and chemical phenomena of samples (Azubuike and Okhamafe, 2012; Morais et al., 2013; Yates, 1972).

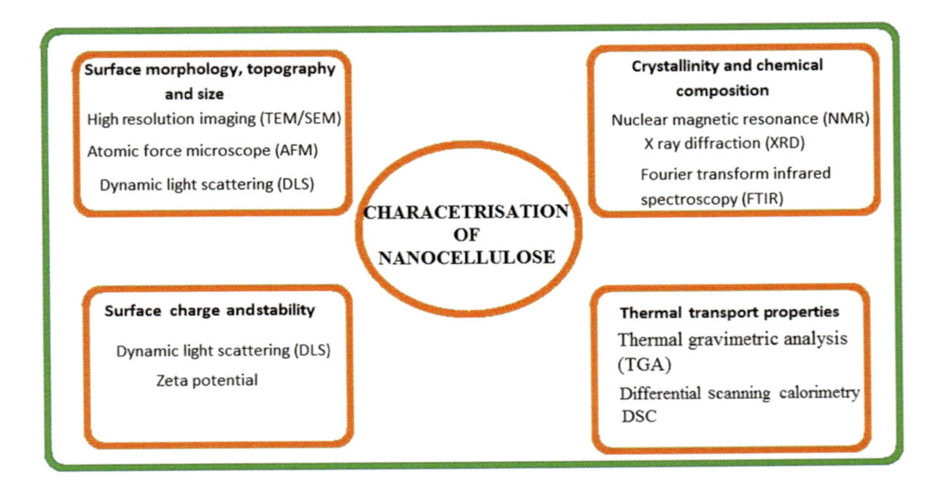

FIGURE 3.4 Varied characterisation tools used for deciphering the nanocellulose materials.

Thermo-physical characteristics of the samples can also be obtained from thermoanalytical techniques (Ioelovich and Luksa, 1990; Kaimin and Ioelovich, 1981; Mochalov et al., 1982).

3.4 APPLICATION OF NANOCELLULOSE IN FOOD PACKAGING

Food packing science is a part of science which studies the characteristics of packaging materials, food packaging needs, and the packaging system using concepts from four key branches of science (Yam and Takhistov, 2016). Material science and food science have both made significant contributions to the creation of food packaging. Understanding the mechanical, physical, and chemical properties of various materials and their composites requires knowledge of material science (Etcheverry and Barbosa, 2012). Food science is critical for understanding food degradation kinetics and determining longevity of foods.

Packing is crucial in the food supply chain. Food packaging currently makes up the majority of the entire packaging industry (85%). As reported by Huang et al., the worldwide packaging market grew from a slightly more than a $42 billion value in 2014 to a value of nearly $48 billion in 2020 (Huang et al., 2019). The prerequisites of packing systems for new, frozen, dehydrated, warm, or aseptically handled items are dictated by intrinsic properties, extrinsic properties, and the necessary time span for usability of the food item. The physical and mechanical characteristics of the packaging materials, too, are crucial during processing, packaging processes, and supply chain handling (Wang et al., 2018).

Plastics are the most often used packaging materials because of their merits of being economical, lightweight, easily processed, and having excellent mechanical and barrier characteristics (Sangroniz et al., 2019). The majority of packaging materials includes synthetic plastics derived from either fossil fuels or various synthetic polymers. Even though the plastic carries many advantages, in terms of environment safety they pose a high risk. Their accumulation on our planet is becoming a matter of great concern to the whole world. Incineration is one of the methods of reducing plastic waste build-up, but it is costly and risky, releasing carbon dioxide and toxic chemicals into the environment (Vilarinho et al., 2017). The non-biodegradable nature of plastic-based materials makes them a less acceptable material for food packaging in the desirable eco-friendly world. Taking into consideration the ever-increasing plastic-based pollution, the European Union's Directive (EU) 2018/852 has set a base recycling objective for all packaging waste and plastic at 70 and 55%, respectively, by the end of 2030 (The European Parliament and the Council of European Union, 2018).

Development of food packaging materials involves three major steps; surface cleaning/etching, functionalising/activation, and deposition/coating (Pankaj et al., 2014). Surface cleaning involves elimination of undesired adulterants and impurities from the surface of packing materials (Cruz et al., 2010). The next step is the surface/polymer modification, which is performed by functionalising the surface of the material with a desired functional group (Vasile et al., 2013). Surface/polymer modification is done to increase the wettability, sealability, printability, and dye-uptake adhesion capability of the packaging material. Surface deposition, often known as coating, alters the substrate by depositing a thin layer on the packing material.

The rapidly growing packaging needs of the modern life style, coupled with environmental safety concern, has put pressure on researchers to find materials which meet both demands, within an affordable cost. Biopolymers are the class of materials able to satisfy both of these concerns by offering recyclability along with a biodegradable nature. Biopolymers can be classified into many groups: polymers straightforwardly extricated from biomass, polymers made through chemical synthesis using renewable bio-based monomers, and polymers created by microorganisms. Although these polymers have a natural origin, they are not acceptable from an application perspective due to their poor intrinsic properties.

Biopolymers have shortcomings, such as poor mechanical properties, they cannot withstand high processing conditions and high temperature, and they have poor barrier properties (high gas and vapour permeability). To bring them to the packaging market and make them a competitor for petrochemical-based films, it was necessary that structural changes should be made to improve their mechanical and barrier properties. Integrating nanomaterial in the evolution of packing material is considered to be one of the key factors for improving the shortcomings of biopolymers. Nanotechnology has the potential to have an impact on the packaging industry by slowing oxidation and regulating moisture migration, microbiological development, respiration rates, and volatile tastes and smells of packaging materials (Brody et al., 2008). In nanotechnology, the term "nanocomposite" is commonly used for a material containing nanosized filler, which could be particles, tubes, or fibres and platelets intercalated in a polymer matrix. Incorporation of nanosized material in polymer makes it mechanically stronger and more robust without compromising the flexibility. Nanocomposite-based food packaging materials possess high mechanical strength, are light weight, and have better heat resistance and barrier ability against O_2, CO_2, UV radiation, moisture, and volatiles. They also help reduce the role of preservatives, which are utilised for extending longevity of food commodities by creating a barrier against air and enzymes.

In addition to nanocomposites, the term *bionanocomposite* is often seen in the literature, referring to material containing a nanocomposite embedded in natural polymer. Bionanocomposite is considered to be the top material for the packaging industry, as it has the ability to control the respiratory exchanges, thus offering material to the food packaging industry that can significantly extend shelf life. There are numerous natural materials that are commonly used with nanocomposite materials to synthesis films. Films synthesised with these natural materials not only exhibit the characteristics of fundamentally ideal packaging films but they can also safely enhance the serviceable life of food commodities. Figure 3.5 indicates the advantages of using bionanocomposites in food packaging materials. They can upgrade the standard of fresh, frozen, and processed meat, as well as poultry, and seafood products.

Cellulose is a linear polymer of anhydroglucose, and it is the most bounteous naturally occurring polymer on our planet. Due to its hydrophilic nature, insolubility, and crystalline structure, cellulose-based film preparation is considered to be tedious task. In order to prepare cellulose films, the material is made miscible in a blend of acid, such as NaOH and CS_2, and then further in a sulphuric acid solution. The use of sulphuric acid as a hydrolysing agent allows implantation of anionic sulphate ester groups through the process of esterification. In this process the sulphuric acid reacts

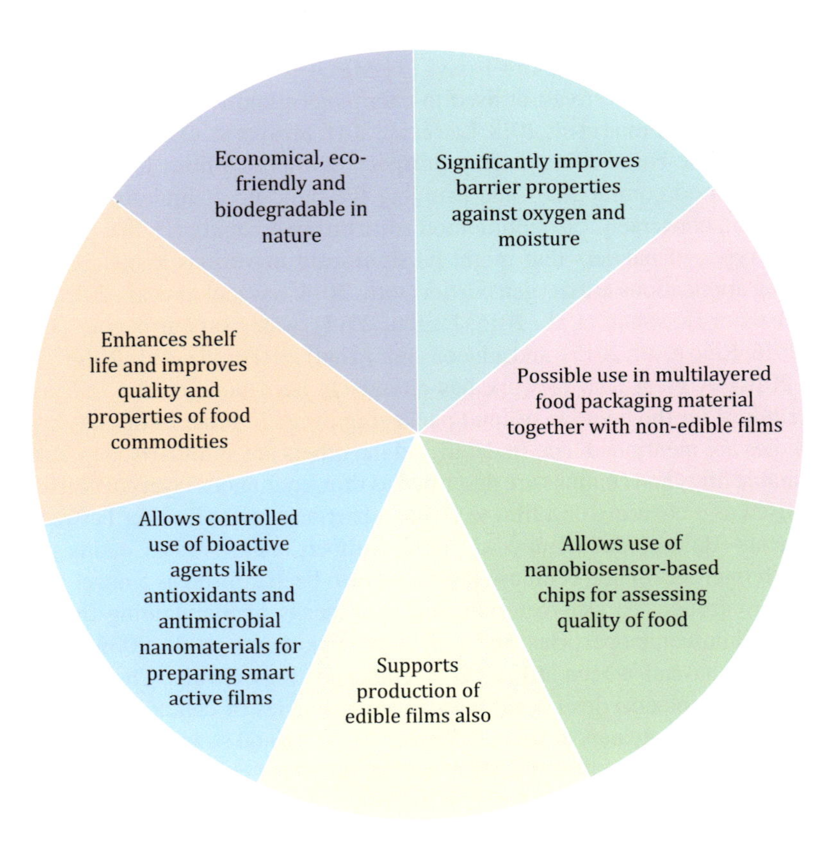

FIGURE 3.5 Advantages and role of bionanocomposites material in food packaging films.

with cellulose through the surface hydroxyl groups (Fortunati et al., 2014). The films produced are hydrophilic, or in other words are moisture sensitive, and are further covered with nitrocellulose wax or poly-vinylidene chloride. Due to the ample availability of cellulose and the economical extraction of it from natural resources, it is considered as a potential base material for synthesis of composite films. Plant cellulose, such as paper and board; Cellophane™; and modified cellulose derivatives, such as cellulose acetate, methylcellulose (MC), hydroxypropyl cellulose (HPC), hydroxypropyl methylcellulose (HPMC), and carboxymethylcellulose (CMC), have a long history of use in food packaging (Paunonen, 2013).

Nanocellulose is an emerging reliable and renewable green substrate for food packaging applications because of its unique characteristics, such as adjustable surface chemistry, barrier qualities, mechanical strength, crystallinity, biodegradability and non-toxicity, and high aspect ratio (Ferrer et al., 2017; Lacroix et al., 2014). The nanocellulose parameters, such as crystallinity, degree of polymerisation (DP), fibre diameter, and length, are influenced by the extraction/production technique, which are important in defining the mechanical and barrier properties (Soykeabkaew et al., 2017). Charreau et al. (2013) looked at the history of nanocellulose patents and identified three key moments: a) 1960s – the nanocellulose process was developed, b) 1980s – it was used as a food ingredient, and c) 2000s – edible nanoemulsions and nanopackaging were developed. Assilane grafting; acetylation; alkylation; esterification; 2,2,6,6-tetramethylpiperidin-1-oxyl (TEMPO) oxidation; carboxymethylation; and the nanoparticles were all utilised to alter nanocellulose in order to achieve the planned approach (Frone et al., 2018; Lee et al., 2011; Shao et al., 2017; Tsai et al., 2018).

Nanocellulose-based films display prospects and possibilities for extending the shelf life of food commodities and avoiding the build-up of undesirable odour or impurities or, conversely, preventing food scents from escaping (Hubbe et al., 2017). The four types of barriers that might be significant in various nanocellulose film packaging applications are oxygen barrier (Iotti, 2014; Savadekar et al., 2012), oil and grease barrier (Kisonan et al., 2015; Raghu, 2015), water vapour barrier (Lundahl et al., 2016; Rojo et al., 2015), and aqueous liquid barrier (Pereda et al., 2014; Shimizu et al., 2016). A great deal of work has effectively been done on the utilisation of nanocellulose for the creation of food packaging films. In order to discuss briefly, a few studies are mentioned here in detail and the others are summarised in Table 3.1.

Palatable films and coatings are described as thin, continuous layers of edible material utilised as a covering or a film to create a barrier to mass transfer between food components (Balasubramaniam et al., 1997; Guilbert et al., 1997). Coating polymer films with microfibrillated cellulose layers has also been utilised as a novel approach to making excellent barrier materials and as a means of maintaining the benefits of both cellulose nanoparticles and polymers (Fukuzumi et al., 2009). Cellulose nanoparticles have also been utilised as an edible coating on fruits. The few reported applications are blueberries coated with aqueous slurries of cellulose nanoparticles and calcium carbonate nanoparticles (Zhao et al., 2014), and strawberries coated with a combination of 1% chitosan and 5% nanocllulose (Dong et al., 2015).

Arrieta et al. (2015) found that the multifunctional PLA-PHB/cellulose nanocrystal films have better thermal stability in comparison to polylactic acid (PLA)–polyhydroxybutyrate (PHB) blends. Also, their thermogravimetric analysis (TGA)

TABLE 3.1
List of Work Performed on Synthesis of Nanocellulose-Based Food Packaging Films

Cellulose Derivative	Functional Group	Raw Source	Extraction Method of Cellulose	Method Used to Prepare Nanofilms	References
Cellulose nanofibers	Nisin	Mixture of Spruce (60%) and pine (40%)	TEMPO-oxidation	Grafting method	Saini et al., 2016
Cellulose nanofibres	Chitosan	----------	----------	------------	Sundaram et al., 2016
Cellulose nanofibres	Polylactic acid	Jute fibre	HNO_3-$NaNO_2$ oxidation	Solvent casting	Kumar et al., 2019
Cellulose nanocrystals	Plastisized starch	Wood	----------	------------	Svagan et al., 2009
Cellulose nanocrystals	Alginate	Microcrystalline cellulose	H_2SO_4 hydrolysis	Solution casting method	Abdollahi et al., 2013
Cellulose nanofibers	Polyvinyl alcohol/ polypyrrole	Kraft wood pulp	TEMPO-oxidation	Chemical polymerization	Bideau et al., 2016
Cellulose nanocrystals	Chitosan nanoparticles/ polyvinyl alcohol	Mango waste	Acid hydrolysis	Solvent casting	Dey et al., 2021
Cellulose nanocrystals methyl ester	poly(3-hydroxybutyrate-co-3-hydroxyvalerate)	Microcrystalline cellulose	HCOOH/ HCl hydrolysis	Solution casting method	Yu et al., 2014
Cellulose nanocrystals/ graphene oxide	poly(3-hydroxybutyrate-co-3-hydroxyvalerate)	Microcrystalline cellulose	citric acid /hydrochloric acid hydrolysis	Grafting	Li et al., 2019
Cellulose nanofibers	Tannin films	Kraft pulp	High shear mixing	----------	Missio et al., 2018
Cellulose nanofibers	Silver nanoparticles	Wood pulp	----------	Reduction with $NaBH_4$	Yu et al., 2019
Cellulose nanocrystals	Starch	Sugarcane bagasse	Sulphuric acid hydrolysis	Solution casting	Slavutsky and Bertuzzi, 2014
Cellulose nanofibers	Polylactic acid	Wheat straw	Acid hydrolysis	Grafting, coating	Song et al., 2014
Cellulose nanocrystals	Polyethylene terephthalate and oriented polyamide	Cotton linters	Sulphuric acid hydrolysis	Coating	Li et al., 2013

revealed that the plasticised PLA–PHB sample deteriorated in two steps, in comparison to the single-step process involved in reinforced cellulose nanocrystals due to its improved interaction between all components in the nanocomposites. Their finding indicates that the multifunctional plasticised PLA–PHB blends reinforced with modified cellulose nanocrystal are a potential food packaging material. Yano et al. (2014) developed another patented technique that uses nanocellulose as a stabiliser. They utilised nanocellulose to help frozen sweets keep their form for longer time duration. Hydroxypropylmethyl cellulose (HPMC) has been discovered to be extremely compatible with chitosan in creating packaging that has both antibacterial and gas barrier characteristics (Imran et al., 2010; Sebti et al., 2007).

Three-layered film structures prepared with cellulose nanofibrils stacked over polyethylene terephthalate and extruded using low-density polyethylene (LDPE) were reported to exhibit better performance over commercially available ethylene vinyl alcohol multilayered barrier film in terms of barrier performance for use in packaging (Vartiainen et al., 2014). When compared to the pure starch film, bamboo generated nanocellulose/starch film that had the lowest penetration of water vapour and lowest oxygen permeability, with reductions of around 27.6 and 32.9%, respectively. The lower permeabilities were attributed to better crystallinity of the bamboo-derived nanocellulose compared to sisal and cotton-derived nanocellulose (Ahankari et al., 2020).

Benzyl acetic acid was grafted on the surface of the carbon nanofibres to reduce the release of active chemicals even further. Benzyl acetic acid is an absorbent that has long been utilised to enhance scent barrier characteristics, and it appears to be a viable choice for reducing active ingredient release. Using methyl hydroxide (MeOH) 95% as a food simulant and carvacrol as an active ingredient, the performance of hybrid composites as active packaging was tested. In comparison to neat polylactic acid (PLA), the hybrid PLA/CNF/C30B 5% exhibited a 90% reduction in oxygen transmission rate and a 74% reduction in water vapour transmission rate (Trifola et al., 2015). To develop films with enhanced barrier characteristics against water vapour movement, composite materials based on PLA and reinforced with bentonite nanoparticles (up to 4% w/w) nanocellulose were produced. Solution casting method was employed to synthesise polyvinyl alcohol/nanocellulose/silver nanocomposite films. The addition of nanoparticles of nanocellulose and silver improved the mechanical and thermal characteristics of the nanocomposite film. It also decreased the water vapour transmission rate, and antibacterial activity was enhanced against gram-positive and gram-negative bacteria. These characteristics made this film suitable for food packaging applications (Sarwara et al., 2017). De Moura et al. utilised a different type of cellulose to integrate silver nanoparticles into a hydroxypropyl methylcellulose (HPMC) matrix for food packaging applications. Antibacterial activities of the matrix were shown to be effective against *E. coli* and *S. aureus* (De Moura et al., 2012). The ternary nanocomposite of poly(3-hydroxybutyrate-*co*-3-hydroxyvalerate) (PHBV)/graphene oxide/cellulose nanocrystals with enhanced thermal and mechanical properties can also be used in food packaging applications (Li et al., 2019). The presence of graphene oxide significantly improves both the thermal and mechanical properties.

3.5 CONCLUSIONS

Bionanocompsites are gaining tremendous attention in the food technology sector for applications related to safe packaging of food commodities, with the additional benefit of extended shelf life. Numbers of natural polymers such as starch, cellulose, etc. are being used in this area. Cellulose is one of the most plentiful natural polymers and can offer the best features as a building block of packaging material. There are number of natural sources for extraction of cellulose, including agricultural waste. The cellulose and nanocellulose obtained from agricultural waste are gaining much attention as the packaging prepared from these base materials will be eco-friendly and recyclable, and it will provide the additional benefit of circular bioeconomy. Nanocellulose in any of its forms is capable of making durable packaging films (both edible and non-edible) that not only keep food fresh but also extend its shelf life without use of any preservatives.

REFERENCES

Abdollahi, M., Alboofetileh, M., Behrooz, M., and Miraki, R. 2013. Reducing water sensitivity of alginate bio-nanocomposite film using cellulose nanoparticles. *International Journal of Biological Macromolecules* 54: 166–173.

Agarwal, U. P., Reiner, R. S., and Ralph, S. A. 2010. Cellulose i crystallinity determination using FT raman spectroscopy. *Cellulose* 17(4): 721–733.

Ahankari, S. S., Subhedar, A. R., Bhadauria, S. S., and Dufresne, A. 2020. Nanocellulose in food packaging: A review. *Carbohydrate Polymers* 255: 117479.

Amara, C., Mahdi, A. E., Medimagh, R., and Khwaldia, K. 2021. Nanocellulose-based composites for packaging applications. *Current Opinion in Green and Sustainable Chemistry* 31: 100512.

Ang, S., Haritos, V., and Batchelor, W. 2020. Cellulose nanofibers from recycled and virgin wood pulp: A comparative study of fiber development. *Carbohydrate Polymers* 234: 115900.

Araki, J., Wada, M., and Kuga, S. 2001. Steric stabilization of a cellulose microcrystal suspension by poly(ethylene glycol) grafting. *Langmuir* 17: 21–27.

Arrieta, M. P., Fortunati, E., Dominici, F., Rayon, E., Lopez, J., and Kenny, J. M. 2015. Bionanocomposite films based on plasticized PLA-PHB/cellulose nanocrystal blends. *Carbohydrate Polymers* 121: 265–275.

Azubuike, C. P., and Okhamafe, A. O. 2012. Physicochemical, spectroscopic and thermal properties of microcrystalline cellulose derived from corn cobs. *International Journal of Recycling of Organic Waste in Agriculture* 1(9): 1–7.

Balasubramaniam, V. M., Chinnan, M. S., Mallikarjunan, P., and Philips, R. D. 1997. The effect of edible film on oil uptake and moisture retention of deep-fat fried poultry product. *Journal of Food Process Engineering* 20(1): 17–29.

Bharimalla, A., Deshmukh, S., Patil, P., and Nadanathangam, V. 2017. Micro/nano-fibrillated cellulose from cotton linters as strength additive in unbleached kraft paper: Experimental, semi-empirical, and mechanistic studies. *Bio Resources* 12(3): 5682–5696.

Bideau, B., Bras, J., Saini, S., Daneault, C., and Loranger, E. 2016. Mechanical and antibacterial properties of a nanocellulose-polypyrrole multilayer composite. *Materials Science and Engineering, C* 69: 977–984.

Bonini, C., Heux, L., Cavaille, J. Y., Lindner, P., Dewhurst, C., and Terech, P. 2002. Rodlike cellulose whiskers coated with surfactant: A small-angle neutron scattering characterization. *Langmuir* 18: 3311–3314.

Brinchi, L., Cotana, F., Fortunati, E., and Kenny, J. M. 2013. Production of nanocrystalline cellulose from lignocellulosic biomass: Technology and applications. *Carbohydrate Polymers* 94(1): 154–169.

Brody, A. I., Bugusu, B., Jung, H. H., Sand, C. K., and McHugh, T. H. 2008. Innovative food packaging solutions. *Journal of Food Science* 73(8): R107–R116.

Brown, R. M. Jr. 1989. Bacterial cellulose. In Kennedy, Philips, Williams (eds.), *Cellulose: Structural and Functional Aspects.* 145–151. Chichester: Ellis Horwood.

Cash, M. J., Chan, A. N., and Conner, H. T., et al. 1999. US Patent 6:602 994.

Cavaille, J. Y., Chanzy, H., Fleury, E., and Sassi, J. F. 1997. US Patent 6:117 545.

Charreau, H., Forestí, M. L., and Vazquez, A. 2013. Nanocellulose patents trends: A comprehensive review on patents on cellulose nanocrystals, microfibrillated and bacterial cellulose. *Recent Patents on Nanotechnology* 7(1): 56–80.

Cook, J. G. 2001. *Handbook of Textile Fibres.* Natural Fibres, Vol. 1. Cambridge: Woodhead.

Chen, Y. W., Lee, H. V., Juan, J. C., and Phang, S. M. 2016. Production of new cellulose nanomaterial from red algae marine biomass *gelidium elegans. Carbohydrate Polymers* 151: 1210–1219.

Cruz, S. A., Zanin, M., Nascente, P. A., and Bica de Moraes, M. A. 2010. Superficial modification in recycled PET by plasma etching for food packaging. *Journal of Applied Polymer Science* 115: 2728–2733.

De Aguiar, J., Bondancia, T. J., Claro, P. I. C., Mattoso, L. H. C., Farinas, C. S., and Marconcini, J. M. 2020. Enzymatic deconstruction of sugarcane bagasse and straw to obtain cellulose nanomaterials. *ACS Sustainable Chemistry and Engineering* 8: 2287–2299.

De Moura, M. R., Mattoso, L. H. C., and Zucolotto, V. 2012. Development of cellulose-based bactericidal nanocomposites containing silver nanoparticles and their use as active food packaging. *Journal of Food Engineering* 109: 520–524.

Dey, D., Dharini, V., and Periyar, S. S., et al. 2021. Physical, antifungal, and biodegradable properties of cellulose nanocrystals and chitosan nanoparticles for food packaging application. *Materials Today* 38(2):860–869.

Dong, F., Li, S. J., Liu, Z. M., and Zhu, K. X. 2015. Improvement of quality and shelf life of strawberry with nanocellulose/chitosan composite coatings. *Bangladesh Journal of Botany* 44:709–717.

Dufresne, A., Dupeyre, D., and Vignon, M. R. 2000. Cellulose microfibrils from potato tuber cells: Processing and characterization of starch-cellulose microfibril composites. *Journal of Applied Polymer Science* 76(14): 2080–2092.

Dungani, R., Abdul Khalil, H. P. S., and Aprilia, N. A., et al. 2017. 3 - Bionanomaterial from agricultural waste and its application. *Cellulose-Reinforced Nanofibre Composites,* 45–88. Woodhead Publishing Series in Composites Science and Engineering, Elsevier, Malaysia. https://www.sciencedirect.com/science/article/pii/B9780081009574000036

Etcheverry, M., and Barbosa, S. E. 2012. Glass fiber reinforced polypropylene mechanical properties enhancement by adhesion improvement. *Materials* 5(12): 1084–1113.

Fang, Z., Li, B., and Liu, Y., et al. 2020. Critical role of degree of polymerization of cellulose in super-strong nanocellulose films. *Matter* 2(4):1000–1014.

Ferrer, A., Pal, L., and Hubbe, M. 2017. Nanocellulose in packaging: Advances in barrier layer technologies. *Industrial Crops and Products* 95: 574–582.

Fortunati, E., Luzi, F., and Puglia, D., et al. 2014. Investigation of thermo-mechanical, chemical and degradative properties of PLA-limonene films reinforced with cellulose nanocrystals extracted from Phormium tenax leaves. *European Polymer Journal* 56:77–91.

Frone, A. N., Panaitescu, D. M., and Chiulan, I., et al. 2018. Surface treatment of bacterial cellulose in mild, eco-friendly conditions. *Coatings* 8:221.

Fukuzumi, H., Saito, T., Iwata, T., Kumamoto, Y., and Isogai, A. 2009. Transparent and high gas barrier films of cellulose nanofibers prepared by TEMPO-mediated oxidation. *Biomacromolecules* 10(1): 162–165.

Gbenga, B. L., Gbemi, Q. W., and Oluyemisi, B. 2013. Evaluation of cellulose obtained from maize husk as compressed tablet excipient. *Der Pharmacia Lettre* 5(5): 12–17.

George, J., and Sabapathi, S. N. 2015. Cellulose nanocrystals: Synthesis, functional properties and applications. *Nanotechnology Science and Applications* 8: 45–54.

George, J., Ramana, K. V., Sabapathy, S. N., and Bawa, A. S. 2005. Physicomechanical properties of chemically treated bacterial (acetobacterxylinum) cellulose membrane. *World Journal of Microbiology and Biotechnology* 21: 1323–1327.

Giri, S., Dutta, P., Kumarasammy, D., and Giri, T. P. 2021. Chapter 1- Natural polysaccharides: Types, basic structure and suitability for forming hydrogels. *Plant and Algal Hydrogels for Drug Delivery and Regenerative Medicine*, 1–35, Elsevier, India.

Gousse, C., Chanzy, H., Excoffier, G., Soubeyrand, L., and Fleury, E. 2002. Stable suspensions of partially silylated cellulose whiskers in organic solvents. *Polymer* 43: 2645–2651.

Grunert, M., and Winter, W. T. 2002. Nanocomposites of cellulose acetate butyrate reinforced with cellulose nanocrystals. *Journal of Environmental Polymer Degradation* 10: 27–30.

Guilbert, S., Cuq, B., and Gontard, N. 1997. Recent innovations in edible and/or biodegradable packaging materials. *Food Additives and Contaminants* 14(6): 741–751.

Habibi, Y., Lucia, L. A., and Rojas, O. J. 2010. Cellulose nanocrystals: Chemistry, self-assembly and applications. *Chemical Reviews* 110: 3479–3500.

Habibi, Y. 2014. Key advances in the chemical modification of nanocelluloses. *Chemical Society Reviews* 43(5): 1519–1542.

Helbert, W., Nishiyama, Y., Okano, T., and Sugiyama, J. 1998. Molecular imaging of halocynthia papillosa cellulose. *Journal of Structural Biology* 124: 42–50.

Herrick, F. W., Casebier, R. L., Hamilton, J. K., and Sandberg, K. R. 1983. Microfibrillated cellulose: Morphology and accessibility. *Journal of Applied Polymer Science. Applied Polymer Symposium* 37: 797–813.

Heux, L., Chauve, G., and Bonini, C. 2000. Nonflocculating and chiral-nematic self-ordering of cellulose microcrystals suspensions in nonpolar solvents. *Langmuir* 16: 8210–8212.

Hsieh, Y. L. 2007. Chemical structure and properties of cotton. In Gordan, S., and Hsieh, Y. L. (ed.), *Cotton: Science and Technology*. Cambridge: Woodhead.

Huang, J., Dufresne, A., and Lin, N. 2019. *Nanocellulose: From Fundamentals to Advanced Materials*, Hoboken, NJ: John Wiley & Sons.

Hubbe, M. A., Ferrer, A., and Tyagi, P., et al. 2017. Nanocellulose in thin films, coatings and plies for packaging applications: A review. *Bioresources* 12(1): 2143–2233.

Ibrahim, M. M., El-Zawawy, W. K., Juttke, Y., Koschella, A., and Heinze, T. 2013. Cellulose and microcrystalline cellulose from rice straw and banana plant waste: Preparation and characterization. *Cellulose* 20: 2403–2416.

Imran, M., El-Fahmy, S., Revol-Junelles, A. M., and Desobry, S. 2010. Cellulose derivative based active coatings: Effects of nisin and plasticizer on physico-chemical and antimicrobial properties of hydroxypropyl methylcellulose films. *Carbohydrate Polymers* 81: 219–225.

Imran, M., Anwar, Z., Irshad, M., Asad, M. J., and Ashfaq, H. 2016. Cellulase production from species of fungi and bacteria from agricultural wastes and its utilization in industry: A review. *Advances in Enzyme Research* 4: 44–55.

Ioelovich, M., and Luksa, R. 1990. Change in the crystalline structure of cellulose in the process of thermal treatment. *Materials Science* 3:18.

Iotti, M. 2014. Aqueous coating composition useful as a coating layer, and as an oxygen barrier or smoothing layer, comprises nanocellulose, and at least one cationic surfactant. US Pat. US2015225590-A1.

John, M. J., and Thomas, S. 2008. Biofibres and biocomposites. *Carbohydrate Polymers* 71: 343–364.

Kadla, J. F., and Gilbert, R. D. 2000. Cellulose structure: A review. *Cellulose Chemistry and Technology* 34: 197–216.

Kaimin, I., and Ioelovich, M. 1981. *Methods of Cellulose Investigation*, Riga: Science.

Kim, D. Y., Nishiyama, Y., and Shigenori, K. 2002. Surface acetylation of bacterial cellulose. *Cellulose* 9: 361–367.

Kisonan, V., Prakobna, K., and Xu, C. L., et al 2015. Composite films of nanofibrillated cellulose o-acetyl galactoglucomannan (GGM) coated with succinic esters of GGM showing potential as barrier material in food packaging. *Journal of Materials Science* 50(8):3189–3199.

Klemm, D., Heublein, B., Fink, H. P., and Andreas, B. 2005. Cellulose: Fascinating biopolymer and sustainable raw material. *Angewandte Chemie International Edition* 44(22): 3358–3393.

Klemm, D., Schumann, D., Kramer, F., Hebler, N., Hornung, M., and Marsch, S. 2006. Nanocelluloses as innovative polymers in research and application. *Advances in Polymer Science* 205: 49–96.

Koh, J. 2011. Dyeing of cellulosic fibres. *Handbook of Textile and Industry Dyeing* 2: 129–146.

Lacroix, M., Criado, P., Fraschini, C., and Salmieri, S. 2014. Modification of nanocrystalline cellulose for bioactive loaded films. *Journal of Research Updates in Polymer Science* 3(2): 122–135.

Lee, K. Y., Quero, F., Blaker, J. J., Hill, C. A. S., Eichhorn, S. J., and Bismarck, A. 2011. Surface only modification of bacterial cellulose nanofibres with organic acids. *Cellulose* 18: 595–605.

Lee, S., Abraham, A., Lim, A. S., Choi, O., Seo, J. G., and Sang, B. I. 2021. Characterisation of bacterial nanocellulose and nanostructured carbon produced from crude glycerol by *komagataeibacter sucrofermentans*. *Biosource Technology* 342: 125918.

Li, F., Yu, H. Y., and Wang, Y. Y., et al. 2019. Natural biodegradable poly(3-hydroxybutyrate-*co*-3-hydroxyvalerate) nanocomposites with multifunctional cellulose Nanocrystals/Graphene oxide hybrids for high-performance food packaging. *Journal of Agriculture and Food Chemistry* 67(39):10954–10967.

Li, F., Biagioni, P., Bollani, M., Maccagnan, A., and Piergiovanni, L. 2013. Multi-functional coating of cellulose nanocrystals for flexible packaging applications. *Cellulose* 20: 2491–2504.

Linton, S. M. 2020. Review: The structure and function of cellulase (endo-β-1,4-glucanase) and hemicellulase (β-1,3-glucanase and endo-β-1,4-mannase) enzymes in invertebrates that consume materials ranging from microbes, algae to leaf litter. *Comparative Biochemistry and Physiology Part B: Biochemistry and Molecular Biology* 240: 110354.

Liu, C., Li, B., and Du, H., et al. 2016. Properties of nanocellulose isolated from corncob residue using sulphuric acid, formic acid, oxidative and mechanical methods. *Carbohydrate Polymers* 151:716–724.

Liu, Z., He, M., Ma, G., Yang, G., and Chen, J. 2019. Preparation and characterization of cellulose nanocrystals from wheat straw and corn stalk. *Journal of Korea Technical Association of the Pulp and Paper Industry* 51(2): 40–48.

Lu, P., and Hsieh, Y. L. 2010. Preparation and properties of cellulose nanocrystals: Rods, spheres, and network. *Carbohydrate Polymers* 82: 331–333.

Lundahl, M. J., Cunha, A. G., and Rojo, E., et al. 2016. Strength and water interactions of cellulose i filaments wet-spun from cellulose nanofibril hydrogels. *Scientific Reports* 6:30695.

Malyushevskaya, A. P., Malyushevskii, P. P., and Yushchishina, A. N. 2021. Extraction of cellulose from flax fiber by electric discharge cavitation. *Surface Engineering and Applied Electrochemistry* 57: 228–232.

Manian, A. P., Cordin, M., and Pham, T. 2021. Extraction of cellulose fibers from flax and hemp: A review. *Cellulose* 28: 8275–8294.

Marchessault, R. H., Morehead, F. F., and Koch, J. M. 1961. Some hydrodynamics properties of neutral suspensions of cellulose crystallites as related to size and shape. *Journal of Colloid Science* 16(4): 327–344.

Masaoka, S., Ohe, T., and Sakota, N. 1993. Production of cellulose from glucose by *acetobacter xylinum*. *Journal of Fermentation Bioengineering* 75: 18–22.

Michelin, M., Gomes, D. G., Romani, A., Polizeli, T. M., and Teixera, J. A. 2020. Nanocellulose production: Exploring the enzymatic route and residues of pulp and paper industry. *Molecules* 24: 3411.

Mihranyan, A. 2011. Cellulose From cladophorales green algae: From environmental problem to high-tech composite materials. *Journal of Applied Polymer Science* 119: 2449–2460.

Missio, A. L., Mattos, B. D., and Ferreira, D. F., et al. 2018. Nanocellulose-tannin films: From trees to sustainable active packaging. *Journal of Cleaner Production* 184:143–151.

Mochalov, A., Chlustova, T., Ioelovich, M., and Kaimin, I. 1982. Effect of the crystallinity degree on the heat capacity of cellulose. *Wood Chemistry* 4: 66–68.

Moon, R. J., Martini, A., Nairn, J., Simonsen, J., and Youngblood, J. 2011. Cellulose nanomaterials review: Structure, properties and nanocomposites. *Chemical Society Reviews* 40: 3941–3994.

Morais, J. P. S., Rosa, M. F., Filho, M. M. S., Nascimento, L. D., Nascimento, D. M., and Cassales, A. R. 2013. Extraction and characterization of nanocellulose structures from raw cotton linter. *Carbohydrate Polymers* 91: 229–235.

Moran, J. I., Alvarez, V. A., Cyras, V. P., and Vazquez, A. 2008. Extraction of cellulose and preparation of nanocellulose from sisal fibres. *Cellulose* 15: 149–159.

Nair, S. S., Zhu, J., Deng, Y., and Ragauskas, A. J. 2014. High performance green barriers based on nanocellulose. *Sustainable Chemical Processes* 2: 23.

Nang An, V., Chi Nhan, H., Tap, T. D., Thanh Van, T. T., Viet, P. V., and Hieu, L. V. 2020. Extraction of high crystalline nanocellulose from biorenewable sources of Vietnamese agricultural wastes. *Journal of Polymers and the Environment* 28(6): 1465–1474.

Newman, R. H., and Hemmingson, J. A. 1995. Carbon-13 NMR distinction between categories of molecular order and disorder in cellulose. *Cellulose* 2: 95–110.

Okahisa, Y., and Sakata, H. 2019. Effects of growth stage of Bamboo on the production of cellulose nanofibers. *Fibers and Polymers* 20: 1641–1648.

Oliveira, F. B., Bras, J., Pimenta, M. T. B., Silva Curvelo, A. A., and Belgacem, M. N. 2016. Production of cellulose nanocrystals from sugarcane bagasse fibres and pith. *Industrial Crops and Products* 93: 48–57.

Pankaj, S., Bueno-Ferrer, C., and Misra, N. N., et al. 2014. Applications of cold plasma technology in food packaging. *Trends in Food Science and Technology* 35(1): 5–17.

Paunonen, S. 2013. Strength and barrier enhancements of cellophane and cellulose derivative films: A review. *Bioresources* 8: 3098–3121.

Pereda, M., Dufresne, A., Aranguren, M. I., and Marcovich, N. E. 2014. Polyelectrolyte films based on chitosan/olive oil and reinforced with cellulose nanocrystals. *Carbohydrate Polymers* 101: 1018–1026.

Perumal, A. B., Nambiar, R. B., Sellamuthu, P. S., Sadiku, E. R., Li, X., and He, Y. 2021. Extraction of cellulose nanocrystals from areca waste and its application in eco-friendly biocomposite film. *Chemosphere* 287(2): 132084.

Raghunathan, D. 2013. Production of microbial cellulose from the new bacterial strain isolated from temple wash waters. *International Journal of Current Microbiology and Applied Sciences* 2(12): 275–290.

Raghu, S. 2015. Formation of nanocellulose composite used for forming packaging material, involves blending co polymer which is reaction product of hydrophilic monomer, cellulose-reactive monomer and amphiphobic monomer, with nanocellulose. US Pat. US20155072581-A1.

Rahman, M. M., Afrin, S., Haque, P., Islam, M. M., Islam, M. S., and Gafur, M. A. 2014. Preparation and characterization of jute cellulose crystals-reinforced poly(L-lactic acid) biocomposite for biomedical applications. *International Journal of Chemical Engineering*. Article ID 842147, 2014: 7. https://doi.org/10.1155/2014/842147

Rojo, E., Peresin, M. S., and Sampson, W. W., et al. 2015. Comprehensive elucidation of the effect of residual ligninon the physical, barrier, mechanical and surface properties of nanocellulose films. *Green Chemistry* 17:1853–1866.

Rosa, M. F., Medeiros, E. S., and Malmonge, J. A., et al. 2010. Cellulose nanowhiskers from coconut husk fibres: Effect of preparation conditions on their thermal and morphological behaviour. *Carbohydrate Polymers* 81(1):83–92.

Ross, C. J., and Sowden, L. C., 1985. The three-dimensional morphology of aggregates of native cotton cellulose microfibrils. *International Journal of Biological Macromolecules*, 7(4): 214–218.

Roy, D., Semsarilar, M., Gythrie, J. T., and Perrier, S. 2009. Cellulose modification by polymer grafting: A review. *Chemical Society Reviews* 38: 2046–2064.

Saini, S., Sillard, C., Belgacem, M. N., and Bras, J. 2016. Nisin anchored cellulose nanofibers for long term antimicrobial active food packaging. *RSC Advances* 6: 12422–12430.

Sangroniz, A., Zhu, J. B., Tang, X., Etxeberria, A., Chen, E. Y. X., and Sardon, H. 2019. Packaging materials with desired mechanical and barrier properties and full chemical recyclability. *Nature Communications* 10: 1–7.

Sarwara, M. S., Niazia, M. B. K., Jahana, Z., Ahmadb, T., and Hussaina, A. 2017. Preparation and characterization of PVA/nanocellulose/Ag nanocomposite films for antimicrobial food packaging. *Carbohydrate Polymers* 84: 453–464.

Sassi, J. F., and Chanzy, H. 1995. Ultrastructural aspects of the acetylation of cellulose. *Cellulose* 2: 111–127.

Savadekar, N. R., Karande, V. S., Vigneshwaran, N., Bharimalla, A. K., and Mhaske, S. T. 2012. Preparation of nano cellulose fibres and its application i kappa-carrageenan based film. *International Journal of Biological Macromolecules* 51(5): 1008–1013.

Scherrer, P. 1918. Bestimmung der grösse und der inneren struktur von kolloidteilchen mittels röntgenstrahlen. *Kolloidchemie Ein Lehrbuch*, 387–409.

Sebti, I., Chollet, E., Degraeve, P., Noel, C., and Peyrol, E. 2007. Water sensitivity, antimicrobial, and physicochemical analyses of edible films based on HPMC and/or chitosan. *Journal of Agricultural and Food Chemistry* 55: 693–699.

Shao, W., Wu, J., Liu, H., Ye, S., Jiang, L., and Liu, X. 2017. Novel bioactive surface functionalization of bacterial cellulose membrane. *Carbohydrate Polymers* 178: 270–276.

Shimizu, M., Saito, T., and Isogai, A. 2016. Water-resistant and high oxygen-barrier nanocellulose films with interfibrillar cross-linkages formed through multivalent metal ions. *Journal of Membrane Science* 500: 1–7.

Sjostrom, E. 1993. *Wood Chemistry: Fundamentals and Applications*. 2nd ed. California: Academic Press Inc.

Song, Z., Xiao, H., and Zhao, Y. 2014. Hydrophobic-modified nano-cellulose fiber/PLA biodegradable composites for lowering water vapor transmission rate (WVTR) of paper. *Carbohydrate Polymers* 111: 442–448.

Soykeabkaew, N., Tawichai, N., Thanomsilp, C., and Suwantong, O. 2017. Nanocellulose-reinforced 'green' composite materials. *Walailak Journal* 14: 353–368.

Sugiyama, J., Persson, J., and Chanzy, H. 1991. Combined infrared and electron diffraction study of the polymorphism of native celluloses. *Macromolecules* 24: 2461–2466.

Sulaiman, S., Mokhtar, M. N., Naim, M. N., Bahariddin, A. S., Mohd Salleh, M. A., and sulaiman, A. 2015. Study on the preparation of cellulose nanofibre (CNF) from kenaf bast fibre for enzyme immobilization application. *Sains Malaysiana* 44(11): 1541–1550.

Sundaram, J., Pant, J., Goudie, M. J., Mani, S., and Handa, H. J. 2016. Antimicrobial and physicochemical characterization of biodegradable, nitric oxide-releasing nanocellulose–chitosan packaging membranes. *Journal of Agricultural and Food Chemistry* 64(25): 5260–5266.

Slavutsky, A. M., and Bertuzzi, M. A. 2014. Water barrier properties of starch films reinforced with cellulose nanocrystals obtained from sugarcane bagasse. *Carbohydrate Polymers* 110: 53–61.

Svagan, A. J., Hedenqvist, M. S., and Berglund, L. 2009. Reduced water vapour sorption in cellulose nanocomposites with starch matrix. *Composites Science and Technology* 69: 500–506.

The European Parliament and the Council of the European Union. 2018. Directive (EU) 2018/852 of the European Parliament and of Council of 30 May 2018 amending Directive 94/62/EC on packaging and packaging waste. *Official Journal of the European Union*, L150/14.

Trifola, J., Garciab, A., and Sillardb, C., et al. 2015. Hybrid PLA based nanocellulose/clay composites for controlled release. *Cost Action FP1105 San Sebastian.*

Tsai, Y. H., Yang, Y. N., Ho, Y. C., Tsai, M. L., and Mi, F. L. 2018. Drug release and antioxidant/antibacterial activities of silymarin-zein nanoparticle/bacterial cellulose nanofiber composite films. *Carbohydrate Polymers* 180: 286–296.

Turbak, A. F., Snyder, F. W., and Sandberg, K. R. 1983. Microfibrillated cellulose, a new cellulose product: Properties, uses, and commercial potential. *Journal of Applied Polymer Science. Applied Polymer Symposium* 37: 815–827.

Ullah, M. W., Manan, S., Kirprono, S., Islam, M. U., and Yang, G. 2019. Synthesis, structure, and properties of bacterial cellulose. In J. Huang, A. Dufresne, and N. Lin (eds.), *Nanocellulose: From Fundamentals to Advanced Materials* (pp. 81–113). Wiley, VCH Verlag GmbH & Co. KGaA. https://doi.org/10.1002/9783527807437.ch4

Usmani, M. A., Khan, I., Gazal, U., Haafiz, M. K., and Bhat, A. H. 2018. Interplay of polymer bionanocomposites and significance of ionic liquids for heavy metal removal. *Polymer-Based Nanocomposites for Energy and Environmental Applications*(pp. 441–463).

Vartiainen, J., Kaijunen, T., Nykanen, H., Maim, T., and Tammelin, T. 2014. Improving multilayer packaging performance with nanocellulose barrier layer. *TAPPI PLACE Conference*, Vol. 2. 763–790.

Vasile, C., Stoleru, E., Sdrobis Irimia, A., Pricope, G., Ioanid, G. E., and Darie, R. 2013. Plasma assisted functionalization of synthetic and natural polymers to obtain new bioactive food packaging materials in ionizing radiation and plasma discharge mediating covalent linking of stratified composites materials for food packaging. In A. Safrany (ed.), *Proceedings of the Co-Ordinated Project: Application of Radiation Technology in the Development of Advanced Packaging Materials for Food Products*; Vienna, Austria. 22–26 April 2013; [(accessed on 19 June 2017)]. pp. 100–110.

Velazquez, V. F., Cordovo-Perez, G. E., and Silahua-Pavon, A. A., et al. 2020. Cellulose obtained from banana plant waste for catalytic production of 5-HMF: Effect of grinding on the cellulose properties. *Fuel* 265:116857.

Vilarinho, F., Silva, A. S., Vaz, M. F., and Farinha, J. P. 2017. Nanocellulose: A benefit for green food packaging. *Critical Reviews in Food Science and Nutrition* 58(9): 1526–1537.

Wan BaderulHisan, W. S., and Mohd Amin, I. N. 2017. Extraction of cellulose from sawdust by using ionic liquid. *International Journal of Engineering and Technology* 9(5): 3869–3873.

Wang, J., Gardner, D. J., Stark, N. M., Bousfield, D. W., Tajvidi, M., and Cai, Z. 2018. Moisture and oxygen barrier properties of cellulose nanomaterial-based films. *ACS Sustainable Chemistry and Engineering* 6: 49–70.

Ward, B. 2015. Chapter 11 - Bacterial energy metabolism. *Molecular Medical Microbiology* (Second Edition) 1: 201–233.

Yam, K. L., and Takhistov, P. 2016. Sustainable packaging technology to improve food safety. *IBM Journal of Research and Development* 60(5/6). 9:1–9:7. https://doi.org/10.1147/JRD.2016.2597018

Yano, H., Abe, K., Nakatani, T., Kase, Y., Kikkawa, S., and Onishi, Y. 2014. Frozen dessert and frozen dessert material. US 20140342075 A1 2014.

Yates, B. 1972. *Thermal Expansion*. New York: Plenum.

Yuan, H., Nishiyama, Y., Wada, M., and Kuga, S. 2006. Surface acylation of cellulose whiskers by drying aqueous emulsion. *Biomacromolecules* 7: 696–700.

Yu, H., Yan, C., and Yao, J. 2014. Fully biodegradable food packaging materials based on functionalized cellulose nanocrystals/poly(3-hydroxybutyrate-co-3-hydroxyvalerate) nanocomposites. *RSC Advances* 4(104): 59792–59802.

Yu, Z., Wang, W., Kong, F., Lin, M., and Mustapha, A. 2019. Cellulose nanofibril/silver nanoparticle composite as an active food packaging system and its toxicity to human colon cells. *International Journal of Biological Macromolecules* 129: 887–894.

Zhanga, J., Elderb, T. J., Puc, Y., and Ragauskas, A. J. 2007. Facile synthesis of spherical cellulose nanoparticles. *Carbohydrate Polymers* 69: 608–610.

Zhao, Y., and Li, J. 2014. Excellent chemical and material cellulose from tunicates: Diversity in cellulose production yield and chemical and morphological structures from different tunicate species. *Cellulose* 21: 3427–3441.

Zhao, Y., Simonsen, J., Cavender, G., Jung, J., and Fuchigami, L. H. 2014. Nano-cellulose coatings to prevent damage in foodstuffs.

Zheng, L., Chaofei, Z., Dang, Z., Zhang, H., Yi, X., and Liu, C. 2012. Preparation of cellulose derived from corn stalk and its application for cadmium ion adsorption from aqueous solution. *Carbohydrate Polymers* 90(2): 1008–1015.

4 Nanomaterials-Based Biosensors for Packaging Application

Farhan Saeed, Muhammad Afzaal, Bushra Niaz, Yasir Abbas Shah, and Atka Afzal
Department of Food Sciences, Government College University, Faisalabad, Pakistan

CONTENTS

4.1 INTRODUCTION

Over the last twenty years, Nano sized substances and gadgets have received a lot of attention. Nanoparticles, or substances with measurements less than 100 nanometers, are well known for providing greater surface area, improved reactions, and superior mechanical and optoelectronic attributes, allowing considerable advancements in disciplines such as electronic equipment, pharmaceutics, food products, beauty products, and energy devices, among others. The speedy evolution of nanoparticles has paved the way for the creation of innovative sensors and food wrapping technologies, tackling long-standing difficulties in the food industry such as extending storage duration, reducing trash, determining food safety, and improving food quality (Kargozar and Mozafari, 2018). Food safety has ever been a major international healthcare issue, particularly in underdeveloped regions. Several methodologies

DOI: 10.1201/9781003207641-5

have been devised to recognize diverse harmful ingredients in food to tackle food safety issues. Nanomaterials-based biomarkers, for example, offer the possibility of achieving speedy, delicate, productive, and versatile tracking, combating conventional optimization techniques' constraints and regulations, such as complex specimen pre-treatment, lengthy identification times, and reliance on highly priced apparatus and well trained personnel (Lv et al., 2018).

Food safety regulations such as BRC, FSSC 22000, IFS, and HACCP are ever more frequently exploited in the agricultural industry around the world (Aung and Chang, 2014). Worldwide food safety accidents, on the other hand, continue to happen on a regular basis, causing serious health, monetary, and even social concerns. For example, in 2008, China experienced a memorable food safety crisis when melamine was discovered in baby milk (powdered milk), resulting in over 290,000 babies suffering from acute healthcare complications such as urinary system stones (Chen, 2009). Every one of these occurrences serves as reminders of the need and necessity of ensuring food safety. Nanotechnology's rapid advancement has revolutionized many elements of food sciences and the food business, resulting in increased capital and marketplace dominance. Nanocomposites and nanostructures, including nanoscale sensors, nanoscale emulsions, nanoscale insecticides, and nanoscale capsules, have recently been developed with the goal of bringing new uses to the food business. In the food sector, nanotechnology can be employed to create containers with improved heating, structural, and safety features. Indeed, nano sensors integrated in food packaging technologies are employed to notify customers when products are about to expire. Figure 4.1 depicts the importance of nanobiotechnology in major food business domains.

The advanced food sector relies heavily on packaging. The development in quality from manufacturing to consuming, on the other hand, is the foremost beneficial

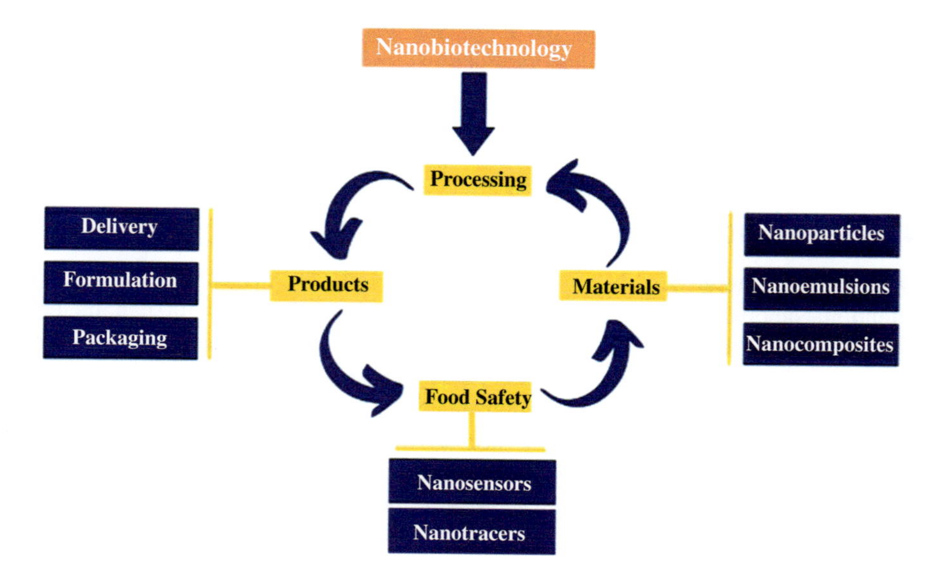

FIGURE 4.1 Nanotechnology's involvement in the agricultural company's numerous areas.

activity in the keeping of food goods (Zhang and Chen, 2019). Nanotechnology's developments when it comes to foodstuff packing hold new potential for increasing the sustainability of foodstuff packing. Owing to their capability to enclose active substances and improve performance, durability, and digestibility, many nanomaterials are manufactured and utilized in the marketplace (Abraham, 2016). Because of their desirable functionality and uses in food packaging, nanofillers are also employed in food packaging. Nanofillers, in biosensing, not only protect food from external variables, but they also incorporate features into packaging materials, opening a slew of fresh opportunities in the foodstuff packaging business. Food nanotechnology has recently been used in commercial operations ranging from smart packing to the fabrication of participatory foodstuff. functional packing is a current concept for food packaging in which the packed material state adjusts to enhance the aesthetic appeal and safety of food goods while also extending life span by retaining product quality (Singh et al., 2017).

Food wastage caused by microbial illnesses are reduced because to innovative food packaging and nano-based solutions. Antimicrobial properties of several nanomaterials including silvery nanostructures, carbon nanostructures, fullerene, titanium dioxide, Mg oxide, Zn oxide analogues, and zero - valent iron have been demonstrated (Salgado et al., 2019). Additionally, distinct nanoparticles have subsequently been designed for detecting hazardous toxins in food goods. An embedded electronic tongue in sophisticated food packaging has a network of nano detectors that are especially delicate to gases emitted from food wastage; this provides a distinct and visual indication showing whether or not the foodstuff is fresh, employing a color-changing detector band (Neethirajan and Jayas, 2011). Magnetic oxide nanostructures are also implemented in the foodstuff business for the quick identification of food-borne diseases (Patra et al., 2018). Owing of its excellent detecting capability, nano sensors can detect germs, harmful compounds, and pollutants in a variety of foods, which is beneficial to food safety (He and Hwang, 2016). Nanotechnology developments in the food sector foster creativity in a variety of food aspects, notably mouthfeel, aroma, aesthetic qualities, product color intensity, and product reliability duration.

4.2 NANOTECHNOLOGY-BASED SENSORS AND BIOSENSORS FOR PATHOGENS AND TOXINS

Microscopic life forms (bacterium, fungus, and protozoa) are everywhere and perform a significant function in nutrition cycles, spontaneous decay, and biotechnology. A few of these, meanwhile, are toxic and infectious. Legionnaires illness is caused by bacteria such as *Legionella pneumophila* attacking macrophages from humans (Swanson and Hammer, 2000), while food poisoning is caused by *Campylobacter jejuni* and *Salmonella typhimurium* (Mor-Mur and Yuste, 2010). To counteract human defenses, some bacteria create chemicals that are released into the environment and are blamed for the signs seen in bacterial diseases. Pathogens and microbial toxins must be detected for health and safety purposes. Nanoparticles have been widely employed to construct nano sensors for quick identification of microbes using toxic metabolites and microbiological colonies as targeted analytes attributed to their distinctive physical and chemical characteristics.

Many technological advancements in microorganism identification are possible because of nanotechnology. Furthermore, the use of nanomaterials as sensors in combination with affinity ligands, antibodies, and current innovative detection methods has boosted the effectiveness of rapid identification of numerous toxins (Valdés et al., 2009). The photonic (optical sensors) or electrical (electrochemical sensors) capabilities of nanomaterials are typically used to determine foodborne microorganisms and poisons (Pérez-López and Merkoçi, 2011). Because of their structure and makeup, metallic nanostructures made of gold or silver exhibit a variety of photonic and electrical characteristics (Nath et al., 2008). Other nanostructures, including luminescent carbon - derived materials and quantum bits have both been employed in a variety of purposes, such as DNA identification and the creation of immunoassays for bacterial and toxins identification (Baptista et al., 2006; Edgar et al., 2006). Nano sensors and electrolytic devices relying on new nanostructures such as carbon nanotubes (single and multiwalled), metallic nanostructures (silver, gold, platinum, copper, and zinc), and super - paramagnetic nanoparticles are now being employed to predict different poisons in food (Rai et al., 2012; Shafiq et al., 2020). Table 4.1 lists different nanoparticles, their characteristics, and their functions in biosensors for microbe identification. The following are some of the most often adopted nano systems for pathogens and toxin identification.

Because of their various surface functions and distinctive features, gold nanoparticles (AuNPs) have been extensively utilized in nano sensors. GNPs have been widely utilized in the manufacturing of biosensors. Gold nanoparticles in various configurations, including nanowires, Nanopatterns, nanospheres, and so on, have been utilized in biosensing applications (Kaittanis et al., 2010). AuNPs' biological tolerability, remarkable conductivity capabilities, and excellent surface to volume ratio are just a few of the properties that make it a promising nanoparticle (Guo and Wang, 2007). The redox action of gold NPs is an intriguing property that improves the responsiveness of nano sensors based on electrochemistry in the identification of harmful microorganisms. The addition of gold NPs to nano sensors based on electrochemistry in combination with ssDNA homologous to the DNA of microbes under investigation boosts their affinity to DNA-gold NPs on the top of the receiver and increases the responsiveness of the nano sensor (Kumar et al., 2020). In another experiment, gold nanoparticles were conjugated with redox catalysts that accurately oxidize/reduce the sample as the reaction's substance. The adsorbed enzyme gold NPs gradually boost the current signals and verify identification after adhering to the sample (Sperling et al., 2008). Food infected with *Aspergillus flavus* and *Aspergillus parasiticus* contains aflatoxins, a category of poisonous and cancerous chemicals. Aflatoxin B1 was detected using gold nanoparticles bifunctional with anti-aflatoxin antigens. Aflatoxin M1 has also been detected in milk samples using superparamagnetic crystals carrying anti-aflatoxin M1 antigens and nanomachines made of gold (Sharma et al., 2010; Xiulan et al., 2005).

Silver nanoparticles (AgNPs) show intriguing traits such as exceptional optical features, good electrocatalytic capacity, and huge surface area, and hence pique attention in the domain of biosensing. Because of their antibacterial qualities, they have been extensively applied in medical and biotechnology. Several investigations have shown that AgNPs can limit the proliferation of harmful microbes

TABLE 4.1
Nanocomposite, Their Characteristics, and Their Applications in Microbe Identification Nano Sensors

Nanostructures	Components	Intrinsic Characteristics	Functions of Biomarkers	References
Metallic Nanostructures	Gold, silver	• Localized surface plasmon resonance (LSPR) • Nano particles agglomeration or changes in the environmental dielectric constant cause observable colour changes. • Increasing the strength of the electric and magnetic environment • Raman scattering intensity	• Colorimetric probe • FRET acceptor, quench fluorescence • Fluorescence enhancer • SERS probe	Petryayeva and Krull, 2011; Sepúlveda et al., 2009; Sutarlie et al., 2014; Wang et al., 2013; Zhao et al., 2008
Quantum dots	Combination of semiconductor materials from periodic group 12-16, 13-15, or 14-16.	• Tunable fluorescence, based on the substance and dimension • Organic fluorophores have lower photocatalytic properties and quantum efficiency. • Multiple access with a confined emission spectrum	• Fluorescent probe • FRET donor	Esteve-Turrillas and Abad-Fuentes, 2013; Kim et al., 2013; Vasudevan et al., 2015
Carbon nanotubes	Carbon	• High conductivity • Fluorescence quenching • Peroxidase catalytic activities	• Nanoelectrode • FRET acceptor, quench fluorescenc • Fluorescent probe	Kumar et al., 2015; Pérez-López and Merkoçi, 2012; Yang et al., 2010
Graphene	Carbon	• High conductivity • Fluorescence quenching • Peroxidase catalytic activities • Photoluminescence • Enhance Raman signals	• Nanoelectrode • Fluorescence quencher, FRET acceptor • Fluorescent probe • SERS	Garg et al., 2015; Ge et al., 2015; Kumar et al., 2015
Silica Nanostructures	Silica	• Deposition of fluorophores with significant permeability and contact area	• Fluorescent probe	Burns et al., 2006a; Burns et al., 2006b

including *Staphylococcus aureus, Streptococcus mutants, Streptococcus pyogenes, Escherichia coli, and Proteus vulgaris* (Abbaszadegan et al., 2015; Gordienko et al., 2019). Kalele et al. (2006) Rabbit immunoglobulin G [IgG] was coupled with silver nanostructures in order to identify *E. coli* in the region of 5-109 by observing SPR band shifts in the presence of *E. coli* cells (Kalele et al., 2006). One of the very effective ways for identifying pathogenic bacteria is surface-enhanced Raman spectroscopy combined with silver nano sensors (Fang et al., 2017). Nano sensors with various nanoparticles, such as nanofibers, nano colloids, graphene oxide, carbon nanotubes, plamonic gold, and magnetic beads, are commonly implemented for the identification of microbes found in food in addition to silver (Baranwal et al., 2016; Holzinger et al., 2014).

Quantum Dots (QDs) are 1–10 nm in size inorganic nanocrystals with unusual optical features such as wide activation, narrower size emission wavelength, strong photochemical durability, and low photobleaching. They have been extensively employed in the fabrication of optical biosensors to identify ions, chemical components, pharmacological substances, and biomolecules like nucleic acids, proteins, amino acids, enzymes, carbohydrates, and neurotransmitters, mostly as replacements to fluorophores (Pandit et al., 2016). Qdots can be made utilizing several tactics, including atomic stream epitaxy, and engraving with an electron gun, but interfacial fabrication is the most prevalent way. Using QDs as fluorophores, multiple researches have proved the identification of foodborne microorganisms. Multiplex identification of *E. coli* and *Salmonella enteritidis* was achieved using QDs and immunomagnetic separation (IMS) (Dudak and Boyaci, 2009). A magnet was used to condense crystals that are conductive and QDs lacquered with matching antibodies and linked to targeted antigens. With an operating range of 5 × 102 to 5 × 104 CFU/mL for *E. coli* and 4 × 102 to 4 × 105 CFU/mL for *S. enteritidis*, the fluorescence intensities created by trapping various proportions of bacteria were examined. There have also been reports using QDs in order to identify gastrointestinal illness at the same time germs in items of foods (Wang et al., 2011; Wang et al., 2015).

Carbon nanotubes (CNTs) one among the more widely employed NMs in nano sensors, diagnostic tools, cellular tracing, tissues regeneration and marking, and pharmaceutical and biomolecule deliveries for the past decade. Single-walled CNTs, double-walled CNTs, and multi-walled CNTs are empty cylindrical tubes made up of one, two, or more contiguous graphite layered topped with fullerenic horizons. They have distinct architectures, great electrical and mechanical characteristics, excellent heat conductance, high chemical durability, outstanding electrocatalyst performance, reduced interface contamination, low voltage unbalance, and a high aspect ratio (surface to volume) (Pandit et al., 2016). CNT-based biosensors and analytics have been used in medicine, industry, environmental control, and food quality screening for extremely sensitive sample identification. CNTs have also been employed for pathogen identification signal amplification (Chunglok et al., 2011). Guner et al. used a pencil graphite electrode enhanced with MWCNT, chitosan, polypyrrole (PPy), and AuNPs to produce an electrochemical immunosensor for the identification of E. col. The hybridized bio nanocomposite was immobilized with an anti-*E. coli* monoclonal antibody. The large interface area of MWCNT on the electrode provided greater room for antibody immobilization. By enhancing the electrochemical signaling and antibody adsorption capability, AuNPs improved identification accuracy. The diagnostic limit for *E. coli* was 3 101 to 3 107 cfu/mL. For bacteria enumeration, the combined nano sensor provided a very delicate, selective, and reproducible system (Güner et al., 2017).

Furthermore, nanotechnology facilitates the use of minimal cost nano sensors in foodstuff packaging to recognize numerous harmful microbes found in diverse foods (Ranjan et al., 2014). Among the highly effective nano biosensors, the nano bioluminescent spray, provides a visible illumination for rapid identification of pathogenic strains in various food items. This spray contains a variety of magnetic nanostructures that interact with viruses in food to give a visible color that can be clearly identified (Arshak et al., 2007).

Magnetic nanoparticles: Foodborne infections have been detected quickly using materials like iron oxide (Fe_3O_4). Magnetic nanomaterials are usually treated with bispecific compounds that allow nanomaterials to attach to sample matrix (e.g., antibodies, bacterial cells, proteins, DNA). When an outer magnetic field was introduced, the intended bacteria turned magnetic, allowing them to be readily isolated from the test solution. The use of carbon-based nanoparticles in biosensors is gaining popularity because they may be employed to rapidly indicate the existence of toxins in food, agricultural, and environmental settings. Because of their remarkable sensitivity, lower detecting limit, and larger linear detecting range, graphene-based electrochemical biosensors offer a lot of potential for food toxin surveillance in food products (Malhotra et al., 2015). Furthermore, the nanostructures implementation in the identification of pathogens and poisons has showed a lot of promise (Manhas et al., 2021). Some nanoparticles work as antimicrobial. The antimicrobial mechanism of nano-particles is shown in Figure 4.2.

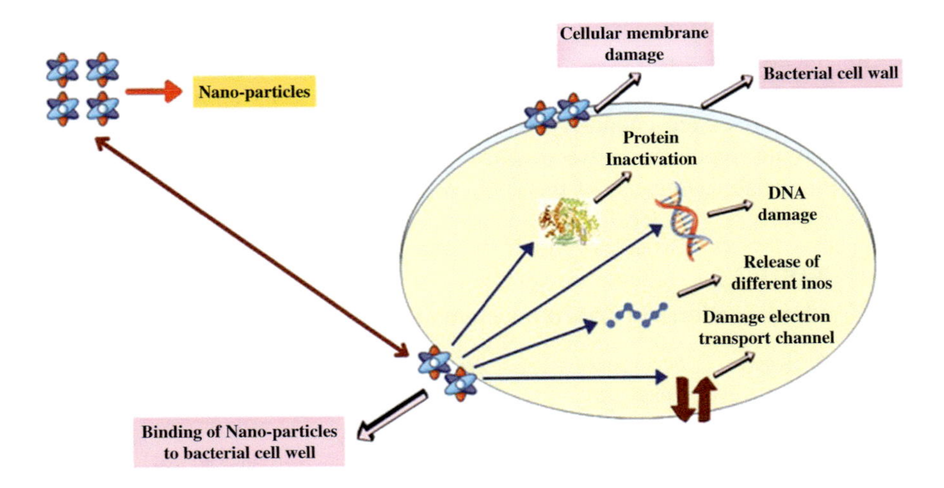

FIGURE 4.2 The antimicrobial mechanism of nanoparticles.

4.3 NANOTECHNOLOGY-BASED SENSORS AND BIOSENSORS FOR HEAVY METALS

Because of their widespread application and inability to degrade in the atmosphere, heavy metals are a significant source of contamination. Heavy metals such as Hg^{2+}, Cd^{2+}, and Pb^{2+} offer major wellness risks, even when tiny amounts of these ions are found in food more than the regulatory limit. Several approaches for detecting heavy metal ions, such as inductively associated plasma mass spectroscopy (ICPMS) and atomic absorption/emission spectroscopy (AAS/AES), have been suggested. Conventional techniques, on the other hand, necessitate either large machinery or time-consuming preparation procedures. Biosensors based on nanoparticles can identify heavy metals in a quick, low-cost, and label-free manner (Lv et al., 2018). Nanoparticles have been meticulously developed and produced for heavy metal identification and elimination with the rise of nanotechnology, offering multiple benefits over earlier approaches. In this paper, nanotechnology-based sensors and biosensors for heavy metal identification are reported. Table 4.2 shows the hazardous elements with biomarkers

Carbon nanoparticles like carbon nanotubes (CNTs), carbon nanosheets, and graphene have been investigated as electrode substrates for heavy metal identification (Shao et al., 2010). For the identification of chemical pollutants for example, heavy metallic ions, alkali and alkaline earth metallic ions and others, a number of electrochemical and photometric tests based on AuNPs have been published (Bülbül et al., 2015). Various sensing methods rely on silver NPs (AgNPs) optical characteristics have been published. The color shift from yellow to brown in scattered and consolidated AgNPs can be linked to a targeted molecule's concentrations variation. Various AgNP-based assays for metal ion identification have been designed depending on this premise (Yoosaf et al., 2007).

Colorimetric AuNP detectors for heavy metals are very appealing since they can be viewed with the human eye (Li et al., 2011). Chen et al. constructed a photometric detecting system using parchment microfluidics diagnostic instruments for Hg^{2+} identification by cooperation of T-Hg^{2+}-T by exploiting the chemicals and photovoltaic features of AuNPs, taking benefit of the affinity among thymine (T) and Hg^{2+}. AuNPs were coupled with oligonucleotide patterns, which diminished the electrostatic repulsion between nearby AuNPs by lowering the zeta potential of AuNP interfaces. Because of the photostability, excellent efficiency in terms of

TABLE 4.2
Identification of Hazardous Elements with Biomarkers

Analytes	Nano Sensors	Nanostructures	References
Hg^{2+}	Colorimetric	AuNPs	Chen et al., 2014
Hg^{2+}	Fluorometric	AuNPs and QDs	Li et al., 2011
Cd^{2+}	Colorimetric	AuNPs	Yin et al., 2011
Pb^{2+}	Electrochemical	DNA and Pt@Pd nanocages	Zhao et al., 2017

quantization, and optically reconfigurable dimension features of QDs, a fluorescent sensor made of QD/DNA/AuNPs assembly for Hg^{2+} identification was constructed leveraging these characteristics (Chen et al., 2014). The inclusion of Hg^{2+} in samples caused the development of T-Hg^{2+}-T and the development of a connection among QDs and AuNPs via T-Hg^{2+}-T contact, resulted in interface energy transfer among QDs and AuNPs and the scorching of QDs. In combined samples, this nano sensor had an outstanding detection limit (LOD, 0.4 ppb) and strong partiality for Hg^{2+} (Li et al., 2011).

Yin et al. (2011) designed a Cd^{2+} detecting surface-enhanced Raman scattering (SERS) sensor based on the interparticle plasmonic interaction formed by Cd^{2+}-induced selective self-aggregation of AuNPs (Yin et al., 2011). Zhao et al. created a robust accelerated biosensor that detects Pb^{2+} selectively via catalytic hairpin construction and synergistic amplification (Zhao et al., 2017). Baxter et al. recommended a convenient and inexpensive method for making gold conductors by implementing manufactured AuNPs stabilized with 4-(dimethylamino) pyridine to filter paper and utilized this detectable gateway Cu ions (Liana et al., 2013). Some biosensors offer a reduced limit of detection and a more delicate way of quantifying heavy metals while being easy, time taking and premium. However, many nano sensors are not yet open for citizen and marketing implementations; the primary rationale is that biosensors' effectiveness is appropriate for the purpose of researching and experimental constraints due to their elements. Even though, in addition to nano sensors, other detecting processes have been established to evaluate the threshold of heavy metals, there is still a long way to evolve shrunken yet intelligent gadgets.

4.4 NANOTECHNOLOGY-BASED SENSORS AND BIOSENSORS FOR CHEMICALS AND PESTICIDES RESIDUE

The identification of dangerous chemicals and pesticides has piqued the interest of both industry and academics over the last three decades. Pesticides are toxic substances that are commonly detected in water, land, fruits, veggies, and other farm goods. These pesticides may be available in massive quantities and at dangerous degrees, posing an environmental damage and dangerous concern. Even little quantities of exposure can have major consequences for human health (Hashwan et al., 2020). In modern agricultural operations, pesticides are routinely employed to avoid and manage pests and weeds in order to enhance harvest output (Songa and Okonkwo, 2016). Despite the application of pesticides boosts food yield, the accumulation of pesticide remnants in food, water, and the surroundings causes substantial food contamination, posing a health risk to humans and causing the ecosystem to collapse (Kantiani et al., 2010). Available analytical technologies such as mass spectrometry (MS), high-performance liquid chromatography (HPLS), and gas chromatography have been utilized to recognize and monitor pesticides (GC) (Fu et al., 2013; Songa and Okonkwo, 2016). Although these strategies have showed efficient trace identification with great specificity and outstanding repeatability, they do have certain downsides, such as time intensive, expensive equipment, and a lengthy sample preparatory procedure, which are barriers to on-the-spot and real-time identification (Zhang et al., 2010).

Metallic nanomaterials are used to make nano sensors that can identify the prevalence of chemical substances or insecticides. The invention of nano sensors for identifying melamine is one case of the exploitation of metallic nanomaterials. Melamine polymers, which are utilized to make household products, ornamental laminates, adhesives, acrylics, and lacquers, are made from it. It was utilized to generate a deceptively high protein level in animal feed and food product studies because of the elevated nitrogen concentration [66 percent of mass]. Since the massive toxicity of infants in China because of the inclusion of melamine in powdered milk, severe melamine content in food restrictions have been implemented. On the basis of energized transformation associated to the fluorescence between nanomaterials, Wu et al. (2015) proposed a nano sensor for sensing the prevalence of melamine (Wu et al., 2015).

Metallic nanomaterials surely facilitate the examination of soap remnants in food products, which are undesirable due to stringent food safety regulations. Kumar et al. (2016) created a device for detecting milk harmful ingredients using nanomaterials. The sensor operates by maintaining gold nanoparticles on the surface by electrostatic repulsion between negatively charged citrate ions, prohibiting them from aggregating. The introduction of activator to gold nanoparticles neutralizes the surface charge and promotes clustering, as seen by a shift in hue from red to purple in the mixture. In the existence of anionic detergent and HCl, the clustering of nanoparticles was slowed, and the solution stayed red (Kumar et al., 2016). Zheng et al. created a sensor for 4-nonylphenol in milk and its packaging polymers using graphene-Au nanoparticles. The scientists' research revealed that 4-nonylphenol was identified in a variety of foodstuffs, including veggies, fruits, cereals, and beverages, with the packing material being the most likely source. As a result, it was critical to create a methodology for detecting 4-nonylphenol in the product and its packaging that was both efficient and quick. The electrochemical sensor was made by depositing a poly [p-aminothiophenol] layer on a graphene-Au nanoparticle-modified electrode. Higher sensitivity and selectivity were among the characteristics of the newly designed sensor (Zheng et al., 2018). Yu et al. designed electrochemical biosensors premised on carbon nanotubes to identify organophosphorus insecticides (OPs) (Yu et al., 2015). Table 4.3 shows the biosensors available for detecting pesticide metabolites.

TABLE 4.3
Biomarkers Available for the Identification of Insecticide Metabolites

Analytes	Biosensors	Nanomaterials	References
Organophosphorus pesticide	Electrochemical	CNTs	(Yu et al., 2015)
Organophosphorus pesticides	Electrochemical	AuNPs and graphene oxide	(Yang et al., 2014)
Organophosphate insecticide	Electrochemical	MNPs	(Dominguez et al., 2015)
Glyphosate	Colorimetric	AuNPs	(Tan et al., 2017)
Methyl parathion	Fluorimetric	CdTe QDs	(Chouhan et al., 2010)
Paraoxon-ethyl	Optical	Carbon dots	(Chang et al., 2017)

4.5 NANOTECHNOLOGY-BASED SENSORS AND BIOSENSORS FOR VETERINARY DRUGS

Veterinary medications are intended to address a wide range of ailments in livestock. Veterinary medications are an important part of advanced animal husbandry and food production, but their metabolites can linger in animal-derived foods, posing a risk to food safety. The administration of veterinary medications in creatures that produce food can finale in leftovers in animal-derived products, posing a health risk to consumers. The most common cause of drug residues is incorrect drug use and failure to adhere to the separation duration. The formation of antibacterial drug tolerance, hypersensitivity response, carcinogenic effects, genotoxicity, teratogenic effects, and alteration of intestine natural flora are the Consequences for Population Healthcare of drug leftovers (Ture et al., 2019).

In subsequent decades, a number of forms of biosensors have been established as an alternate method for screening veterinary medications in animal-derived food products. Biosensor technology is a bio-sensitive technology that turns biological matter concentrations into a detectable signal. For the analysis of veterinary medication residues, many sensors are available. The use of biosensors should be determined by the characteristics of the residues and sensors. Tetracyclines in poultry muscle samples were determined using a luminous bacterial biosensor (Pikkemaat et al., 2010). A colorimetric aptasensor premised on enzyme was created by integrating magnetic Loop-DNA probes with catalytic hairpin assembly (CHA)- targeted reprocessing with assistance amplification for the measurement of antibiotic residual in milk (Luan et al., 2017).

An apta-nanosensor was employed to identify the existence of ampicillin in food items. Apta-nano sensors are biosensors that use a dual fluorescence calorimetric approach to measure temperature (Patel et al., 2020). To identify ampicillin from other antibiotics, it employs an aptamer designed on AuNP solution. AuNPs are released when a single-stranded DNA aptamer termed AMP17 encounters ampicillin. These AuNPs clump together and react with salt, turning the reddish-purple solution purple. This technique is a precise and delicate option for detecting the antibiotic ampicillin in food products (Kuswandi et al., 2017). Septicemia, pneumonia, intestinal illnesses, and urinary system infectious diseases are all treated with kanamycin, a popular antibiotic employed in veterinary medicine (Yazdian-Robati et al., 2017). Song et al. indicated the manufacturing of AuNP-based aptasensor to identify kanamycin. Once more, it is a calorimetric method premised nano sensor in which the accumulation of NPs when responded with salt tends to cause the transform in color in the existence of kanamycin (Song et al., 2011). The identification of kanamycin using calorimetric-based approaches is straightforward and premium, based on published research, because the change in color may be seen with the human eye (Patel et al., 2020; Niu et al., 2014; Zhou et al., 2015).

Tetracycline antibiotics are employed in veterinary and aquaculture medicines for a variety of reasons. Antibiotic leftovers have been discovered in land, water, and other habitats. Tetracycline can be detected using an MNP-based liquid chromatography technique. The effectiveness of this approach for detecting tetracycline is 86.2 4000 g/kg (Cristea et al., 2017). Nanomaterials' exceptional sensitivity as excellent detectors of different chemical and biological pollutants in the food industry has made these flexible modified molecular structures an indispensable image in the expanding food business.

4.6 NANOTECHNOLOGY-BASED SENSORS AND BIOSENSORS FOR ADULTERANTS

Adulterants have become more prevalent in food goods offered to customers as a result of the abrupt increase in supply for food to meet the ever-increasing population. Food adulteration is a huge concern all over the world that requires to be addressed immediately. As people become more conscious of food safety and quality, new technologies and procedures for detecting food adulterants are developed. Although sensitive, traditional experimental methods are costly, involve substantial creation of a specimen, experimental apparatus, and individualized instruction, and have a low productivity. Food nanotechnology is an emerging invention in nanoscience and nanotechnology, which are this generation's emerging horizons. Adulterants in food can sometimes cause chemical risks, so it's important to keep an eye out for them.

Melamine (1,3,5-triazine-2,4,6-triamine) is a professionally produced organic molecule made from cyanic acid released from urea following the chemical reaction. In animals, it can induce long-term toxic effects and bladder stones (Kuswandi et al., 2017). Several pet fatalities have been reported as a result of food adulteration induced by the prevalence of melamine. In most cases, the measurement of nitrogen portion is used to determine protein level in food. As a result, the introduction of melamine, which contains 66 percent nitrogen, can boost the protein composition of milk products. Prolonged use, on the other hand, can lead to the development of melamine cyanurate nanocrystals that are hydrophobic. Animals' kidneys and reproductive organs are severely harmed as a result of this adulteration (Kim et al., 2010). Testing methods based on current instrumentation and nano biosensors are employed to detect melamine in food, according to the literature review (Rovina and Siddiquee, 2015). Sensors made of cadmium telluride (CdTe), quantum dots (QDs), gold NPs (AuNPs), and single-walled carbon nanotubes (SWCNTs) have all been identified as promising possibilities for quick melamine detection in food (Li and Rothberg, 2004). In a different technique, urea is discovered in milk as an adulteration molecule like melamine, resulting in a falsely increased proportion of proteins. One urea-specific biosensor is based on urease-based enzymatic identification (Ramesh et al., 2015).

The determination of biomarkers is applied to identify manuka honey, a honey from New Zealand. Depending on a competing immunoassay, Kato et al. devised an asymmetrical flowing test for the determination of leptosperin, a glycoside. Before the honey specimen is placed to the asymmetrical flowing strip, anti-leptosperin monoclonal antibodies coupled with gold nanostructures are introduced. The antibody-leptosperin intricate, as well as single antibodies, move to the test line (leptosperin-conjugated BSA), where only single antibodies are collected and detected by the gold nanoparticles' reddish-violet color (Kato et al., 2016). Because of their remarkable affinity, broad accessibility, and versatility to employ in a variety of configurations, antibodies have been widely employed in biosensor technology. Biosensors were utilized to check for adulteration of fresh milk from a specific animal with raw milk from different dairy cows. To establish easy, rapid, direct, and inhibitory biosensor immunoassays (BIA) for detecting cow's milk in the milk of ewes and goats, monoclonal antibodies (MAb) were generated against bovine k-casein and employed in photonic biosensor that is programmed (Haasnoot et al., 2004; Jha et al., 2016).

They have shown to be reliable diagnostic methods for detecting various adulterants and pollutants in milk and milk powder by computerized immunochemical identification (Haasnoot et al., 2006). The identification of cow's milk in ewe's and goat's milk, as well as the identification of bovine rennet whey powder in milk powder, was tested using an inhibitory immunoassay for bovine K-casein. By analyzing changes in bulk responses, the sensor proved also beneficial in detecting fake water infusions to milk.

Food safety is critical to public health and, as a result, to society. Adulteration in particular goods, such as pork, processed beef, meat, and beef jerky, requires worldwide concern. The shift in hue from pinkish red to gray-purple can be used to identify adulteration in pig, poultry meat, and beef using a calorimetric approach based on AuNPs (Ali et al., 2012).

4.7 APPLICATIONS OF NANOTECHNOLOGY IN BIODEGRADABLE FOOD PACKAGING

Food safety issues are evolving at an unpredictably fast level in the twenty-first century (Weiskopf et al., 2020). Surprisingly, more than a third of all food manufactured internationally is destroyed. Because the human stomach weighs roughly 1.3 billion tonnes, this yearly global food-as-waste value has a direct impact on it. Surprisingly, food waste has an economic influence on the worldwide economy, with a value of about $1 trillion. Considering these facts, it is clear that large quantities of food are not consumed by people. This has a significant detrimental impact on the atmosphere, sociological standing, and the economy. Another concerning problem is the continuous intake of tainted food products with extremely less nutritional status. This intake eventually gives rise to numerous disease epidemics and jeopardizes every nation's food security. As a result, to address the issues in the food industry and/or food safety, food-borne diseases or spoilage require thorough screening and monitoring (Kumar et al., 2021). Through the 20th century, nanobiotechnology has gotten a lot of interest and has positioned itself as a discipline with a lot of implications (Ball et al., 2019; Zahin et al., 2020).

Non-biodegradable petroleum-based plastic polymer polymers make up most components utilized in the packaging industry today. As a result, non-biodegradable food packaging materials pose a severe environmental threat worldwide. As a result, the adoption of bio-based packaging materials, such as edible and biodegradable materials made from renewable resources, could help to alleviate the waste problem by lowering packaging waste and extending shelf life, both of which would improve food quality. In this regard, it is feasible to ensure commodity authenticity and durability by using the appropriate materials and packaging technology. Direct inclusion into food commodities as well as their inclusion into food packaging materials and use in food manufacturing are the three main options for nano-technological interventions in food packaging.

The consumer's attitude toward and acceptance of newly presented technologies and their applications determines the industrialization, successful implementation, and responses to various uses of nanotechnology (Kim et al., 2014). Sophisticated nanomaterial supplemented polymers will assist to multiply the advantages connected with current polymers, while also resolving ecological consequences, to achieve efficient food packaging standards. The new packaging material will help to prevent any serious interactions between packaging and food matrices, as well as their impact on consumer health, waste material reduction, enhanced biodegradability, obstacles to vapors and photons, as well as Leaks of carbon dioxide (Sharma et al., 2017).

4.7.1 Starch

Over the years, starch has been thoroughly scrutinized as a potential substance for wrapping of foodstuff. Because of its cyclic accessibility from numerous plants, surplus productivity relevant to recent requirements, and low cost, starch is a viable natural resource and regenerative resource. Corn is the most common source of starch for bioplastics, although more recent global research is looking into the use of starches from potato, wheat, rice, barley, oat, and soy in bioplastics (Kuswandi, 2016).

The application of starch in conjunction with other ingredients to mitigate the shortcomings of this organic polymer has paved to a surge in its use in a variety of businesses, particularly the packaging industry (Sadeghizadeh-Yazdi et al., 2019). The addition of various nanoparticles such as TiO_2, ZnO nanoparticles, graphene, and poly (methyl methacrylate-co-acrylamide) in the polymer structure can also increase their mechanical, UV, and moisture resistance characteristics (Goudarzi et al., 2017; Jayakumar et al., 2019). The addition of various nanoparticles such as TiO_2, ZnO nanoparticles, graphene, and poly (methyl methacrylate-co-acrylamide) in the polymer structure can also increase their mechanical, UV, and moisture resistance characteristics.

The influence of varying concentrations of graphene oxide on the performance of the starch/PVA/graphene oxide composite film was investigated. The addition of graphene oxide (at an ideal concentration of 2 mg/mL) to the composite film can increase its mechanical characteristics, transmittance, and water vapor permeability (Wu et al., 2019). By integrating clove essential oil (15–30 percent (w/w)) and graphene oxide nanosheets (1 percent (w/w)) into PLA, antimicrobial nano-packaging films against *E. coli* and *S. aureus* were established, with the optical and anti-UV attributes of the film effected by the inclusion of graphene oxide nanosheets and essential oil (Arfat et al., 2018).

Owing to their distinctive functional features, starch nanoparticles (SNPs) or nanocrystals are excellent for food applications, particularly biodegradable food packaging. They're made from starch using acid or enzymatic hydrolysis, as well as numerous mechanical techniques (Kim et al., 2015).

Starch-based films could be used to package perishable foods (such as fruits and vegetables, snacks, or dry products). However, effective mechanical, oxygen, and moisture protection are required in these applications. TPS (thermoplastic starch) is frequently insufficient to meet all these needs. Because of its hydrophilicity, its performance varies during and after processing as the water content changes. The use of clay as a filler has been reported to alleviate this disadvantage. Clay has been employed as a potential filler to improve the characteristics of TPS in such applications (Chen and Evans, 2005; Yoon and Deng, 2006).

4.7.2 Chitosan

Chitin is the 2nd most prevalent biopolymer on the planet after cellulose, and it is the building component of crustaceans, insects, and fungi. Deacetylation of chitin yields chitosan, a deacetylated derivative of chitin. Due to the involvement of amino groups fundamental for the polymer's different properties, it is a functionally flexible biopolymer (Priyadarshi and Rhim, 2020). It has a long history of commercial processes, but one of the most recent is as a biodegradable antimicrobial food packaging material. Plasticizers and cross-linkers have been used to boost the capabilities of chitosan, as have fillers such as nanoparticles, fibers, and whiskers, as well as combining the polymer with natural extracts and essential oils, as well as other natural and synthetic polymers.

The addition of zinc oxide and gallic acid to chitosan films significantly enhanced mechanical and physical attributes such as oxygen and water vapor permeability, swelling, water solubility, and UV-vis light transmission. In comparison to basic chitosan, the modified chitosan-based composite has substantial antibacterial and antioxidant activities (Yadav et al., 2021). The preservation stability of raw meat was determined by incorporating ZnO nanoparticles and linseed oil into a chitosan/potato protein-based polymer. The inclusion of ZnO nanoparticles increased the transparency and tensile strength of the films, while the introduction of linseed oil enhanced the composite film's elastic behavior. In addition, biopolymer films had outstanding moisture barrier properties (Wang et al., 2020). Using the solution casting process, a nanocomposite film with chitosan, gelatin, and polyethylene glycol as host materials and silver nanoparticles was explored. To increase the mechanical characteristics and limit visible light penetration, AgNPs were inserted. Film was found to be an effective antibacterial and biodegradable food packaging material in this investigation (Kumar et al., 2018).

4.7.3 Cellulose

Cellulose is a naturalistic streamlined organic molecule made up of d-glucopyranose units joined by b-1,4-glycosidic linkages that can be derived from wood, cotton, leaves, and a variety of other sources (Johar et al., 2012). Cellulose is a biodegradable and environmentally acceptable substance that has been successfully employed as a food preservation packaging material (Ghaderi et al., 2014). Pure cellulose, on the other hand, lacks inherent antibacterial qualities, which can be advantageous in the production of antimicrobial resources. cellulose nanocomposites with antibacterial characteristics due to the integration of inorganic nanoparticles are an example of such materials (Jia et al., 2012).

Cellulose is also employed as a supportive element for some nanomaterials, allowing the available surface of the nanoparticles to enlarge, resulting in improved action. Furthermore, adsorbed silver nanoparticles on cellulose fibers have high antibacterial activity against *S. aureus* and *E. coli*, with antimicrobial activity of up to 99.99 percent due to connections between oxygen (from cellulose) and silver (Jung et al., 2009). Silver, gold, and platinum nanoparticles were produced and encased in a cellulose gel using the hydrothermal reduction process, then dried using supercritical CO_2. The generated aerogels had exceptional porosity, surface area, transmittance, mechanical strength, and moderate thermal stability, among other characteristics (Cai et al., 2009).

Earlier studies looked at the qualities of the created nanocellulose material by looking at the crystallinity index, which reduced in comparison to microcrystalline cellulose, indicating that it might be used as a biodegradable composite film enhancer (Huang et al., 2018). Another study found that combining cellulose nanocrystals with starch-based nanocomposite films allows D-limonene permeability to be controlled (Liu et al., 2018). Furthermore, cellulose was applied to verify the antibacterial capability (*E. coli* and *S. aureus*) (Fahmy et al., 2020).

4.7.4 Proteins

Many proteins, together with as gelatin, keratin, and casein, have intriguing polymer characteristics like flexural, strength characteristics, and tensile modulus, as well as outstanding structural qualities like hardness, resilience, and flexibility. As a result, these proteins can be used to create new biodegradable polymers for a variety of commercialized uses. Protein-based biodegradable polymers have a growing number of potential applications in food and non-food packaging, as well as biomaterials for reconstructive surgery, tissue engineering, and other medical procedures. As a result, the polymer reinforcement can be made from a protein-based polymer. Blending protein polymers with other protein and/or non-protein molecules can improve their mechanical characteristics. Blending technology allows us to create next-generation biodegradable polymer/plastics that will eventually replace traditional plastics on the marketplace (Mangaraj et al., 2019).

The application of nanometric particles in combination with natural proteins has the potential to improve mechanical, barrier, and water resistance qualities, among other things. Nano clays are frequently employed as fillers in a variety of plastics to minimize thermal properties and liquid and gas permeation, as well as reinforcements. Commercially, nano clays are used to increase barrier qualities, which will have a vast range of applications in the food packaging industry. Clay-based composites have already been approved by the US Food and Drug Administration for use in meals, medications, drinks, and biomedical devices (Alexandre and Dubois, 2000). The most common multilayer silicate clay is montmorillonite (MMT). Both solution intercalation and synthesis of soy protein/MMT nanocomposites were possible (Tian et al., 2021). Soy protein nanomaterials have also been documented to be made using other types of layered silicates, such as rectorite (Yu et al., 2007).

These protein coverings could extend their storage life and expand their usefulness by introducing antibacterial nanomaterials such as titanium dioxide (TiO_2), silver, and others. In Soy protein isolate (SPI)/titanium dioxide composite films, anti-bacterial efficacy against *E. coli* and *S. aureus* has been demonstrated (Wang et al., 2014).

4.8 CONCLUSION

The rapid growth of nanotechnology has transformed many aspects of food science and the food industry, leading in increasing capital and market dominance. The advancements in nanotechnology in the field of food packaging show great promise for enhancing the sustainability of food packaging. Many nanoparticles are manufactured and utilized in the marketplace because of their abilities to enclose active substances and enhance quality, durability, and digestibility. Nano-biosensors research aims to develop innovative technologies that can be beneficial for food safety and quality. For biosensors, new nanomaterials and nanostructures must be explored. Nanomaterials' remarkable sensitivity as excellent sensors of various chemical and biological contaminants in the food sector has made these dynamic synthetic molecular frameworks an important representative in the expanding food business. Government authorities must evaluate if nanomaterial migration and exposure are safe because it is critical to assess the consequences of nanomaterials on environmental and human impact. The greater the number of goods and products that contain nanomaterials, the stricter safety evaluations will have to be.

REFERENCES

Abbaszadegan, A., Ghahramani, Y., Gholami, A., Hemmateenejad, B., Dorostkar, S., Nabavizadeh, M., and Sharghi, H. 2015. The effect of charge at the surface of silver nanoparticles on antimicrobial activity against gram-positive and gram-negative bacteria: A preliminary study. *Journal of Nanomaterials* 2015.

Abraham, A. 2016. *Understanding the Effect of Phytochemical Coated Silver Nanoparticles on Mammalian Cells and the Protein Interactions with the Surface Corona of These Nanoparticles*. RMIT University.

Alexandre, M., and Dubois, P. 2000. Polymer-layered silicate nanocomposites: Preparation, properties and uses of a new class of materials. *Materials Science and Engineering: R: Reports* 28: 1–63.

Ali, M., Hashim, U., Mustafa, S., Che Man, Y., and Islam, K. N. 2012. Gold nanoparticle sensor for the visual detection of pork adulteration in meatball formulation. *Journal of Nanomaterials* 2012: 1–7.

Arfat, Y. A., Ahmed, J., Ejaz, M., and Mullah, M. 2018. Polylactide/graphene oxide nanosheets/clove essential oil composite films for potential food packaging applications. *International Journal of Biological Macromolecules* 107: 194–203.

Arshak, K., Adley, C., Moore, E., Cunniffe, C., Campion, M., and Harris, J. 2007. Characterisation of polymer nanocomposite sensors for quantification of bacterial cultures. *Sensors and Actuators B: Chemical* 126: 226–231.

Aung, M. M., and Chang, Y. S. 2014. Traceability in a food supply chain: Safety and quality perspectives. *Food Control* 39: 172–184.

Ball, A. S., Patil, S., and Soni, S. 2019. Introduction into nanotechnology and microbiology. *Methods in Microbiology*. Elsevier.

Baptista, P. V., Koziol-Montewka, M., Paluch-Oles, J., Doria, G., and Franco, R. 2006. Gold-nanoparticle-probe–based assay for rapid and direct detection of mycobacterium tuberculosis DNA in clinical samples. *Clinical Chemistry* 52: 1433–1434.

Baranwal, A., Mahato, K., Srivastava, A., Maurya, P. K., and Chandra, P. 2016. Phytofabricated metallic nanoparticles and their clinical applications. *RSC Advances* 6: 105996–106010.

Bülbül, G., Hayat, A., and Andreescu, S. 2015. Portable nanoparticle-based sensors for food safety assessment. *Sensors* 15: 30736–30758.

Burns, A., Ow, H., and Wiesner, U. 2006a. Fluorescent core–shell silica nanoparticles: Towards "Lab on a particle" architectures for nanobiotechnology. *Chemical Society Reviews* 35: 1028–1042.

Burns, A., Sengupta, P., Zedayko, T., Baird, B., and Wiesner, U. 2006b. Core/shell fluorescent silica nanoparticles for chemical sensing: Towards single-particle laboratories. *Small* 2: 723–726.

Cai, J., Kimura, S., Wada, M., and Kuga, S. 2009. Nanoporous cellulose as metal nanoparticles support. *Biomacromolecules* 10: 87–94.

Chang, M. M. F., Ginjom, I. R., and Ng, S. M. 2017. Single-shot 'turn-off' optical probe for rapid detection of paraoxon-ethyl pesticide on vegetable utilising fluorescence carbon dots. *Sensors and Actuators B: Chemical* 242: 1050–1056.

Chen, B., and Evans, J. R. 2005. Thermoplastic starch–clay nanocomposites and their characteristics. *Carbohydrate Polymers* 61: 455–463.

Chen, G.-H., Chen, W.-Y., Yen, Y.-C., Wang, C.-W., Chang, H.-T., and Chen, C.-F. 2014. Detection of mercury (II) ions using colorimetric gold nanoparticles on paper-based analytical devices. *Analytical Chemistry* 86: 6843–6849.

Chen, J.-S. 2009. *A Worldwide Food Safety Concern in 2008—melamine-Contaminated Infant Formula in China Caused Urinary Tract Stone in 290 000 Children in China.* LWW.

Chouhan, R. S., Vinayaka, A. C., and Thakur, M. S. 2010. Thiol-stabilized luminescent CdTe quantum dot as biological fluorescent probe for sensitive detection of methyl parathion by a fluoroimmunochromatographic technique. *Analytical and Bioanalytical Chemistry* 397: 1467–1475.

Chunglok, W., Wuragil, D. K., Oaew, S., Somasundrum, M., and Surareungchai, W. 2011. Immunoassay based on carbon nanotubes-enhanced ELISA for Salmonella enterica serovar typhimurium. *Biosensors and Bioelectronics* 26: 3584–3589.

Cristea, C., Tertis, M., and Galatus, R. 2017. Magnetic nanoparticles for antibiotics detection. *Nanomaterials* 7: 119.

Dominguez, R. B., Alonso, G. A., Muñoz, R., Hayat, A., and Marty, J.-L. 2015. Design of a novel magnetic particles based electrochemical biosensor for organophosphate insecticide detection in flow injection analysis. *Sensors and Actuators B: Chemical* 208: 491–496.

Dudak, F. C., and Boyaci, I. H. 2009. Multiplex detection of *Escherichia coli* and *Salmonella enteritidis* by using quantum dot-labeled antibodies. *Journal of Rapid Methods & Automation in Microbiology* 17: 315–327.

Edgar, R., Mckinstry, M., Hwang, J., Oppenheim, A. B., Fekete, R. A., Giulian, G., Merril, C., Nagashima, K., and Adhya, S. 2006. High-sensitivity bacterial detection using biotin-tagged phage and quantum-dot nanocomplexes. *Proceedings of the National Academy of Sciences* 103: 4841–4845.

Esteve-Turrillas, F. A., and Abad-Fuentes, A. 2013. Applications of quantum dots as probes in immunosensing of small-sized analytes. *Biosensors and Bioelectronics* 41: 12–29.

Fahmy, H. M., Eldin, R. E. S., Serea, E. S. A., Gomaa, N. M., Aboelmagd, G. M., Salem, S. A., Elsayed, Z. A., Edrees, A., Shams-Eldin, E., and Shalan, A. E. 2020. Advances in nanotechnology and antibacterial properties of biodegradable food packaging materials. *RSC Advances* 10: 20467–20484.

Fang, Z., Zhao, Y., Warner, R. D., and Johnson, S. K. 2017. Active and intelligent packaging in meat industry. *Trends in Food Science & Technology* 61: 60–71.

Fu, G., Chen, W., Yue, X., and Jiang, X. 2013. Highly sensitive colorimetric detection of organophosphate pesticides using copper catalyzed click chemistry. *Talanta* 103: 110–115.

Garg, B., Bisht, T., and Ling, Y.-C. 2015. Graphene-based nanomaterials as efficient peroxidase mimetic catalysts for biosensing applications: An overview. *Molecules* 20: 14155–14190.

Ge, S., Lan, F., Yu, F., and Yu, J. 2015. Applications of graphene and related nanomaterials in analytical chemistry. *New Journal of Chemistry* 39: 2380–2395.

Ghaderi, M., Mousavi, M., Yousefi, H., and Labbafi, M. 2014. All-cellulose nanocomposite film made from bagasse cellulose nanofibers for food packaging application. *Carbohydrate Polymers* 104: 59–65.

Gordienko, M. G., Palchikova, V. V., Kalenov, S. V., Belov, A. A., Lyasnikova, V. N., Poberezhniy, D. Y., Chibisova, A. V., Sorokin, V. V., and Skladnev, D. A. 2019. Antimicrobial activity of silver salt and silver nanoparticles in different forms against microorganisms of different taxonomic groups. *Journal of Hazardous Materials* 378: 120754.

Goudarzi, V., Shahabi-Ghahfarrokhi, I., and Babaei-Ghazvini, A. 2017. Preparation of eco-friendly UV-protective food packaging material by starch/TiO2 bio-nanocomposite: Characterization. *International Journal of Biological Macromolecules* 95: 306–313.

Güner, A., Çevik, E., Şenel, M., and Alpsoy, L. 2017. An electrochemical immunosensor for sensitive detection of Escherichia coli O157: H7 by using chitosan, MWCNT, polypyrrole with gold nanoparticles hybrid sensing platform. *Food Chemistry* 229: 358–365.

Guo, S., and Wang, E. 2007. Synthesis and electrochemical applications of gold nanoparticles. *Analytica Chimica Acta* 598: 181–192.

Haasnoot, W., Marchesini, G. R., and Koopal, K. 2006. Spreeta-based biosensor immunoassays to detect fraudulent adulteration in milk and milk powder. *Journal of AOAC International* 89: 849–855.

Haasnoot, W., Smits, N. G., Kemmers-Voncken, A. E., and Bremer, M. G. 2004. Fast biosensor immunoassays for the detection of cows' milk in the milk of ewes and goats. *Journal of Dairy Research* 71: 322–329.

Hashwan, S. S. B., Khir, M. H. B. M., Al-Douri, Y., and Ahmed, A. Y. 2020. Recent progress in the development of biosensors for chemicals and pesticides detection. *IEEE Access* 8: 82514–82527.

He, X., and Hwang, H.-M. 2016. Nanotechnology in food science: Functionality, applicability, and safety assessment. *Journal of Food and Drug Analysis* 24: 671–681.

Holzinger, M., Le Goff, A., and Cosnier, S. 2014. Nanomaterials for biosensing applications: A review. *Frontiers in Chemistry* 2: 63.

Huang, Y., Mei, L., Chen, X., and Wang, Q. 2018. Recent developments in food packaging based on nanomaterials. *Nanomaterials* 8: 830.

Jayakumar, A., Heera, K., Sumi, T., Joseph, M., Mathew, S., Praveen, G., Nair, I. C., and Radhakrishnan, E. 2019. Starch-PVA composite films with zinc-oxide nanoparticles and phytochemicals as intelligent pH sensing wraps for food packaging application. *International Journal of Biological Macromolecules* 136: 395–403.

Jha, S. N., Jaiswal, P., Grewal, M. K., Gupta, M., and Bhardwaj, R. 2016. Detection of adulterants and contaminants in liquid foods—a review. *Critical Reviews in Food Science and Nutrition* 56: 1662–1684.

Jia, B., Mei, Y., Cheng, L., Zhou, J., and Zhang, L. 2012. Preparation of copper nanoparticles coated cellulose films with antibacterial properties through one-step reduction. *ACS Applied Materials & Interfaces* 4: 2897–2902.

Johar, N., Ahmad, I., and Dufresne, A. 2012. Extraction, preparation and characterization of cellulose fibres and nanocrystals from rice husk. *Industrial Crops and Products* 37: 93–99.

Jung, R., Kim, Y., Kim, H.-S., and Jin, H.-J. 2009. Antimicrobial properties of hydrated cellulose membranes with silver nanoparticles. *Journal of Biomaterials Science, Polymer Edition* 20: 311–324.

Kaittanis, C., Santra, S., and Perez, J. M. 2010. Emerging nanotechnology-based strategies for the identification of microbial pathogenesis. *Advanced Drug Delivery Reviews* 62: 408–423.

Kalele, S. A., Kundu, A. A., Gosavi, S. W., Deobagkar, D. N., Deobagkar, D. D., and Kulkarni, S. K. 2006. Rapid detection of Escherichia coli by using antibody-conjugated silver nanoshells. *Small* 2: 335–338.

Kantiani, L., Llorca, M., Sanchís, J., Farré, M., and Barceló, D. 2010. Emerging food contaminants: A review. *Analytical and Bioanalytical Chemistry* 398: 2413–2427.

Kargozar, S., and Mozafari, M. 2018. Nanotechnology and nanomedicine: Start small, think big. *Materials Today: Proceedings* 5: 15492–15500.

Kato, Y., Araki, Y., Juri, M., Ishisaka, A., Nitta, Y., Niwa, T., Kitamoto, N., and Takimoto, Y. 2016. Competitive immunochromatographic assay for leptosperin as a plausible authentication marker of manuka honey. *Food Chemistry* 194: 362–365.

Kim, C.-W., Yun, J.-W., Bae, I.-H., Lee, J.-S., Kang, H.-J., Joo, K.-M., Jeong, H.-J., Chung, J.-H., Park, Y.-H., and Lim, K.-M. 2010. Determination of spatial distribution of melamine– cyanuric acid crystals in rat kidney tissue by histology and imaging matrix-assisted laser Desorption/Ionization quadrupole time-of-flight mass spectrometry. *Chemical Research in Toxicology* 23: 220–227.

Kim, H.-Y., Park, S. S., and Lim, S.-T. 2015. Preparation, characterization and utilization of starch nanoparticles. *Colloids and Surfaces B: Biointerfaces* 126: 607–620.

Kim, J. Y., Voznyy, O., Zhitomirsky, D., and Sargent, E. H. 2013. 25th anniversary article: Colloidal quantum dot materials and devices: A quarter-century of advances. *Advanced Materials* 25: 4986–5010.

Kim, Y.-R., Lee, E. J., Park, S. H., Kwon, H. J., An, S. S. A., Son, S. W., Seo, Y. R., Pie, J.-E., Yoon, M., and Kim, J. H. 2014. Comparative analysis of nanotechnology awareness in consumers and experts in South Korea. *International Journal of Nanomedicine* 9: 21.

Kumar, A., Choudhary, A., Kaur, H., Mehta, S., and Husen, A. 2021. Metal-based nanoparticles, sensors, and their multifaceted application in food packaging. *Journal of Nanobiotechnology* 19: 1–25.

Kumar, H., Kuča, K., Bhatia, S. K., Saini, K., Kaushal, A., Verma, R., Bhalla, T. C., and Kumar, D. 2020. Applications of nanotechnology in sensor-based detection of foodborne pathogens. *Sensors* 20: 1966.

Kumar, P., Kumar, P., Manhas, S., and Navani, N. K. 2016. A simple method for detection of anionic detergents in milk using unmodified gold nanoparticles. *Sensors and Actuators B: Chemical* 233: 157–161.

Kumar, S., Ahlawat, W., Kumar, R., and Dilbaghi, N. 2015. Graphene, carbon nanotubes, zinc oxide and gold as elite nanomaterials for fabrication of biosensors for healthcare. *Biosensors and Bioelectronics* 70: 498–503.

Kumar, S., Shukla, A., Baul, P. P., Mitra, A., and Halder, D. 2018. Biodegradable hybrid nanocomposites of chitosan/gelatin and silver nanoparticles for active food packaging applications. *Food Packaging and Shelf Life* 16: 178–184.

Kuswandi, B. 2016. Nanotechnology in food packaging. *Nanoscience in Food and Agriculture 1*. Springer.

Kuswandi, B., Futra, D., and Heng, L. 2017. Nanosensors for the detection of food contaminants. *Nanotechnology Applications in Food*. Elsevier.

Li, H., and Rothberg, L. J. 2004. Label-free colorimetric detection of specific sequences in genomic DNA amplified by the polymerase chain reaction. *Journal of the American Chemical Society* 126: 10958–10961.

Li, M., Wang, Q., Shi, X., Hornak, L. A., and Wu, N. 2011. Detection of mercury (II) by quantum dot/DNA/gold nanoparticle ensemble based nanosensor via nanometal surface energy transfer. *Analytical Chemistry* 83: 7061–7065.

Liana, D. D., Raguse, B., Wieczorek, L., Baxter, G. R., Chuah, K., Gooding, J. J., and Chow, E. 2013. Sintered gold nanoparticles as an electrode material for paper-based electrochemical sensors. *RSC Advances* 3: 8683–8691.

Liu, S., Li, X., Chen, L., Li, L., Li, B., and Zhu, J. 2018. Tunable d-limonene permeability in starch-based nanocomposite films reinforced by cellulose nanocrystals. *Journal of Agricultural and Food Chemistry* 66: 979–987.

Luan, Q., Gan, N., Cao, Y., and Li, T. 2017. Mimicking an enzyme-based colorimetric aptasensor for antibiotic residue detection in milk combining magnetic loop-DNA probes and CHA-assisted target recycling amplification. *Journal of Agricultural and Food Chemistry* 65: 5731–5740.

Lv, M., Liu, Y., Geng, J., Kou, X., Xin, Z., and Yang, D. 2018. Engineering nanomaterials-based biosensors for food safety detection. *Biosensors and Bioelectronics* 106: 122–128.

Malhotra, B. D., Srivastava, S., and Augustine, S. 2015. Biosensors for food toxin detection: Carbon nanotubes and graphene. *MRS Online Proceedings Library (OPL)*, 1725: 24–34.

Mangaraj, S., Yadav, A., Bal, L. M., Dash, S., and Mahanti, N. K. 2019. Application of biodegradable polymers in food packaging industry: A comprehensive review. *Journal of Packaging Technology and Research* 3: 77–96.

Manhas, P. K., Quintela, I. A., and Wu, V. C. 2021. Enhanced detection of major pathogens and toxins in poultry and livestock with zoonotic risks using nanomaterials-based diagnostics. *Frontiers in Veterinary Science* 8: 602.

Mor-Mur, M., and Yuste, J. 2010. Emerging bacterial pathogens in meat And poultry: An overview. *Food and Bioprocess Technology* 3: 24–35.

Nath, S., Kaittanis, C., Tinkham, A., and Perez, J. M. 2008. Dextran-coated gold nanoparticles for the assessment of antimicrobial susceptibility. *Analytical Chemistry* 80: 1033–1038.

Neethirajan, S., and Jayas, D. S. 2011. Nanotechnology for the food and bioprocessing industries. *Food and Bioprocess Technology* 4: 39–47.

Niu, S., Lv, Z., Liu, J., Bai, W., Yang, S., and Chen, A. 2014. Colorimetric aptasensor using unmodified gold nanoparticles for homogeneous multiplex detection. *PLoS One* 9: e109263.

Pandit, S., Dasgupta, D., Dewan, N., and Prince, A. 2016. Nanotechnology based biosensors and its application. *The Pharma Innovation* 5: 18.

Patel, G., Pillai, V., Bhatt, P., and Mohammad, S. 2020. Application of nanosensors in the food industry. *Nanosensors for Smart Cities*. Elsevier.

Patra, J. K., Shin, H.-S., and Paramithiotis, S. 2018. Application of nanotechnology in food science and food microbiology. *Frontiers in Microbiology* 9: 714.

Pérez-López, B., and Merkoçi, A. 2011. Nanomaterials based biosensors for food analysis applications. *Trends in Food Science & Technology* 22: 625–639.

Pérez-López, B., and Merkoçi, A. 2012. Carbon nanotubes and graphene in analytical sciences. *Microchimica Acta* 179: 1–16.

Petryayeva, E., and Krull, U. J. 2011. Localized surface plasmon resonance: Nanostructures, bioassays and biosensing—A review. *Analytica Chimica Acta* 706: 8–24.

Pikkemaat, M. G., Rapallini, M. L., Karp, M. T., and Elferink, J. A. 2010. Application of a luminescent bacterial biosensor for the detection of tetracyclines in routine analysis of poultry muscle samples. *Food Additives and Contaminants* 27: 1112–1117.

Priyadarshi, R., and Rhim, J.-W. 2020. Chitosan-based biodegradable functional films for food packaging applications. *Innovative Food Science & Emerging Technologies* 62: 102346.

Rai, M., Gade, A., Gaikwad, S., Marcato, P. D., and Durán, N. 2012. Biomedical applications of nanobiosensors: The state-of-The-art. *Journal of the Brazilian Chemical Society* 23: 14–24.

Ramesh, R., Puhazhendi, P., Kumar, J., Gowthaman, M. K., D'souza, S. F., and Kamini, N. R. 2015. Potentiometric biosensor for determination of urea in milk using immobilized arthrobacter creatinolyticus urease. *Materials Science and Engineering: C* 49: 786–792.

Ranjan, S., Dasgupta, N., Chakraborty, A. R., Samuel, S. M., Ramalingam, C., Shanker, R., and Kumar, A. 2014. Nanoscience and nanotechnologies in food industries: Opportunities and research trends. *Journal of Nanoparticle Research* 16: 1–23.

Rovina, K., and Siddiquee, S. 2015. A review of recent advances in melamine detection techniques. *Journal of Food Composition and Analysis* 43: 25–38.

Sadeghizadeh-Yazdi, J., Habibi, M., Kamali, A. A., and Banaei, M. 2019. Application of edible and biodegradable starch-based films in food packaging: A systematic review and meta-analysis. *Current Research in Nutrition and Food Science Journal* 7: 624–637.

Salgado, P. R., Di Giorgio, L., Musso, Y. S., and Mauri, A. N. 2019. Bioactive packaging: Combining nanotechnologies with packaging for improved food functionality. *Nanomaterials for Food Applications*. Elsevier.

Sepúlveda, B., Angelomé, P. C., Lechuga, L. M., and Liz-Marzán, L. M. 2009. LSPR-based nanobiosensors. *Nano Today* 4: 244–251.

Shafiq, M., Anjum, S., Hano, C., Anjum, I., and Abbasi, B. H. 2020. An overview of the applications of nanomaterials and nanodevices in the food industry. *Foods* 9: 148.

Shao, Y., Wang, J., Wu, H., Liu, J., Aksay, I. A., and Lin, Y. 2010. Graphene based electrochemical sensors and biosensors: A review. *Electroanalysis: An International Journal Devoted to Fundamental and Practical Aspects of Electroanalysis* 22: 1027–1036.

Sharma, A., Matharu, Z., Sumana, G., Solanki, P. R., Kim, C., and Malhotra, B. 2010. Antibody immobilized cysteamine functionalized-gold nanoparticles for aflatoxin detection. *Thin Solid Films* 519: 1213–1218.

Sharma, C., Dhiman, R., Rokana, N., and Panwar, H. 2017. Nanotechnology: An untapped Resource for food packaging. *Frontiers in Microbiology* 8: 1735.

Singh, T., Shukla, S., Kumar, P., Wahla, V., Bajpai, V. K., and Rather, I. A. 2017. Application of nanotechnology in food science: Perception and overview. *Frontiers in Microbiology* 8: 1501.

Song, K.-M., Cho, M., Jo, H., Min, K., Jeon, S. H., Kim, T., Han, M. S., Ku, J. K., and Ban, C. 2011. Gold nanoparticle-based colorimetric detection of kanamycin using a DNA aptamer. *Analytical Biochemistry* 415: 175–181.

Songa, E. A., and Okonkwo, J. O. 2016. Recent approaches to improving selectivity and sensitivity of enzyme-based biosensors for organophosphorus pesticides: A review. *Talanta* 155: 289–304.

Sperling, R. A., Gil, P. R., Zhang, F., Zanella, M., and Parak, W. J. 2008. Biological applications of gold nanoparticles. *Chemical Society Reviews* 37: 1896–1908.

Sutarlie, L., Aung, K. M. M., Lim, M. G. L., Lukman, S., Cheung, E., and Su, X. 2014. Studying protein–DNA complexes using gold nanoparticles by exploiting particle aggregation, refractive index change, and fluorescence quenching and enhancement principles. *Plasmonics* 9: 753–763.

Swanson, M., and Hammer, B. 2000. Legionella pneumophila pathogenesis: A fateful journey from amoebae to macrophages. *Annual Reviews in Microbiology* 54: 567–613.

Tan, M. J., Hong, Z.-Y., Chang, M.-H., Liu, C.-C., Cheng, H.-F., Loh, X. J., Chen, C.-H., Liao, C.-D., and Kong, K. V. 2017. Metal carbonyl-gold nanoparticle conjugates for highly sensitive SERS detection of organophosphorus pesticides. *Biosensors and Bioelectronics* 96: 167–172.

Tian, H., Weng, Y., Kumar, R., Rani, P., and Guo, G. 2021. Protein-Based biodegradable polymer: From sources to innovative sustainable materials for packaging applications. *Bio-based Packaging: Material, Environmental and Economic Aspects*, 51–67.

Ture, M., Fentie, T., & Regassa, B. (2019). Veterinary drug residue: the risk, public health significance and its management. *Journal of Dairy & Veterinary Science*, 13(2), 001–011.

Valdés, M. G., González, A. C. V., Calzón, J. A. G., and Díaz-García, M. E. 2009. Analytical nanotechnology for food analysis. *Microchimica Acta* 166: 1–19.

Vasudevan, D., Gaddam, R. R., Trinchi, A., and Cole, I. 2015. Core–shell quantum dots: Properties and applications. *Journal of Alloys and Compounds* 636: 395–404.

Wang, B., Wang, Q., Cai, Z., and Ma, M. 2015. Simultaneous, rapid and sensitive detection of three food-borne pathogenic bacteria using multicolor quantum dot probes based on multiplex fluoroimmunoassay in food samples. *LWT-Food Science and Technology* 61: 368–376.

Wang, C., Chang, T., Dong, S., Zhang, D., Ma, C., Chen, S., and Li, H. 2020. Biopolymer films based on chitosan/potato protein/linseed oil/ZnO NPs to maintain the storage quality of raw meat. *Food Chemistry* 332: 127375.

Wang, H., Li, Y., Wang, A., and Slavik, M. 2011. Rapid, sensitive, and simultaneous detection of three foodborne pathogens using magnetic nanobead–based immunoseparation and quantum dot–based multiplex immunoassay. *Journal of Food Protection* 74: 2039–2047.

Wang, Y., Yan, B., and Chen, L. 2013. SERS tags: Novel optical nanoprobes for bioanalysis. *Chemical Reviews* 113: 1391–1428.

Wang, Z., Zhang, N., Wang, H.-Y., Sui, S.-Y., Sun, X.-X., and Ma, Z.-S. 2014. The effects of ultrasonic/microwave assisted treatment on the properties of soy protein isolate/titanium dioxide films. *LWT-Food Science and Technology* 57: 548–555.

Weiskopf, S. R., Rubenstein, M. A., Crozier, L. G., Gaichas, S., Griffis, R., Halofsky, J. E., Hyde, K. J., Morelli, T. L., Morisette, J. T., and Muñoz, R. C. 2020. Climate change effects on biodiversity, ecosystems, ecosystem services, and natural resource management in the United States. *Science of the Total Environment* 733: 137782.

Wu, Q., Long, Q., Li, H., Zhang, Y., and Yao, S. 2015. An upconversion fluorescence resonance energy transfer nanosensor for one step detection of melamine in raw milk. *Talanta* 136: 47–53.

Wu, Z., Huang, Y., Xiao, L., Lin, D., Yang, Y., Wang, H., Yang, Y., Wu, D., Chen, H., and Zhang, Q. 2019. Physical properties and structural characterization of starch/polyvinyl alcohol/graphene oxide composite films. *International Journal of Biological Macromolecules* 123: 569–575.

Xiulan, S., Xiaolian, Z., Jian, T., Zhou, J., and Chu, F. 2005. Preparation of gold-labeled antibody probe and its use in immunochromatography assay for detection of aflatoxin B1. *International Journal of Food Microbiology* 99: 185–194.

Yadav, S., Mehrotra, G., and Dutta, P. 2021. Chitosan based ZnO nanoparticles loaded gallic-acid films for active food packaging. *Food Chemistry* 334: 127605.

Yang, W., Ratinac, K. R., Ringer, S. P., Thordarson, P., Gooding, J. J., and Braet, F. 2010. Carbon nanomaterials in biosensors: Should you use nanotubes or graphene? *Angewandte Chemie International Edition* 49: 2114–2138.

Yang, Y., Asiri, A. M., Du, D., and Lin, Y. 2014. Acetylcholinesterase biosensor based on a gold nanoparticle–polypyrrole–reduced graphene oxide nanocomposite modified electrode for the amperometric detection of organophosphorus pesticides. *Analyst* 139: 3055–3060.

Yazdian-Robati, R., Arab, A., Ramezani, M., Abnous, K., and Taghdisi, S. M. 2017. Application of aptamers in treatment and diagnosis of leukemia. *International Journal of Pharmaceutics* 529: 44–54.

Yin, J., Wu, T., Song, J., Zhang, Q., Liu, S., Xu, R., and Duan, H. 2011. SERS-active nanoparticles for sensitive and selective detection of cadmium ion (Cd^{2+}). *Chemistry of Materials* 23: 4756–4764.

Yoon, S. Y., and Deng, Y. 2006. Clay–starch composites and their application in papermaking. *Journal of Applied Polymer Science* 100: 1032–1038.

Yoosaf, K., Ipe, B. I., Suresh, C. H., and Thomas, K. G. 2007. In situ synthesis of metal nanoparticles and selective naked-eye detection of lead ions from aqueous media. *The Journal of Physical Chemistry C* 111: 12839–12847.

Yu, G., Wu, W., Zhao, Q., Wei, X., and Lu, Q. 2015. Efficient immobilization of acetylcholinesterase onto amino functionalized carbon nanotubes for the fabrication of high sensitive organophosphorus pesticides biosensors. *Biosensors and Bioelectronics* 68: 288–294.

Yu, J., Cui, G., Wei, M., and Huang, J. 2007. Facile exfoliation of rectorite nanoplatelets in soy protein matrix and reinforced bionanocomposites thereof. *Journal of Applied Polymer Science* 104: 3367–3377.

Zahin, N., Anwar, R., Tewari, D., Kabir, M., Sajid, A., Mathew, B., Uddin, M., Aleya, L., and Abdel-Daim, M. M. 2020. Nanoparticles and its biomedical applications in health and diseases: Special focus on drug delivery. *Environmental Science and Pollution Research* 27: 19151–19168.

Zhang, H., and Chen, S. 2019. Nanoparticle-based methods for food safety evaluation. *Evaluation Technologies for Food Quality*. Elsevier.

Zhang, X., Mobley, N., Zhang, J., Zheng, X., Lu, L., Ragin, O., and Smith, C. J. 2010. Analysis of agricultural residues on tea using d-SPE sample preparation with GC-NCI-MS and UHPLC-MS/MS. *Journal of Agricultural and Food Chemistry* 58: 11553–11560.

Zhao, J., Jing, P., Xue, S., and Xu, W. 2017. Dendritic structure DNA for specific metal ion biosensor based on catalytic hairpin assembly and a sensitive synergistic amplification strategy. *Biosensors and Bioelectronics* 87: 157–163.

Zhao, W., Brook, M. A., and Li, Y. 2008. Design of gold nanoparticle-based colorimetric biosensing assays. *ChemBioChem* 9: 2363–2371.

Zheng, L., Zhang, C., Ma, J., Hong, S., She, Y., Abd Ei-Aty, A., He, Y., Yu, H., Liu, H., and Wang, J. 2018. Fabrication of a highly sensitive electrochemical sensor based on electropolymerized molecularly imprinted polymer hybrid nanocomposites for the determination of 4-nonylphenol in packaged milk samples. *Analytical Biochemistry* 559: 44–50.

Zhou, N., Luo, J., Zhang, J., You, Y., and Tian, Y. 2015. A label-free electrochemical aptasensor for the detection of kanamycin in milk. *Analytical Methods* 7: 1991–1996.

5 Nanotechnology Derived Antimicrobial Packaging

Iqra Bashir, Qudsiya Ayaz, Haamiyah Sidiq, Aiman Zehra, and Sajad Mohd Wani
Division of Food Science and Technology,
Sher-e- Kashmir University of Agricultural Sciences and Technology, Shalimar Srinagar, J&K, India

CONTENTS

5.1 INTRODUCTION

Packaging is the process of enclosing produced goods in a material to safeguard, store, handle, distribute, and identify each item as it makes its way down the supply chain from raw materials to final consumers (Kuswandi, 2016). Preservation, information, convenience, and containment are the four basic components of food packaging provided by conventional packaging. Almost all of the materials utilized in conventional petroleum-based plastic packaging are non-biodegradable polymers. Consequently, around the globe, non-biodegradable food packaging materials have created a great environmental issue (Kirwan et al., 2011). New packaging options

DOI: 10.1201/9781003207641-6

(like smart and active packaging) are selected for research work to meet the expectations of modern society while also protecting the environment (Majid et al., 2018). Smart packaging updates consumers regarding the kinetic changes that are happening to the food quality or the environment in which food product is kept. To achieve this, packaging is integrated with time-temperature, gas indicators, and biosensors. The issue with this technology is the production of sensitive sensors that need to be safe and easy to interpret while keeping the packaging costs economical (Lloyd et al., 2019; Silvestre et al., 2012; Yucel, 2016). Active packaging is made up of components that facilitate releasing or absorbing substances inside the packaged product or in the environment to increase the food's quality, usability, and safety. Active packaging systems possess carbon dioxide emitters and absorbers of oxygen and moisture, making these systems the subject of attention and application (Echegoyen and Nerín, 2013; Sothornvit, 2019).

Microorganism exposure and oxidation reactions are the two major causes of food deterioration. Because of growth in the global market for fresh foods and the requirement for centralized logistics, there is a need to implement the distribution of fresh foods, which lengthens their transit times. Hence, there is a demand for unique strategies to suppress the development of microbes resulting in considerable scientific and commercial interest in the development of antimicrobial packaging (da Silva Barbosa and dos Santos, 2019). Antimicrobial packaging is a new type of material that prevents microbial proliferation and offers protective qualities to food (Sofi et al., 2018). Antimicrobial packing has three primary aims: i) quality check, ii) safety, and iii) long life span, which are the opposite order of concern of traditional packaging systems (Han, 2003). Antimicrobial packaging material inhibits or stops the growth rate of microorganisms by lengthening the slack time and, thus, decreasing the microbial population (Han, 2003). It also targets microbes by removing the essential requirements for their growth from the products (Singh and Shalini, 2016). The properties of microbial suppression in antimicrobial packing are obtained by adding the active constituents into the packaging material or by proactively utilizing functional polymeric compounds (Sofi et al., 2018). Antimicrobial substances affect microbes differently due to physiological differences among microbes. Certain antimicrobial substances interfere with basic metabolic functions of microorganism, whereas others alter the makeup of cell membranes and cell walls (Sofi et al., 2018). Moisture-content reduction, low and high temperature protection, pH alteration, the addition of competitive microbes, and preservative agents are all examples of factors that inhibit microorganisms in food products. The use of these protections in combination against microbes is called "hurdle effect" (Leistner, 1994). Antimicrobial packaging is a hurdle that protects packaged food from microorganisms and also prevents the overall quality degradation of foods. Antimicrobial polymeric materials can provide antimicrobial effects in three ways: release, absorption, and immobilization (Ahvenainen, 2003). The method of release permits antimicrobial agents to migrate into foods or into space inside the container, thus preventing the development of microbes. In the absorption mode, vital elements, like O_2, CO_2, pH, and moisture, are removed from food systems, preventing the growth of microbes, thereby inhibiting the growth of bacteria, fungi, and molds within food packages. The immobilization technique does not release antimicrobials, but rather the growth of microbes at the contact surface is inhibited, as occurs when immobilized lysozyme and glucose oxidase enzyme are used in polymer packaging materials (Sofi et al., 2018).

The demand for safe, high-quality, and shelf-stable food has become a concern for the food business, and the application of nanotechnology seems to be a potential strategy to solve this problem. Any technology that is specially executed on a nanoscale and can be used in real world situations is referred to as a nanotechnology. Nanotechnology is generally related to manipulation or reorganization at the atomic level and in a size ranging from 1 to 100 nanometers (Bhushan, 2017). It embraces the development and use of physiological, biological, and biochemical systems, as well as organizational qualities spanning individual atoms to sub-micron scales, as well as the incorporation of subsequent structures into complex systems (Daniel and Astruc, 2004; Rao and Cheetham, 2001). Nanotechnology is a branch of knowledge, having a sub-classifications of technology in colloidal science, chemistry, biology, physics, and other scientific fields, encompassing the study of phenomena at the nanoscale (Mansoori and Soelaiman, 2005). Matter has different properties at the nanoscale than it does on a large scale. When the dimensions of matter decrease from a bigger size, the characteristics at first stay the same but then small changes in properties occur. When the size drops below 100 nm, drastic alterations of properties happen. The exclusive physical and chemical properties of nanomaterials can be exploited for commercial applications that benefit society (Bhushan et al., 2014). Nanomaterials or nanostructures show diverse morphologies, such as clusters, nano-fibers, or crystallites. Nanomaterials possess higher surface-to-volume ratios and reveal exclusive physical and chemical attributes, like solubility, toxicity, and strength as well as magnetic, optical, and thermal properties (Bajpai et al., 2018). Nanotechnology has ushered in a new era of industrialization, and both developed and under-developed nations are ready to increase their investment in such technology (Qureshi et al., 2012). Henceforth, nanotechnology presents a broad range of opportunities for the application of its novel properties in structures, materials, or systems in several areas, such as agriculture, food, and medicine, and so on (Singh et al., 2017). Manufacturers are already using nanotechnology in food packaging to develop packaging materials that increase food safety while simultaneously extending the life of foods and beverages (Kuswandi, 2016). With the aid of nanotechnology, industries manufacture new packaging materials that possess novel attributes. Since the formation of nanocomposites from nanomaterials, there have been gains in the development of packaging materials with improved properties such as greater mechanical and barrier properties (Cerqueira et al., 2018). Nano-based "active" and "smart" food packaging give numerous benefits beyond those already existing packaging systems. These packaging systems provide superior packing materials that have advanced mechanical strength, barrier properties, antimicrobial activity, and nano-sensing for detecting harmful microbes and notifying customers regarding the safety of the food (Mihindukulasuriya and Lim, 2014). Nanotechnology opens up new avenues for developing antimicrobial packaging systems and improving the antibacterial activity of active ingredients. Nano-sized antimicrobial compounds may combat unwanted microbes successfully as they have an improved surface activity and also have very large surface area per unit of volume, unlike microscopic and macroscopic equivalents (Padmavathy and Vijayaraghavan, 2008;

Radusin et al., 2016). Nanoparticles (NPs), in contrast to the two mentioned above, can adhere to the bacteria's surface and interact with them precisely. This can result in alterations to the bacteria's structure, interfering with critical cell activities like permeability, forming gaps, decreasing enzymatic activity, and ultimately destroying the microbial cells (Thirumurugan et al., 2013). Different antimicrobial agents have been employed in food technology to deter the spread of pathogenic microbes and the degradation of packaging material. These agents include inorganic, organic, or biologically active substances that hinder the growth of bacteria (da Silva Barbosa and dos Santos, 2019). In this chapter, emphasis is on the nanotechnology-derived antimicrobial packages.

5.2 METAL NANOPARTICLES AS ANTIMICROBIAL AGENTS

Heavy metals of different forms, like colloids, oxides, and salts are antibacterial in nature. These metals can be added to the food-contact polymers or their surfaces to improve the structural and barrier properties and also to extend the shelf life of food products. The widely used metal nanomaterials are silver (Ag), copper (Cu), and gold (Au) (Corrales et al., 2014).

5.2.1 Silver

Since earlier times, silver (Ag) has been employed as a bactericide for food and beverage storage, and the most powerful antimicrobial property of Ag ions and salts are well recognized (Duncan, 2011; Kim et al., 2007; Morones et al., 2005). The most important antimicrobial operations of silver include inhibiting duplication of deoxyribonucleic acid, the intervention of silver metal with essential biological actions such as attaching to the protein and enzyme surfaces, and the production of reactive oxygen species by stimulating oxidative stress (OS) (Duncan, 2011). However, there are ongoing debates about which of these pathways is the most significant (Duncan, 2011; Kim et al., 2007; Morones et al., 2005). However, the antimicrobial properties of Ag ions and salts are limited in their metallic state due to a number of factors, including salt interfering with the antibacterial process, particularly the continuous release of sufficient Ag ions (Kim et al., 2007). The aforementioned restrictions can be resolved with the use of Ag-NPs. Silver nanoparticles are arguably the most researched nanoparticles in medical as well as food packaging applications, and their antibacterial activity is noted within an extensive range of bacteria, algae, fungi, and yeasts, and probably several viruses (Yildirim et al., 2018). Silver NPs have an increased area of contact and a fraction of atoms on the surface; hence, they have a stronger antimicrobial impact than bulk silver (He et al., 2019). Chemically stable Ag-NPs are a particularly appealing alternative as they are integrated into polymer matrices and they allow for controlled release over extended storage durations (Duncan, 2011). Mahdi et al. (2012) studied the antimicrobial efficacy of polyvinyl chloride (PVC) and polyethylene (PE) laminate trays that had been inkjet printed with Ag-NPs. The nanosilver packaging material is shown to dramatically decrease the growth rate of microorganisms (like total bacteria counts of *S. aureus* and *E. coli*) in minced beef. The minced beef in the above mentioned packaging material at refrigerated conditions has an extended shelf life, from two to seven days. In 2016 Azlin-Hasim and his research team demonstrated that Ag-NPs (as solution) have great antimicrobial action toward *B. cereus*, *E. coli*, *microflora*, *S. aureus*, *P. fluorescens*, and the microflora associated with raw chicken, cooked ham, uncooked beef, and chicken (Azlin-Hasim et al., 2016). The results revealed that the gram negative bacteria were usually extra susceptible to Ag-NPs compared with gram positive bacteria. Besides, the microflora drawn out from meat was much more antagonistic toward Ag-NPs than pure bacterial culture. Ahmed et al. (2018) generated silver and copper nanoparticle-loaded films for poultry meat packing. The antibacterial action of the synthesized linear low-density polyethylene (LLDPE) film bearing Ag-Cu NPs was highest toward *Salmonella typhimurium*, *L. Monocytogenes*, and *Campylobacter jejuni*. Chicken sections infected with *C. jejuni* and *S. Typhimurium*, wrapped in nanocomposites, and kept in the refrigerator for 21 days exhibited thorough inhibition. LLDPE film's tensile, thermodynamic, and insulating characteristics were also enhanced by using NPs.

Silver nanoparticles are also employed in mixture with gold NPs and zeolites minerals. Silver/zeolite and silver/gold in conjunction have a stronger antimicrobial property than silver alone (Pereda et al., 2017). Silver zeolites have an effect on the mechanical qualities of packaging and may help to lower polymer degradation rates. The antibacterial ability of Ag-doped zeolites toward *S. aureus* together with *E. coli* was discovered (Fernandez et al., 2010). Mastromatteo et al. (2015) studied the effect of silver in MAP (modified atmosphere packaging) having 50% CO_2 and 50% N_2 to enhance the shelf-life of fiordilatte packed with and without conventional covering fluid. The outcome suggested that the combined result of Ag-NPs in the coating could display suitable protection to expand the marketability of the fiordilatte dairy product beyond the local market. Due to the active antimicrobial action of the packaging against *Pseudomonas spp.*, *Enterobacteriaceae* and *E. coli.*, Ahmed et al. (2018) developed polylactic acid composite films by packing bimetallic Ag-Cu NPs and cinnamon essential oil in a matrix of polymer by a compression molding method. The film was utilized in packing of poultry; the film represented a novel type of active packaging, to limit the growth of decay and harm-causing microbes connected with fresh chicken meat.

5.2.2 Copper

Copper (Cu) is a co-factor for metal protein and enzymes, and it can be found in most foods as ions or salts (Costa et al., 2016; Llorens et al., 2012). Copper has been considered an antibacterial component for centuries as ancient Egyptians used it to sterilize water and disinfect wounds (Borkow and Gabbay, 2009). Copper possesses advantages for active packing as it demonstrates a wide range of antimicrobic properties. Copper ions kill bacteria by quickly receiving and giving electrons and having strong reduction and oxidation potential as well as the capacity to damage microbial cell components (Nan et al., 2008). Copper's antibacterial characteristics are widely known in numerous forms, including oxides of copper, zero-valent Cu, copper in ionic form, and molecular complexes (Zhong et al., 2017). Cu-NPs are preferable over Ag-NPs due to their low price, physical and mechanical durability, and easiness with which they mix with polymers (Costa et al., 2016; Shankar and Rhim, 2014). Researchers have shown that since they have a large specified surface area ratio, reactivity, and optimal discharge properties, Cu-NPs have a high ability to combine with the polymer network and induce biotechnological actions (Anvar et al., 2019). Shankar et al. (2014) showed the antimicrobic action of ecological agar-based bio-nanocomposite films that contain distinct types of Cu salts. All films displayed great antimicrobial action toward both gram positive and gram negative food-borne bacteria (*L. Monocytogenes* and *E. coli*, respectively). Cellulose/Cu composites developed by Llorens et al. (2012) were recorded to be effective against *Saccharomyces cerevisiae* in vitro. Furthermore, the nanocomposite absorbent materials revealed exceptional antibacterial and antifungal action in pineapple and melon juice; load of decay-causing yeasts and molds were reduced to about four log cycles. Bikiaris and Triantafyllidis (2013) investigated the impacts of high density polyethylene (HDPE) and copper-nanofibers (in concentrations of 0.5%, 1%, and 5%). According to the findings, Cu-NPs had considerably improved mechanical properties, Young's modulus, and oxidation resistance abilities, in addition to antimicrobial action toward *Pseudomonas fluorescens BS3*, *Escherichi coli DHSa*, and *Staphylococcus aureus.* Also, nano-copper oxide was incorporated in LDPE to manufacture a nanocomposite with antimicrobic characteristics, and the composite was tested on the packaging of cheese. The results revealed a decline of 4.21 log cfu/g of coliforms over one month of storage under refrigeration (Beigmohammadi et al., 2016). Lomate et al. (2018) created a packaging material for Peda (Indian sweet dairy product) that uses CuO-NPs to enhance the product life span. Cu-NPs were observed to diffuse evenly in the LDPE polymer, improving tensile characteristics. Nanoparticle material seems to have a better antibacterial impact toward gram positive and gram negative bacteria, which are responsible for food contamination.

5.2.3 Gold

Long ago in the west, gold was called nervine because it was thought to be able to cure neurological problems (Hoseinnejad et al., 2018). Due to its special antibacterial effect, gold nanoparticles (Au-NPs) have been regarded as excellent bactericidal options. Typically, gold nanoparticles demonstrate their antimicrobial effects in one of two different ways: The first method involves altering membrane charge, therefore limiting ATP synthetase functions, and lowering ATP content and thereby delaying metabolic activity. The second method involves limiting ribosome subunit formation for tRNA, ensuing in metabolic dysfunction. Au-NPs are ideal compounds for stopping the multiplication of microbes due to their strong potential for standardization (Cui et al., 2012). Au-NPs have been found to be a viable agent in food packaging systems (Hoffmann et al., 2019). Pagno et al. (2015) assessed the biocidal action of Au-NPs in activated biofilms of *Chenopodium quinoa* toward *S. aureus* and *E.coli*. The use of gold-NPs in the bio-film significantly decreased the amount of bacteria (by 99%) while also improving the physical and optical features of the bio-film. Ramakritina et al. (2013) assessed the effects of PVA/Au-NPs composites toward gram negative and gram positive bacteria, like *E. coli* and *S. aureus*, respectively.

5.3 METAL OXIDE NANOPARTICLES AS ANTIMICROBIAL AGENTS

Antimicrobial packaging films can be made from inorganic metal oxides, like titanium dioxide, zinc oxide, or magnesium oxide. Apart from having antimicrobial activity, these metal oxides have mineral elements that are important to humans even when used in low quantities (Sawai and Yoshikawa, 2004). They are utilized as photocatalysts; that is, they collect energy from a light source to generate catalytic activity. UV light leads in the production of highly reactive oxygen species (ROS) that represent and important mechanism of their antimicrobial action (Yildirim et al., 2018). The chemical stability and antimicrobial activity of metal oxides are better than those of organic antimicrobial compounds (Zhang et al., 2010).

5.3.1 Titanium Dioxide

Titanium dioxide (TiO_2) is a harmless antimicrobial agent that can credibly kill bacteria and fungi present on food and packaging surfaces. On exposure to UV light, the production of ROS by TiO_2 nanoparticles is linked to their antimicrobial action. To suppress or minimize microbial development on the surface of solid food, TiO_2 nanoparticles (TiO_2-NPs) are integrated within the matrixes of polymer or applied as a coating on the surface of polymer film (Hernández-Muñoz et al., 2019). TiO_2-NPs weighing 2–5% were inserted into ethylene-vinyl alcohol (EVOH) to manufactured nanocomposites that have auto-sterilizing characteristics. The nanocomposite displayed inhibitory effects toward gram positive and gram negative bacteria, respectively (Cerrada et al., 2008). Also, Razali et al. in 2019 incorporated titanium oxide nanoparticles in gellan gum film and the studies reveal that the film possessed acceptable antimicrobial action toward *S. aureus, Streptococcus, Pseudomonas aeruginosa*, and *E. coli* (Razali et al., 2019). Xing and his co-workers (Xing et al., 2012) checked the antimicrobial action of titanium oxide nanoparticles by employing polyethylene (PE) based films as a composite. The findings revealed that UV irradiation drastically enhanced the antibacterial property to neutralize *E. coli* or *S. aureus*. The outcome also showed that the inhibition ratio for *E. coli* is 89% and for *S. aureus* it is 95%, when compared to TiO_2-PE film that is not irradiated. Also, LDPE films incorporated with TiO_2-NPs (anatase and rutile) prevents the in vitro *Pseudomonas spp.* and yeasts. The films were then used for packaging pears, and the results reveal that antimicrobial activity of the film was improved on exposing to UV-A light (Bodaghi et al., 2013). By combining polyethylene with Ag/TiO_2/kaolin nanopowder, Li et al. (2009) generated packaging films. The developed nano-packaging films have been used successfully to prevent mold formation in strawberries (Yang et al., 2010), Chinese jujube (Li et al., 2009), and Chinese bayberries (Wang et al., 2018). In addition, nano-Ag/TiO_2 was used in high density polyethylene films and for bread packaging. The result showed that the films were effective against yeasts, molds, and bacteria (Mihaly Cozmuta et al., 2015).

5.3.2 Zinc Oxide (ZnO)

Zinc oxide (ZnO) is frequently used in the food industry as a zinc supplement, and it has also been used in the linings of food cans to prevent rotting and preserve color (Yildirim et al., 2018). The antimicrobial action of Zn nanoparticles (NPs) against food-borne pathogens has been investigated carefully, but the definite mechanism of its antibacterial activity remains unidentified. Initially, the primary mechanism of action was thought to be the discharge of antimicrobial ions, then interaction of ZnO-NPs with microbes, followed by bacterial cell destruction and the reactive oxygen species (ROS) being formed on exposure to UV light. Thus, visible light or UV radiation can activate ZnO-NPs and results in the release of ROS, like hydrogen peroxide (H_2O_2), superoxide (O_2• -), or hydroxyl radicals (•OH) (Espitia et al., 2012; Padmavathy and Vijayaraghavan, 2008). In 2009, Jin and his fellow researchers used nano-ZnO in powders, films, capped polyvinyl-prolidone (PVP), and coatings (Jin et al., 2009). They determined that all the developed packaging had antimicrobial effects against *L. monocytogenes* and *Salmonella entiriditis* in liquid egg and bacteria media. In addition, orange juice packed in LDPE that contained Ag+ and ZnO nanoparticles had a lower count of deteriorating microbes (Emamifar et al., 2010). Calcium alginate film was prepared by Akbar and Anal (2014) and the film was coated with ZnO-NPs (approximately 50 nm). The antimicrobial efficacy of film was tested against *S. typhimurium* and *S. aureus* in ready-to-eat chicken meat. Results revealed that the alginate-based film that had ZnO-NP concentration of 3 mg/mL was most suited to and successful when used in active packaging. Challenge testing was done at 8±1°C with poultry meat wrapped in active film, and a reduction of 2 log cycles in the original quantity of inoculation bacteria (*S. aureus* and *S. typhimurium* 10^6–10^7 CFU/mL) was observed within a day. No active *S. aureus* or *S. typhimurium* cells were found after six and eight days of incubation, respectively. Essential oils, which are widely known for their antibacterial properties, have also been tested in conjunction with ZnO-NPs. Arfat et al. (2015) used fish protein isolate/fish skin gelatin films that contain about 3% zinc-oxide nanoparticles and basil leaf essential oil to wrap sea bass slices. In the course of 12 days and at a temperature 4°C, the slices that were wrapped in the prepared, active film showed the lowest growth for LAB (lactic acid bacteria), psychrophilic bacteria, and other decay-causing microbes (*Enterobacteriaceae*, H_2S-producing bacteria, and *Pseudomonas*) in comparison to films that contained only basil leaf essential oil without zinc oxide nanoparticles, or vice versa. Tankhiwale and Bajpai (2012) coated PE film with a starch coating ZnO-NPs on it. The resulting substance had antibacterial activity against *E. coli.*

5.3.3 Magnesium Oxide (MgO)

MgO nanoparticles (MgO-NPs) have also been shown to have an antibacterial effect on vegetative bacteria (Jin and He, 2011; Stoimenov et al., 2002; Zhang et al., 2010) and bacterium spores (Zhang et al., 2010). MgO nanoparticles deform and destroy the microbial cell membrane, allowing the constituents of the cell to flow out, resulting in cell death (Jin and He, 2011). Jin and He (2011) concluded that magnesium oxide nanoparticles are very effective against harmful bacteria, and they achieve log reductions of more than 7 in microbial counts of *E. coli* O157:H7 and *Salmonella Stanley,* respectively. Furthermore, mixed Zn–MgO NPs were integrated into an alginate film and the results showed that film inhibited *L. monocytogenes* proliferation in cold smoked salmon meat (Vizzini et al., 2020).

Silica (SiO_2) and aluminum oxide (Al_2O_3) are other metal oxide nanoparticles that could be used in the creation of antimicrobial food packaging. According to Luo et al. (2015), SiO_2 nanoparticles can be integrated into polymers to provide an active packaging system that is antibacterial and enzyme inhibitory. In addition, a multi-functional nanocomposite made of SiO_2/TiO_2/Ag- NPs were integrated into LDPE that was capable of regulating the carbon dioxide and oxygen levels, scavenging ethylene, and suppressing bacterial development, which helped in prolonging the shelf life of packed mushrooms (Donglu et al., 2016). The antimicrobial properties of Al_2O_3-NPs are significant. Al_2O_3-NPs had a moderate growth limiting effect on *E. coli*, but only at exceptionally high doses, which may be related to electrostatic interaction of charges among particles and the bacterial cells (Sadiq et al., 2009).

5.4 ORGANIC BIOPOLYMER-BASED NANOMATERIALS AS ANTIMICROBIAL AGENTS

Production of biopolymer-based materials was inspired by the desire to replace non-renewable fossil fuels with those from inexhaustible organic sources, such as polysaccharides and proteins. Organic nanoparticles, which are primarily separated into polysaccharide-based and protein-based nanomaterials are used in food packaging to offer a biopolymer matrix for nanocomposites (Huang et al., 2018).

5.4.1 Polysaccharide-Based Nanomaterials

Polysaccharides are macromolecules formed of multiple monosaccharide units that are connected by glycosidic linkages, which can be linear or branched. These biopolymers, also known as glycans, are divided into homo-polysaccharides (monosaccharide units are all alike) and heteropolysaccharides (monosaccharide units are not alike; i.e., made of two or more unlike kinds). The glycosidic linkages could be α or β (1→4, 1→6, 1→3, etc.). Depending on the role of polysaccharides, they may be structural polysaccharides (like cellulose and chitin) or storage polysaccharides (like starch, inulin, and glycogen). Polysaccharides (such as starch, cellulose, alginate, and chitosan) can be used to make packaging materials because of their low cost. The majority of them can be utilized in the production of nanoparticles (Lopez-Lopez et al., 2015).

5.4.1.1 Chitosan

Chitin, a polymer of N-acetyl glucosamine entities is joined by β (1→4) glycoside linkages. It is a natural polysaccharide. Present abundantly, it is the major upholding substance of crustaceans and insects. It has a modest chemical reactivity and is generally water insoluble (Lopez-Lopez et al., 2015). Chitosan is naturally derived from deacetylated chitin (Corrales et al., 2014). Biocompatibility, biodegradability, non-antigenicity, antibacterial activity, and metal complexation are just a few of the outstanding features of chitosan. Its antibacterial activity is due to its polycationic nature, which works in three ways: i) electrostatic contact involving chitosan's protonated amino groups and cell surface negative residues, resulting in plasma membrane breakdown as well as cell wall permissibility; ii) metal chelation, which may result in instability of cell wall or outer covering of gram negative bacteria; and iii) internalized chitosan that interacts with the DNA of a microorganism, thus potentially inhibiting gene expression (Marquez et al., 2013). Also, the entrance of chitosan in the bacterial cell nucleus inhibits the synthesis of RNA and protein. It may also restrict microbial growth by serving as a chelating agent for vital nutrients, or for essential elements that the organism needs in order to reproduce at its usual pace. It is unclear whether the antibacterial properties of chitosan are due to growth limitation of bacterial cells or to its death (Aranaz et al., 2009). The antimicrobial action of chitosan depends on its lower molecular weight, elevated temperature, high degree of deacetylation, high concentration in solution, lesser pH of the medium, and static metal oxides in its matrix. Radiation exposure of chitosan boosts its antibacterial activity (Honarkar and Barikani, 2009). Chitosan nanoparticles are generally made by the reaction of oppositely charged macromolecules. The electrostatic interaction between the amine groups of chitosan and the negatively charged groups of a polyanion, such as tripolyphosphate, is used to make chitosan nanoparticles (Zhao et al., 2011). Chitosan nanoparticles tend to diffuse well in biopolymers, indicating that they could be employed against microbes in biodegradable or edible food packaging (Hoffmann et al., 2019). In one study, tara gum (TG) edible films enclosing bulk chitosan (CS) and chitosan nanoparticles (CNPs) were developed as food packaging (Antoniou et al., 2015). By homogenization, various concentrations (0–15% wt/wt) of CS along with CNPs suspensions (150 mL) were provided at 1% w/v TG and 20% w/v of a plasticizer (glycerol) and the concentration of acetic acid in all of the solutions was the same (0.42% v/v). Growth inhibitory zone (mm^2) of TG edible films holding CNPs and CS over *E.coli* and *S. aureus* was 41.53 and 85.30 at 5% CS/CNP content, as well as 87.32 and 111.71 at 10% CS/CNP content, and 85.80 and 93.41 at 15% CS/CNP content for TG + CNPs type films, respectively. A quick look at the results showed that the CS had improved antimicrobial action over the CNPs. Furthermore, CNPs were highly effective toward *S. aureus* (gram positive) than *E. coli* (gram negative), and the antimicrobial ability of CNPs was reduced at increased levels, due to their agglomeration. Cui et al. (2020) developed biodegradable and active zein films to package foods and the films were loaded with pomegranate peel extract (PE) that was encapsulated in CNPs. The ionic gelation process was used to develop this nanocomposite (CNPs/PE) and the prepared nanocomposite was incorporated into zein film during nanocomposite preparation by solution casting technique. The antimicrobial activity of fabricated zein/CNPs/PE nanocomposite film, before and after the plasma treatment, was considered for fresh pork meat inoculated with *L. monocytogenes* within a storage period. The outcome revealed that the number of *L. monocytogenes* cells rose from 3.40 to 7.38 log cfu/g after storage of 14 days at 4°C for fresh pork meat, while 3.2 log cfu/g of cells were observed for fresh pork meat in zein/CNPs/PE nanocomposite, for a storage period of ten days (4°C). Additionally, nanocomposite films that were treated with plasma revealed a gentle release of PE and potent antibacterial action against *L. monocytogenes* during refrigerated storage (4°C). As a matter of fact, the occurrence of hydrolyzable polyphenols in PE besides the CNPs and their controlled release upon plasma treatment could lyse cell membrane of bacteria, thus imparting efficient antimicrobial attributes. Chitosan nanostructures have also been shown to diffuse well in bio-polymers, suggesting that they could be employed as antibacterial substances in edible or biodegradable food packaging. Several studies focused on the antibacterial effects of nanostructures integrating chitosan and other antimicrobial agents (Peng et al., 2013; Medina et al., 2019). The inclusion of various metal ions also improves chitosan's antibacterial properties (Lopez-Lopez et al., 2015). The zeta potential of CNPs increased considerably with positive charge of ions, and it helps in the increase of stability of NPs and enhances their antimicrobial activity (Arora et al., 2016). The antibacterial efficacy of CNPs loaded with Ag+, Cu^{2+}, Zn^{2+}, and Mn^{2+} was investigated against *S. aureus* and *S. cholerasuis* by Du et al. (2009). Among these metals, CNPs loaded with copper ions had the best antibacterial action. This activity was better for the NPs than free ions and was directly linked to their zeta potential (Du et al., 2009). Shapi'i et al. (2020) discovered that starch/CNP sheets have the capacity to be used as antibacterial packaging. They discovered that cherry tomatoes packed in starch/CNP films had the least mold growth due to CNP's small size, which inhibits the microbial activity. Tripathi et al. (2011) also developed chitosan silver oxide encapsulated nanocomposite film by solution casting that tested as effective toward pathogenic bacteria like *E. coli, S. aureus, Bacillus subtilis*, and *Pseudomonas aeruginosa*. Gull et al. (2021) studied the consequence of nano-chitosan coatings on the post-harvest quality of apricots containing extract of pomegranate peel (PPE) at concentrations 0.5, 0.75, and 1% (w/v) for a storage period of 30 days at 4°C. The results showed that apricot fruit treated with chitosan and 1% PPE had lower decay percentage and had retained the DPPH radical scavenging activity, ascorbic acid, and firmness than untreated apricots.

5.4.1.2 Starch

This natural polysaccharide is made up of amylose and amylopectin (Nešić et al., 2020). Starch is inexhaustible, compostable, and can be easily modified both physically and chemically, and it is inexpensive (Dai et al., 2019). It is generated by various plants and is suitable for the manufacturing of nanoparticles (Le Corre et al., 2010). Any positive charge that occurs on an antimicrobial agent plays a key role in its action against the microbes. Thus, the antimicrobial action of metals that are absorbed on the polysaccharide surface escalates significantly due to the increase in its surface area (Arora et al., 2016). Many antimicrobial drugs are compatible with starch, and functional films prepared from starch have proved effectively efficient against wide number of microorganisms (Mlalila et al., 2018). Starch nanoparticles (SNPs) are defined as particles that possess at least one measurement that is smaller than 1000 nm but are larger than a single molecule. Furthermore, the literature frequently specifies more rigorous size criteria; such as, at least one dimension must not exceed than 300 nm (Sun et al., 2014). The goal of developing starch nanocrystals or NPs is to employ them as filler in polymeric matrices in order to enhance their mechanical and/or barrier properties (Sandhu and Nain, 2017). Nanostructured starch materials are being primarily manufactured by solution casting (Majdzadeh-Ardakani et al., 2010), co-precipitation (Chung et al., 2010), and melt intercalation techniques (Lee et al., 2008). As per Le Corre et al. (2012), starch nanocrystals are crystalline platelets formed when the semi-crystalline morphology of starch molecules is disrupted by acid hydrolysis of their amorphous state. Researchers have demonstrated that hydrolysis is oriented toward amylose, and the SNPs are chemically changed. The resultant starch then has varied concentrations of amylose, which affects the process yield, crystal form, and industrial uses of starch, thereby creating a new avenue for research. Nano-starch-based film was developed with ginger starch; the developed film revealed increased antimicrobial, anti-inflammatory, and anti-analgesic activity because of the presence of components like terpenes, oleoresin, and gingerol (Aisyah et al., 2018). SNPs show effective action against *E. coli*, *Salmonella typhi*, *Candida albicans*, and *Bacillus subtilis* (Chen et al., 2008). Starch stabilized with Ag-NPs were studied for antimicrobial and biofilm inhibition ability toward *S. aureus, P. aeruginosa, Shigella flexneri*, and *Salmonella typhi* (Mohanty et al., 2012).

5.4.1.3 Cellulose

Cellulose is the most plentiful natural resource present on the earth. It is found in trees, plants (like cotton, jute, flax, hemp, sisal, coir, ramie, abaca, and kenaf), marine life (like tunicates), algae (like red, green, grey, and yellow-green), and bacteria. This polysaccharide results from the linear condensation of glucose units joined by β (1→4) glycosidic linkages (Lopez-Lopez et al., 2015). Cellulose nanoparticles (CNPs), together with cellulose nano-fibers (CNFs) and cellulose nano-crystals (CNCs), can be drawn by acid hydrolyzing the cellulose, enzyme treatment, mechanical disintegration, and 2,2,6,6-Tetramethyl-1- piperidinyloxy (TEMPO) mediated oxidation. Due to their large surface area, high aspect ratio, and Young's modulus and also because they are economical, lightweight, inexhaustible and biodegradable, CNPs are being utilized as multifunctional substances in different fields, such as engineering composites, paper-package films, biomedical materials, hydrogels, aerogels, magnetic nano-rods, and super-capacitors (Li et al., 2015). Nano cellulose comes in a diversity of shapes and sizes, including spheres, rods, and whiskers, all of which have a highly crystalline structure (Bondeson et al., 2006; Li et al., 2015). Various sources of CNPs have been discovered, which include banana peel, blue agave leaves, rice straw (Wang et al., 2018), and potato peel (Ramesh and Radhakrishnan, 2019). Eco-friendly bio-nanocomposites can be manufactured by incorporating nanocellulose into biopolymers. Jancy et al. (2020) conducted an investigation in which optimization of CNP-reinforced polyvinyl alcohol (PVA)-based film was done by central composite design. The outcome revealed that the optimized bio-nanocomposite film outperformed the PVA film in tensile properties (7 times), ductility (6 times), antioxidant properties (21 times), and antibacterial properties (11 times). Also, the incorporation of essential oil greatly decreased growth of *E.coli* and, thus, improved the film's activity against microbes. Nanosilver (when loaded in the range of 0.5 wt % to 1 wt %) for polylactic acid/nanocellulose biocomposites generated antimicrobial effects toward *S. aureus* and *E. coli* (Gan and Chow, 2018). Yang et al. (2020) used sugarcane bagasse cellulose nanofibrils (CNFs) and nisin to create a nanocellulose-based hybrid film with antibacterial characteristics. CNFs/nisin hybrid film was utilized as LDPE plastic packaging for ready-to-eat ham. The outcome revealed that during storage period of seven days and a temperature 4°C, *L. monocytogenes* were completely inhibited.

Several more polysaccharides were also used as antimicrobial packaging. Alginate nanoparticles were used in food packaging in order to demonstrate antimicrobial properties. It was revealed that nisin filled with chitosan/alginate NPs were effective against *S. aureus* in raw and pasteurized milk (Zohri et al., 2010). A study by Manzoor et al. (2021) also reported that nanoemulsions made from 2% alginate and carboxymethylcellulose, as well as Tween 80, ascorbic acid, and vanillin at concentrations of 0.5 and 1.0%, were used to coat fresh cut kiwi slices. According to the findings, the created coatings successfully reduced bacteria, yeast, and mold growth during storage. As a result, the shelf life of ready-to-eat kiwi fruit during storage is extended. Similarly, carrageen is also used in antimicrobial packaging. Roy and Rhim (2019) developed carrageenan-based antimicrobial bio-nanocomposite films that were incorporated with ZnO-NPs, stabilized by melanin. They revealed that the films showed a powerful antimicrobial activity toward *E. coli* but showed a little effect toward *L. monocytogenes*.

5.4.2 Protein-Based Nanomaterials

The functional properties of proteins are determined by the composition and essence of amino acids (Shine and Rostom, 2021). The side chains of amino acids are responsible for the interactions between the polypeptide chains, and the stability of the protein structure thus formed depends on the non-covalent interactions (Zubair and Ullah, 2020). Proteins are considered to be a valuable resource of biopolymer that can be used to develop bio-plastics (Silva et al., 2015).

5.4.2.1 Zein-Based Nanomaterials

When compared with other films, zein-based films have low water vapor permeability. With the incorporation of silver nanoclusters, these films have decreased toxicity (Mei et al., 2017). Aytac et al. (2017) studied the zein-thymol-gamma cyclodextrin inclusion of complex nano-fibrous films and the results demonstrated inhibition of the bacterial growth in meat samples. Essential oils have also been encapsulated using zein-based nanomaterials (Luo et al., 2011).

5.4.2.2 Gelatin-Based Nanomaterials

Gelatin has excellent filming properties and various antimicrobial agents have been integrated in this film that inhibits the growth of microorganisms. Ag-NPs were featured in gelatin-based films and this amalgamation successfully inhibited the multiplication of *E. coli* and *L. monocytogenes* (Kanmani and Rhim, 2014). Fusion of essential oils, such as oregano, has also been investigated and the findings reveal bacterial obstruction (Hosseini et al., 2016). Insertion of citric acid into film has also reduced the proliferation of *E. coli* (Uranga et al., 2018). Among the various biopolymers that have been studied so far, gelatin-based films work as a better medium as far as the release of antimicrobial agents is considered (Said and Sarbon, 2019).

5.5 CONCLUSION

The association of nanotechnology with the food industry has led to a number of improvements in the fields of food packing, preservation, and processing systems. This technology offers abundant choice for economic, environmental friendly, biodegradable, and sustainable packaging materials that are attaining more recognition and acknowledgement in efforts to reduce environmental impacts and provide more efficient packaging systems. With the aid of nanotechnology, antimicrobial nanoparticles/nanocomposite films are produced that assist in limiting the microbial populations, providing high quality and safe food products, and extending foods' shelf life. It is noteworthy that with the use nanotechnology in food packaging systems food shortage problems in several parts of the globe could be resolved with fewer difficulties.

REFERENCES

Ahmed, J., Mulla, M., Arfat, Y. A., Bher, A., Jacob, H., and Auras, R. 2018. Compression molded LLDPE films loaded with bimetallic (Ag-Cu) nanoparticles and cinnamon essential oil for chicken meat packaging applications. *LWT-Food Science and Technology* 93: 329–338.

Ahvenainen, R. (Ed.). 2003. *Novel Food Packaging Techniques*. Elsevier, Sawston, United Kingdom.

Aisyah, Y., Irwanda, L. P., Haryani, S., and Safriani, N. 2018. Characterization of corn starch-based edible film incorporated with nutmeg oil nanoemulsion. In *IOP Conference Series: Materials Science and Engineering* 352(1): 012050.

Akbar, A., and Anal, A. K. 2014. Zinc oxide nanoparticles loaded active packaging, a challenge study against salmonella typhimurium and Staphylococcus aureus in ready-to-eat poultry meat. *Food Control* 38: 88–95.

Antoniou, J., Liu, F., Majeed, H., and Zhong, F. 2015. Characterization of tara gum edible films incorporated with bulk chitosan and chitosan nanoparticles: A comparative study. *Food Hydrocolloids* 44: 309–319.

Anvar, A., Haghighat Kajavi, S., Ahari, H., Sharifan, A., Motallebi, A., Kakoolaki, S., and Paidari, S. 2019. Evaluation of the antibacterial effects of Ag-TiO2 nanoparticles and optimization of its migration to sturgeon caviar (Beluga). *Iranian Journal of Fisheries Sciences* 18(4): 954–967.

Aranaz, I., Mengíbar, M., Harris, R., Paños, I., Miralles, B., Acosta, N., and Heras, Á 2009. Functional characterization of chitin and chitosan. *Current Chemical Biology* 3(2): 203–230.

Arfat, Y. A., Benjakul, S., Vongkamjan, K., Sumpavapol, P., and Yarnpakdee, S. 2015. Shelf-life extension of refrigerated sea bass slices wrapped with fish protein isolate/fish skin gelatin-ZnO nanocomposite film incorporated with basil leaf essential oil. *Journal of Food Science and Technology* 52(10): 6182–6193.

Arora, D., Sharma, N., Sharma, V., Abrol, V., Shankar, R., and Jaglan, S. 2016. An update on polysaccharide-based nanomaterials for antimicrobial applications. *Applied Microbiology and Biotechnology* 100(6): 2603–2615.

Aytac, Z., Ipek, S., Durgun, E., Tekinay, T., and Uyar, T. 2017. Antibacterial electrospun zein nanofibrous web encapsulating thymol/cyclodextrin-inclusion complex for food packaging. *Food Chemistry* 233: 117–124.

Azlin-Hasim, S., Cruz-Romero, M. C., Morris, M. A., Padmanabhan, S. C., Cummins, E., and Kerry, J. P. 2016. The potential application of antimicrobial silver polyvinyl chloride nanocomposite films to extend the shelf-life of chicken breast fillets. *Food and Bioprocess Technology* 9(10): 1661–1673.

Bajpai, V. K., Kamle, M., Shukla, S., Mahato, D. K., Chandra, P., Hwang, S. K., and Han, Y. K. 2018. Prospects of using nanotechnology for food preservation, safety, and security. *Journal of Food and Drug Analysis* 26(4): 1201–1214.

Beigmohammadi, F., Peighambardoust, S. H., Hesari, J., Azadmard-Damirchi, S., Peighambardoust, S. J., and Khosrowshahi, N. K. 2016. Antibacterial properties of LDPE nanocomposite films in packaging of UF cheese. *LWT-Food Science and Technology* 65: 106–111.

Bhushan, B. 2017. Introduction to nanotechnology. In *Springer Handbook of Nanotechnology* (pp. 1–19). Springer, Berlin, Heidelberg.

Bhushan, B., Luo, D., Schricker, S. R., Sigmund, W., and Zauscher, S. (Eds.). 2014. *Handbook of Nanomaterials Properties*. Springer Berlin, Heidelberg.

Bikiaris, D. N., and Triantafyllidis, K. S. 2013. HDPE/Cu-nanofiber nanocomposites with enhanced antibacterial and oxygen barrier properties appropriate for food packaging applications. *Materials Letters* 93: 1–4.

Bodaghi, H., Mostofi, Y., Oromiehie, A., Zamani, Z., Ghanbarzadeh, B., Costa, C., and Del Nobile, M. A. 2013. Evaluation of the photocatalytic antimicrobial effects of a TiO_2 nanocomposite food packaging film by in vitro and in vivo tests. *LWT-Food Science and Technology* 50(2): 702–706.

Bondeson, D., Mathew, A., and Oksman, K. 2006. Optimization of the isolation of nanocrystals from microcrystalline cellulose by acid hydrolysis. *Cellulose* 13(2): 171–180.

Borkow, G., and Gabbay, J. 2009. Copper, an ancient remedy returning to fight microbial, fungal and viral infections. *Current Chemical Biology* 3(3): 272–278.

Cerqueira, M. A., Vicente, A. A., and Pastrana, L. M. 2018. Nanotechnology in food packaging: Opportunities and challenges. Nanomaterials for Food Packaging, 1–11.

Cerrada, M. L., Serrano, C., Sánchez-Chaves, M., Fernández, García, M., Fernández-Martín, F., de Andres, A., Fernández, and García, M. 2008. Self-sterilized EVOH-TiO_2 nanocomposites: Interface effects on biocidal properties. *Advanced Functional Materials* 18(13): 1949–1960.

Chen, I. N., Chang, C. C., Ng, C. C., Wang, C. Y., Shyu, Y. T., and Chang, T. L. 2008. Antioxidant and antimicrobial activity of *Zingiberaceae* plants in Taiwan. *Plant Foods for Human Nutrition* 63(1): 15–20.

Chung, Y. L., Ansari, S., Estevez, L., Hayrapetyan, S., Giannelis, E. P., and Lai, H. M. 2010. Preparation and properties of biodegradable starch–clay nanocomposites. *Carbohydrate Polymers* 79(2): 391–396.

Corrales, M., Fernández, A., and Han, J. H. 2014. Antimicrobial packaging systems. In *Innovations in Food Packaging* (pp. 133–170). Academic Press, London, UK.

Costa, C., Conte, A., Alessandro, M., and Nobile, D. 2016. Use of metal nanoparticles for active packaging applications. In *Antimicrobial Food Packaging* (pp. 399–406). Academic Press, London, UK.

Cui, H., Surendhiran, D., Li, C., and Lin, L. 2020. Biodegradable zein active film containing chitosan nanoparticle encapsulated with pomegranate peel extract for food packaging. *Food Packaging and Shelf Life* 24: 100511.

Cui, Y., Zhao, Y., Tian, Y., Zhang, W., Lü, X., and Jiang, X. 2012. The molecular mechanism of action of bactericidal gold nanoparticles on Escherichia coli. *Biomaterials* 33(7): 2327–2333.

da Silva Barbosa, R. F., and dos Santos Rosa, D. 2019. New trends of antimicrobial packaging applying nanotechnology. Importance & Applications of Nanotechnology. MedDocs Publishers LLC, Nevada, USA.

Dai, L., Zhang, J., and Cheng, F. 2019. Effects of starches from different botanical sources and modification methods on physicochemical properties of starch-based edible films. *International Journal of Biological Macromolecules* 132: 897–905.

Daniel, M. C., and Astruc, D. 2004. Gold nanoparticles: Assembly, supramolecular chemistry, quantum-size-related properties, and applications toward biology, catalysis, and nanotechnology. *Chemical Reviews* 104(1): 293–346.

Donglu, F., Wenjian, Y., Kimatu, B. M., Mariga, A. M., Liyan, Z., Xinxin, A., and Qiuhui, H. 2016. Effect of nanocomposite-based packaging on storage stability of mushrooms (*flammulina velutipes*). *Innovative Food Science & Emerging Technologies* 33: 489–497.

Du, W. L., Niu, S. S., Xu, Y. L., Xu, Z. R., and Fan, C. L. 2009. Antibacterial activity of chitosan tripolyphosphate nanoparticles loaded with various metal ions. *Carbohydrate Polymers* 75(3): 385–389.

Duncan, T. V. 2011. Applications of nanotechnology in food packaging and food safety: Barrier materials, antimicrobials and sensors. *Journal of Colloid and Interface Science* 363(1): 1–24.

Echegoyen, Y., and Nerín, C. 2013. Nanoparticle release from nano-silver antimicrobial food containers. *Food and Chemical Toxicology* 62: 16–22.

Emamifar, A., Kadivar, M., Shahedi, M., and Soleimanian-Zad, S. 2010. Evaluation of nanocomposite packaging containing ag and ZnO on shelf life of fresh Orange juice. *Innovative Food Science & Emerging Technologies* 11(4): 742–748.

Espitia, P. J. P., Soares, N. D. F. F., dos Reis Coimbra, J. S., de Andrade, N. J., Cruz, R. S., and Medeiros, E. A. A. 2012. Zinc oxide nanoparticles: Synthesis, antimicrobial activity and food packaging applications. *Food and Bioprocess Technology* 5(5): 1447–1464.

Fernandez, A., Picouet, P., and Lloret, E. 2010. Reduction of the spoilage-related microflora in absorbent pads by silver nanotechnology during modified atmosphere packaging of beef meat. *Journal of Food Protection* 73(12): 2263–2269.

Gan, I., and Chow, W. S. 2018. Antimicrobial poly (lactic acid)/cellulose bio-nanocomposite for food packaging application: A review. *Food Packaging and Shelf Life* 17: 150–161.

Gull, A., Bhat, N., Wani, S. M., Masoodi, F. A., Amin, T., and Ganai, S. A. 2021. Shelf life extension of apricot fruit by application of nanochitosan emulsion coatings containing pomegranate peel extract. *Food Chemistry* 349: 129149.

Han, J. H. 2003. Antimicrobial food packaging. *Novel Food Packaging Techniques* 8: 50–70.

He, X., Deng, H., and Hwang, H. M. 2019. The current application of nanotechnology in food and agriculture. *Journal of Food and Drug Analysis* 27(1): 1–21.

Hernández-Muñoz, P., Cerisuelo, J. P., Domínguez, I., López-Carballo, G., Catalá, R., and Gavara, R. 2019. Nanotechnology in food packaging. In *Nanomaterials for Food Applications* (pp. 205–232). Elsevier, MA, USA.

Hoffmann, T., Peters, D., Angioletti, B., Bertoli, S., Péres, L., Reiter, M., and De Souza, C. 2019. Potentials nanocomposites in food packaging. *Chemical Engineering Transactions* 75: 253–258.

Honarkar, H., and Barikani, M. 2009. Applications of biopolymers i: Chitosan. *Monatshefte Für Chemie-Chemical Monthly* 140(12): 1403–1420.

Hoseinnejad, M., Jafari, S. M., and Katouzian, I. 2018. Inorganic and metal nanoparticles and their antimicrobial activity in food packaging applications. *Critical Reviews in Microbiology* 44(2): 161–181.

Hosseini, S. F., Rezaei, M., Zandi, M., and Farahmandghavi, F. 2016. Development of bioactive fish gelatin/chitosan nanoparticles composite films with antimicrobial properties. *Food Chemistry* 194: 1266–1274.

Huang, Y., Mei, L., Chen, X., and Wang, Q. 2018. Recent developments in food packaging based on nanomaterials. *Nanomaterials* 8(10): 830.

Jancy, S., Shruthy, R., and Preetha, R. 2020. Fabrication of packaging film reinforced with cellulose nanoparticles synthesised from jack fruit non-edible part using response surface methodology. *International Journal of Biological Macromolecules* 142: 63–72.

Jin, T., and He, Y. 2011. Antibacterial activities of magnesium oxide (MgO) nanoparticles against foodborne pathogens. *Journal of Nanoparticle Research* 13(12): 6877–6885.

Jin, T., Sun, D., Su, J. Y., Zhang, H., and Sue, H. J. 2009. Antimicrobial efficacy of zinc oxide quantum dots against *listeria monocytogenes*, *salmonella enteritidis*, and *Escherichia coli* O157: H7. *Journal of Food Science* 74(1): M46–M52.

Kanmani, P., and Rhim, J. W. 2014. Physical, mechanical and antimicrobial properties of gelatin based-active nanocomposite films containing ag-NPs and nanoclay. *Food Hydrocolloids* 35: 644–652.

Kim, J. S., Kuk, E., Yu, K. N., Kim, J. H., Park, S. J., Lee, H. J., and Cho, M. H. 2007. Antimicrobial effects of silver nanoparticles. *Nanomedicine: Nanotechnology, Biology and Medicine* 3(1): 95–101.

Kirwan, M. J., Plant, S., and Strawbridge, J. W. 2011. Plastics in food packaging. *Food and Beverage Packaging Technology*: 157–212.

Kuswandi, B. 2016. Nanotechnology in food packaging. In *Nanoscience in Food and Agriculture 1* (pp. 151–183). Springer, Cham.

Le Corre, D., Bras, J., and Dufresne, A. 2010. Starch nanoparticles: A review. *Biomacromolecules* 11(5): 1139–1153.

Le Corre, D., Bras, J., and Dufresne, A. 2012. Influence of native starch's properties on starch nanocrystals thermal properties. *Carbohydrate Polymers* 87(1): 658–666.

Lee, S. Y., Chen, H., and Hanna, M. A. 2008. Preparation and characterization of tapioca starch–poly (lactic acid) nanocomposite foams by melt intercalation based on clay type. *Industrial Crops and Products* 28(1): 95–106.

Leistner, L. 1994. Further developments in the utilization of hurdle technology for food preservation. *Journal of Food Engineering* 22: 421–432.

Li, H., Li, F., Wang, L., Sheng, J., Xin, Z., Zhao, L., and Hu, Q. 2009. Effect of nano-packing on preservation quality of Chinese jujube (*Ziziphus jujuba* Mill. var. *inermis* (*Bunge*) *Rehd*). *Food Chemistry* 114(2): 547–552.

Li, M. C., Wu, Q., Song, K-., Lee, S., Qing, Y., and Wu, Y. 2015. Cellulose nanoparticles: structure–morphology–rheology relationships. *ACS Sustainable Chemistry & Engineering* 3(5): 821–832.

Llorens, A., Lloret, E., Picouet, P. A., Trbojevich, R., and Fernandez, A. 2012. Metallic-based micro and nanocomposites in food contact materials and active food packaging. *Trends in Food Science & Technology* 24(1): 19–29.

Lloyd, K., Mirosa, M., and Birch, J. 2019. Active and intelligent packaging. In P. Varelis, L. Melton, and F. Shahidi (Eds.), *Encyclopedia of Food Chemistry* (pp. 177–182). Elsevier, MA, USA.

Lomate, G. B., Dandi, B., and Mishra, S. 2018. Development of antimicrobial LDPE/Cu nanocomposite food packaging film for extended shelf life of peda. *Food Packaging and Shelf Life* 16: 211–219.

Lopez-Lopez, E. A., Hernández-Gallegos, M. A., Cornejo-Mazón, M., and Hernández-Sánchez, H. 2015. Polysaccharide-based nanoparticles. In H. Hernandez-Sanchez and G. F. Gutierrez-Lopez (Eds.), *Food Nanoscience and Nanotechnology* (pp. 59–68). Springer, Cham.

Luo, Y., Zhang, B., Whent, M., Yu, L. L., and Wang, Q. 2011. Preparation and characterization of zein/chitosan complex for encapsulation of α-tocopherol, and its in vitro controlled release study. *Colloids and Surfaces B: Biointerfaces* 85(2): 145–152.

Luo, Z., Xu, Y., and Ye, Q. 2015. Effect of nano-SiO_2-LDPE packaging on biochemical, sensory, and microbiological quality of Pacific white shrimp Penaeus vannamei during chilled storage. *Fisheries Science* 81(5): 983–993.

Mahdi, S. S., Vadood, R., and Nourdahr, R. 2012. Study on the antimicrobial effect of nanosilver tray packaging of minced beef at refrigerator temperature. *Global Veterinaria* 9(3): 284–9.

Majdzadeh-Ardakani, K., Navarchian, A. H., and Sadeghi, F. 2010. Optimization of mechanical properties of thermoplastic starch/clay nanocomposites. *Carbohydrate Polymers* 79(3): 547–554.

Majid, I., Nayik, G. A., Dar, S. M., and Nanda, V. 2018. Novel food packaging technologies: Innovations and future prospective. *Journal of the Saudi Society of Agricultural Sciences* 17(4): 454–462.

Mansoori, G. A., and Soelaiman, T. F. 2005. Nanotechnology—an introduction for the standards community. *Journal of ASTM International* 2(6): 1–22.

Manzoor, S., Gull, A., Wani, S. M., Ganaie, T. A., Masoodi, F. A., Bashir, K., and Dar, B. N. 2021. Improving the shelf life of fresh cut kiwi using nanoemulsion coatings with antioxidant and antimicrobial agents. *Food Bioscience* 41: 101015.

Marquez, I. G., Akuaku, J., Cruz, I., Cheetham, J., Golshani, A., and Smith, M. L. 2013. Disruption of protein synthesis as antifungal mode of action by chitosan. *International Journal of Food Microbiology* 164(1): 108–112.

Mastromatteo, M., Conte, A., Lucera, A., Saccotelli, M. A., Buonocore, G. G., Zambrini, A. V., and Del Nobile, M. A. 2015. Packaging solutions to prolong the shelf life of fiordilatte cheese: Bio-based nanocomposite coating and modified atmosphere packaging. *LWT-Food Science and Technology* 60(1): 230–237.

Medina, E., Caro, N., Abugoch, L., Gamboa, A., Díaz-Dosque, M., and Tapia, C. 2019. Chitosan thymol nanoparticles improve the antimicrobial effect and the water vapour barrier of chitosan-quinoa protein films. *Journal of Food Engineering* 240: 191–198.

Mei, L., Teng, Z., Zhu, G., Liu, Y., Zhang, F., Zhang, J., and Wang, Q. 2017. Silver nanocluster-embedded zein films as antimicrobial coating materials for food packaging. *ACS Applied Materials & Interfaces* 9(40): 35297–35304.

Mihaly Cozmuta, A., Peter, A., Mihaly Cozmuta, L., Nicula, C., Crisan, L., Baia, L., and Turila, A. 2015. Active packaging system based on Ag/TiO_2 nanocomposite used for extending the shelf life of bread. Chemical and microbiological investigations. *Packaging Technology and Science* 28(4): 271–284.

Mihindukulasuriya, S. D. F., and Lim, L. T. 2014. Nanotechnology development in food packaging: A review. *Trends in Food Science & Technology* 40(2): 149–167.

Mlalila, N., Hilonga, A., Swai, H., Devlieghere, F., and Ragaert, P. 2018. Antimicrobial packaging based on starch, poly (3-hydroxybutyrate) and poly (lactic-co-glycolide) materials and application challenges. *Trends in Food Science & Technology* 74: 1–11.

Mohanty, S., Mishra, S., Jena, P., Jacob, B., Sarkar, B., and Sonawane, A. 2012. An investigation on the antibacterial, cytotoxic, and antibiofilm efficacy of starch-stabilized silver nanoparticles. *Nanomedicine: Nanotechnology, Biology and Medicine* 8(6): 916–924.

Morones, J. R., Elechiguerra, J. L., Camacho, A., Holt, K., Kouri, J. B., Ramírez, J. T., and Yacaman, M. J. 2005. The bactericidal effect of silver nanoparticles. *Nanotechnology* 16(10): 2346.

Nan, L., Yang, W. C., Liu, Y. Q., Xu, H., Li, Y., Lu, M. Q., and Yang, K. 2008. Antibacterial mechanism of copper-bearing antibacterial stainless steel against *E. coli*. *Journal of Materials Science and Technology -Shenyang* 24(2): 197–20.

Nešić, A., Cabrera-Barjas, G., Dimitrijević-Branković, S., Davidović, S., Radovanović, N., and Delattre, C. 2020. Prospect Of polysaccharide-based materials as advanced food packaging. *Molecules* 25(1): 135.

Padmavathy, N., and Vijayaraghavan, R. 2008. Enhanced bioactivity of ZnO nanoparticles—an antimicrobial study. *Science and Technology of Advanced Materials* 9(3):035004.

Pagno, C. H., Costa, T. M., de Menezes, E. W., Benvenutti, E. V., Hertz, P. F., Matte, C. R., and Flôres, S. H. 2015. Development of active biofilms of quinoa (*Chenopodium quinoa* W.) starch containing gold nanoparticles and evaluation of antimicrobial activity. *Food Chemistry* 173: 755–762.

Peng, Y., Wu, Y., and Li, Y. 2013. Development of tea extracts and chitosan composite films for active packaging materials. *International Journal of Biological Macromolecules* 59: 282–289.

Pereda, M., Marcovich, N. E., and Ansorena, M. R. 2017. Nanotechnology in food packaging applications: barrier materials, antimicrobial agents, sensors, and safety assessment. In L. M. T. Martínez, O. V. Kharissova, and B. I. Kharisov (Eds.), *Handbook of Ecomaterials* (pp. 2035–2056). Springer, Cham, Switzerland.

Qureshi, M. A., Karthikeyan, S., Karthikeyan, P., Khan, P. A., Uprit, S., and Mishra, U. K. 2012. Application of nanotechnology in food and dairy processing: An overview. *Pakistan Journal of Food Sciences* 22(1): 23–31.

Radusin, T. I., Ristić, I. S., Pilić, B. M., and Novaković, A. R. 2016. Antimicrobial nanomaterials for food packaging applications. *Food and Feed Research* 43(2): 119–126.

Ramakritinan, C. M., Kaarunya, E., Shankar, S., and Kumaraguru, A. K. 2013. Antibacterial effects of ag, au and bimetallic (ag-au) nanoparticles synthesized from red algae. Nanoparticles Synthesized from Red Algae. *Solid State Phenomena* 201: 211–230.

Ramesh, S., and Radhakrishnan, P. 2019. Cellulose nanoparticles from agro-industrial waste for the development of active packaging. *Applied Surface Science* 484: 1274–1281.

Rao, C. N. R., and Cheetham, A. K. 2001. Science and technology of nanomaterials: Current status and future prospects. *Journal of Materials Chemistry* 11(12): 2887–2894.

Razali, M. H., Ismail, N. A., and Amin, K. A. M. 2019. Fabrication and characterization of antibacterial titanium dioxide nanorods incorporating gellan gum films. *Journal of Pure and Applied Microbiology* 13: 1909–1916.

Roy, S., and Rhim, J. W. 2019. Carrageenan-based antimicrobial bionanocomposite films incorporated with ZnO nanoparticles stabilized by melanin. *Food Hydrocolloids* 90: 500–507.

Sadiq, I. M., Chowdhury, B., Chandrasekaran, N., and Mukherjee, A. 2009. Antimicrobial sensitivity of *Escherichia coli* to alumina nanoparticles. *Nanomedicine: Nanotechnology, Biology and Medicine* 5(3): 282–286.

Said, N. S., and Sarbon, N. M. 2019. Protein-based active film as antimicrobial food packaging. In I. Var and S. Uzunlu (Eds.), *Active Antimicrobial Food Packaging* (pp. 53–70). IntecOpen, London.

Sandhu, K. S., and Nain, V. 2017. Starch nanoparticles: Their preparation and applications. In *Plant Biotechnology: Recent Advancements and Developments* (pp. 213–232). Springer, Singapore.

Sawai, J., and Yoshikawa, T. 2004. Quantitative evaluation of antifungal activity of metallic oxide powders (MgO, CaO and ZnO) by an indirect conductimetric assay. *Journal of Applied Microbiology* 96(4): 803–809.

Shankar, S., and Rhim, J. W. 2014. Effect of copper salts and reducing agents on characteristics and antimicrobial activity of copper nanoparticles. *Materials Letters* 132: 307–311.

Shankar, S., Teng, X., and Rhim, J. W. 2014. Properties and characterization of agar/CuNP bionanocomposite films prepared with different copper salts and reducing agents. *Carbohydrate Polymers* 114: 484–492.

Shapi'i, R. A., Othman, S. H., Nordin, N., Basha, R. K., and Naim, M. N. 2020. Antimicrobial properties of starch films incorporated with chitosan nanoparticles: In vitro and in vivo evaluation. *Carbohydrate Polymers* 230: 115602.

Shine, B., and Rostom, H. 2021. Basic metabolism: Proteins. *Surgery (Oxford)* 39(1): 1–6.

Silva, K. S., Garcia, C. C., Amado, L. R., and Mauro, M. A. 2015. Effects of edible coatings on convective drying and characteristics of the dried pineapple. *Food and Bioprocess Technology* 8(7): 1465–1475.

Silvestre, C., Duraccio, D., and Cimmino, S. 2012. Progress in polymer science: Food packaging based on polymer nanomaterials. *Cellular Polymers* 31(1): 54–55.

Singh, S., and Shalini, R. 2016. Effect of hurdle technology in food preservation: A review. *Critical Reviews in Food Science and Nutrition* 56(4): 641–649.

Singh, T., Shukla, S., Kumar, P., Wahla, V., Bajpai, V. K., and Rather, I. A. 2017. Application of nanotechnology in food science: Perception and overview. *Frontiers in Microbiology* 8: 1501.

Sofi, S. A., Singh, J., Rafiq, S., Ashraf, U., Dar, B. N., and Nayik, G. A. 2018. A comprehensive review on antimicrobial packaging and its use in food packaging. *Current Nutrition & Food Science* 14(4): 305–312.

Sothornvit, R. 2019. Nanostructured materials for food packaging systems: New functional properties. *Current Opinion in Food Science* 25: 82–87.

Stoimenov, P. K., Klinger, R. L., Marchin, G. L., and Klabunde, K. J. 2002. Metal oxide nanoparticles as bactericidal agents. *Langmuir* 18(17): 6679–6686.

Sun, Q., Li, G., Dai, L., Ji, N., and Xiong, L. 2014. Green preparation and characterisation of waxy maize starch nanoparticles through enzymolysis and recrystallisation. *Food Chemistry* 162: 223–228.

Tankhiwale, R., and Bajpai, S. K. 2012. Preparation, characterization and antibacterial applications of ZnO-nanoparticles coated polyethylene films for food packaging. *Colloids and Surfaces B: Biointerfaces* 90: 16–20.

Thirumurugan, A., Ramachandran, S., and Shiamala Gowri, A. 2013. Combined effect of bacteriocin with gold nanoparticles against food spoiling bacteria-an approach for food packaging material preparation. *International Food Research Journal* 20(4): 1909–1912.

Tripathi, S., Mehrotra, G. K., and Dutta, P. K. 2011. Chitosan–silver oxide nanocomposite film: Preparation and antimicrobial activity. *Bulletin of Materials Science* 34(1): 29–35.

Uranga, J., Puertas, A. I., Etxabide, A., Duenas, M. T., Guerrero, P., and Caba, K. D. L. 2018. Citric acid-incorporated fish gelatin/chitosan composite films. *Food Hydrocolloids* 8: 95–103.

Vizzini, P., Beltrame, E., Zanet, V., Vidic, J., and Manzano, M. 2020. Development and evaluation of qPCR detection method and Zn-MgO/alginate active packaging for controlling *Listeria monocytogenes* contamination in cold-smoked salmon. *Foods* 9(10): 1353.

Wang, Z., Qiao, X., and Sun, K. 2018. Rice straw cellulose nanofibrils reinforced poly (vinyl alcohol) composite films. *Carbohydrate Polymers* 197: 442–450.

Xing, Y., Li, X., Zhang, L., Xu, Q., Che, Z., Li, W., and Li, K. 2012. Effect of TiO_2 nanoparticles on the antibacterial and physical properties of polyethylene-based film. *Progress in Organic Coatings* 73(2–3): 219–224.

Yang, F. M., Li, H. M., Li, F., Xin, Z. H., Zhao, L. Y., Zheng, Y. H., and Hu, Q. H. 2010. Effect of nano-packing on preservation quality of fresh strawberry (*Fragaria ananassa Duch. cv Fengxiang*) during storage at 4 c. *Journal of Food Science* 75(3): C236–C240.

Yang, Y., Liu, H., Wu, M., Ma, J., and Lu, P. 2020. Bio-based antimicrobial packaging from sugarcane bagasse nanocellulose/nisin hybrid films. *International Journal of Biological Macromolecules* 161: 627–635.

Yildirim, S., Röcker, B., Pettersen, M. K., Nilsen, Nygaard, J., Ayhan, Z., Rutkaite, R., and Coma, V. 2018. Active packaging applications for food. *Comprehensive Reviews in Food Science and Food Safety* 17(1): 165–199.

Yucel, U. 2016. Intelligent packaging. In *Reference Module in Food Science*. Elsevier, MA, USA.

Zhang, L., Jiang, Y., Ding, Y., Daskalakis, N., Jeuken, L., Povey, M., and York, D. W. 2010. Mechanistic investigation into antibacterial behavior of suspensions of ZnO nanoparticles against *E. coli*. *Journal of Nanoparticle Research* 12(5): 1625–1636.

Zhao, L. M., Shi, L. E., Zhang, Z. L., Chen, J. M., Shi, D. D., Yang, J., and Tang, Z. X. 2011. Preparation and application of chitosan nanoparticles and nanofibers. *Brazilian Journal of Chemical Engineering* 28: 353–362.

Zhong, T., Oporto, G. S., and Jaczynski, J. 2017. Antimicrobial food packaging with cellulose-copper nanoparticles embedded in thermoplastic resins. In *Food Preservation* (pp. 671–702). Academic Press.

Zohri, M., Alavidjeh, M. S., Haririan, I., Ardestani, M. S., Ebrahimi, S. E. S., Sani, H. T., and Sadjadi, S. K. 2010. A comparative study between the antibacterial effect of nisin and nisin-loaded chitosan/alginate nanoparticles on the growth of *Staphylococcus aureus* in raw and pasteurized milk samples. *Probiotics and Antimicrobial Proteins* 2(4): 258–266.

Zubair, M., and Ullah, A. 2020. Recent advances in protein derived bionanocomposites for food packaging applications. *Critical Reviews in Food Science and Nutrition* 60(3): 406–434.

6 Nanofibrils- and Nanorods-Based Nanofillers for Food Packaging Application

A. J. Cenobio-Galindo[1], I. Hernández-Soto[1], E. Pérez-Soto[1], L. González-Montiel[1], R. G. Campos-Montiel[1], and Gulzar Ahmad Nayik[2]

[1]Instituto de Ciencias Agropecuarias, Universidad Autónoma del Estado de Hidalgo, Tulancingo, Hidalgo, México

[2]Department of Food Science & Technology, Govt. Degree College Shopian, J&K, India

CONTENTS

DOI: 10.1201/9781003207641-7

6.1 INTRODUCTION

In the area of food processing, packaging means more than simple protection, since the main objective of food packing is becoming the barrier that maintains food characteristics and quality during all subsequent stages of production, including storage and transport, protecting the food from adverse conditions, which may appear as chemical contaminants, microorganisms, and environmental conditions, among others. It is for this reason that the constant development of components that become auxiliary to maintaining the properties of food is essential (Rhim et al., 2013). For this reason, various nanomaterials have been developed to perfect the attributes of biopolymers intended for food packaging.

Nanomaterials can be used in the packaging industry in a variety of ways. It can improve some properties of packaging, such as mechanical properties, and it can also provide other benefits, such as actively interacting with the packaging to release certain antimicrobials or antioxidant compounds (Othman, 2014). Specifically, nanofiller materials added in small amounts (5%) in food packaging applications, generally, improved the mechanical and barrier properties during processing and storage of products that it protects (Rhim et al., 2013).

We can classify nanofillers according to their nature: organic and inorganic. In the inorganic we can find particles of silica or titanium dioxide, among others (Saba et al., 2014). Among organic nanofillers, the most important and by far the most used is cellulose, a growing trend because when introduced it helps generate improved properties without neglecting environmental benefits (Sun et al., 2018). This chapter approaches the development of nanofibrils and nanorods and their importance in food packaging technology as an emerging alternative to help ensure the quality of food during its useful life.

6.2 CELLULOSE NANOFIBRILS AND THEIR APPLICATION AS NANOFILLERS

It is not possible to begin this chapter without mentioning that the use of natural fibers has received special attention in recent times, especially due to its characteristics, among which its biodegradable nature, low cost, low density, and high resistance stand out (Zimmermann et al., 2004). The different layers of the cell wall of cellulose are made up of nanofibrils, which provide the resistance that the plant vascular system needs (Emons and Mulder, 2000). Depending on the processes of separating and retrieving the nanofibrils, their diameter can range between 2 and 4 nm and their length could be several microns (Dinand et al., 1999), making them an excellent raw material to use when talking about nanofillers. This biopolymer forms fibrils linking approximately 100 cellulose chains, in amorphous and crystalline forms. This is fundamental because that is where the predominant characteristics, such as the degree of crystallinity, are determined (Gan and Chow, 2018). The production of cellulose nanocomposites is carried out through three main processes: high pressure homogenization, microfluidization, and micro-grinding, each method leading to a different result (Lee et al., 2017). This material can be referred to by different names; for example, microfibrillated cellulose, nanofibrillated cellulose, or cellulose nanofibrils (CNF), the latter being the most used currently, and an aqueous suspension (Dufresne, 2017).

The development of CNF has taken on great relevance, as it has the advantages of being a cheap, biodegradable material and its raw materials can also be considered "waste," giving it value as an agro-industrial by product. The application of nanofillers in biopolymers has attracted the attention of various studies because of the attributes of being a biodegradable material that can be extracted from a wide diversity of raw materials.

6.3 DEVELOPMENT AND APPLICATION OF NANOFILLERS USING NANOFIBRILS EMBEDDED IN CARBOHYDRATE-BASED MATRICES

6.3.1 Starch

Packaging based on natural polymers is currently considered a viable alternative to petroleum-derived plastics, with biopolymers being one of the most used for its versatility, due to starch. Starch is widely studied in the food area as a promising ingredient for the improvement of packaging (Cenobio-Galindo et al., 2019). Recently, interest in the incorporation of these nano-sized compounds has further expanded the development and use of this type of packaging. However, the system must be designed well as the fiber-matrix interactions and the affinity of the cellulose with starch are the critical points in determining the properties of the system, as the best distribution of nanofibrils, the best interfacial area, and an excellent fiber network could result in improvements in the properties (Eslami et al., 2021). Investigations carried out by Balakrishnan et al. (2018) on starch have isolated CNF from pineapple leaves, using them to reinforce barrier properties of films made with this biopolymer. This study found that the 25 nm CNF obtained presented good dispersion during film production, which was verified by polarized optical microscope as a key point for its final properties (Table 6.1). It is noteworthy that when a 4% filler is used, it is not possible to observe desired properties in the film (Zhang et al., 2007).

Among natural biopolymers, starch is the most interesting raw material because of its characteristics of versatility, low price, availability, and biodegradability, which make it extremely attractive for food packaging innovation. Of its properties, native starch is the most outstanding, since with proper processing it can be turned into thermoplastic starch with a plasticizer in the filmogenic solution, because the starch granules lose their structure, generating a characteristic material (Avérous and Halley, 2009; Hietala et al., 2013). With the above and knowledge of its limited mechanical properties, Kargarzadeh et al. (2017) investigated the addition of nanocomposites from rice husk, an agricultural residue that is normally burned, producing a high impact on the environment. The aggregation of this filler improved the mechanical properties, such as traction and thermal properties, and it was possible to reduce water absorption, making the nanofiller and the starch matrix highly compatible.

TABLE 6.1
Application of Cellulose Nanofibrils Used as Nanofillers in Carbohydrate-Based Films

Matrix	CNF Source	Effect	Reference
Starch packaging	Pineapple leaves	Strengthening of mechanical and barrier properties	Balakrishnan et al., 2018
Native corn starch foam packaging	Commercial	The reinforced films had better thermal stability and degradation temperature than the usual films	Ghanbari et al., 2018
Cassava starch films	Rice husk	Improvement in opacity, permeability to water vapor and in general mechanical properties	Nascimento et al., 2016
Tapioca starch films	Ramio (*Boehmeria nivea*)	Improvement in thermal and mechanical properties	Syafri et al., 2018
Chitosan films with oregano essential oil	bleached bagasse	Inhibition of microbial growth and improvement of mechanical properties	Chen et al., 2020
Refined and semi-refined carrageenan films	Commercial	Improvement in barrier, mechanical and thermal properties	Sedayu et al., 2020
Agar films	Paper mulberry (*Broussonetia papyrifera*)	Improves barrier and mechanical properties	Reddy and Rhim, 2014

CNF: Cellulose nanofibrils

Up to 85–90% of all biodegradable products on the market are made of starch, which is why improvements are constantly being sought in order to avoid its most common problems, such as low thermal stability (Glenn et al., 2011) and poor reinforcement, as it is fragile due to inter and intramolecular hydrogen bonding between the starch chains (Balakrishnan et al., 2018). Other variants of film, such as foaming, offer advantages, such as obtaining light materials, better thermal insulation, and the possibility of improving the cost of materials. Generally, this process is carried out by extrusion, or molding. Ghanbari et al. (2018) carried out a study to evaluate the effect of reinforcing the foam packaging made with native corn starch along with commercial CNF at different concentrations (0, 0.5, 1.0 and 1.5%) (Table 6.1), evaluating mechanical, thermal, and barrier properties, and observing with scanning electron microscopy that with added CNF there was a better dimensional homogeneity—in other words, the alveoli formed were more defined. Its thermal properties showed that the reinforced films have better thermal stability and degradation temperature than normal films, and that the density and water absorption were decreased.

Currently, there is special interest in agro-industrial residues that have high cellulose content. One of these is rice husk, which is produced in considerable quantities, estimated to be up to 500 million tons per year. This shell is the main by-product of grinding, and it allows new inputs, such as CNF that creates interlocking structures with nano size (Khalil et al., 2014; Soltani et al., 2015). In this context, Nascimento et al. (2016) obtained CNF from this residue to incorporate it (2.5% w/w) into cassava starch films to determine the effect on biodegradable films (Table 6.1). Their research placed special emphasis on mechanical and barrier properties, finding that the films made with CNF were more homogeneous and not rough when compared with those that were not added, attributing this behavior to the excellent dispersion and incorporation in the polymeric matrix. This resulted in a better adhesion between the materials, in addition to improving the opacity, the permeability to water vapor, and the mechanical properties in general.

Starch has been the subject of multiple investigations due to its versatility, and it is attractive because it is obtained mainly from agro-industrial waste or from non-conventional sources. Research carried out by Syafri et al. (2018) used ramie, a plant family of the urticaceae, as a source of CNF (Table 6.1) to reinforce tapioca starch films and observe the effect on its properties. The nanofibrils that the authors obtained depended on the duration the sample was subjected to ultrasound, finding sizes that ranged from 34 to 193 nm. When this was added to the tapioca matrix, no agglomeration was detected. The results showed an improvement in the mechanical and thermal properties as the thermal stability increases in proportion to increases in the nanofiller concentration, due to the adhesion effect between this filler and the main matrix (Prachayawarakorn et al., 2013).

It should be observed that a film made solely with gelatinized or retrograded starch has little resistance to wet materials and will easily collapse in water. Soni et al. (2020) applied an innovative method, using retrograde starch in different concentrations to evaluate the effect that this retrograde tapioca starch causes in cellulose reinforced films. They found that with the retrograded starch there was less swelling by water and improvement in the mechanical properties, confirming that the retrograded starch granules, like CNF, act as fillers, providing higher levels of mechanical resistance and improved crystallinity of the films, which is a determining factor for successful food packaging.

6.3.2 Chitosan

One of the raw materials with a great future in the packaging industry is chitosan as it is considered ideal for making films due to its conformation. It is a linear polymer made up of D-glucosamine (β1-4) in greater quantity, derived from chitin, non-toxic, and above all a biodegradable raw material (Moradi et al., 2012). Rachtanapun et al. (2021) mention that it is a product that forms films with good barrier properties compared with films made with other biopolymers, and that it also provides antimicrobial properties. With the above, Azeredo et al. (2010) evaluated the effect of incorporating CNF as a nanofiller, in multiple concentrations, on the mechanical, barrier, and thermal attributes of chitosan films, because although they present good properties, they cannot yet be compared to synthetic polymers. The authors found that, according to the AFM images carried out, CNF presented excellent dispersion in the matrix, improving the elongation at break and Young's modulus, even when compared with synthetic plastics, and that there was reduction in the permeability to water. These good results were found when 15% of CNF was added.

Chen et al. (2020) elaborated "active" films of chitosan with oregano essential oil reinforced with CNF obtained from bleached bagasse, achieving an inhibition up to 99% of the growth of some gram negative microorganisms, such as *Escherichia coli* and *Listeria monocytogenes* (Table 6.1). Also, with the addition of CNF, the network formed between the chitosan and the filler became stronger, with an increase in the attributes. These films have great potential to significantly improve packaging characteristics.

6.3.3 Carrageenan

In constant search for the development and application of polymers, there is growing interest in marine resources, with carrageenan among them because of its use in a diversity of food products. Packaging based on this polymer offers good properties. However, the hydrophilic nature of this raw material is a recurring problem and fillers to reduce these problems are being sought (Rajendran et al., 2012). Seeking the improvement of these materials, Sedayu et al. (2020) investigated reinforcing films of refined and semi-refined carrageenan with commercial CNF, evaluating the permeability and thermal properties with varying the filling concentration (Table 6.1). The authors found that, in general terms, the addition improved the properties of the films, especially for water permeability. One of the most noticeable differences between the films was in opacity, being higher in films with unrefined carrageenan due to the presence of various residual components. It should be noted that the opacity increased for both films when the filler concentration was higher.

Hamid et al. (2019) developed a biodegradable active packaging based on semi-refined carrageenan, with a dual purpose: first, to increase the mechanical attributes with the addition of CNF (from 0 to 13%), and to incorporate α-tocopherol as a natural antioxidant. The authors reported that the addition of CNF did improve the mechanical properties, increasing elongation and tensile strength, opacity, permeability, among other characteristics, depending on concentration. Also, with the addition of an antioxidant to the matrix, an extra protective effect is achieved for the food against undesirable aspects, such as oxidative degeneration.

6.3.4 Agar

A novel material for the development of food packaging is agar. This ingredient is a generally recognized as safe (GRAS) product by the US Food and Drug Administration (US-FDA), and currently its main uses have been in the biotechnology area (Mostafavi and Zaeim, 2020). Chemically, agar is composed of agaropectin, the part that does not gel, and agarose, the fraction that gels giving the film-forming capacity. This occurs through the structure of hydrogen bonds between agarose molecules forming a interphase of double helices strengthened through aqueous molecules. Because of this it acquires the characteristic of storing a significant concentration of liquid inside (Nieto and Akins, 2010). Reddy and Rhim (2014) characterized agar films reinforced with nanocellulose from paper mulberry in different concentrations (1, 3, 5 and 10%) (Table 6.1), determining the effect this filler had on the properties. Finding that their hypothesis was true, the addition of this nanofiller served to significantly increase the barrier and mechanical attributes, mainly due to the strong interaction that nanocellulose induces with the matrix. Likewise, this fill, when it is well distributed, generates a complicated route for the dispersion of vapor, increasing the length in this route, generally resulting in a decrease in permeability (Rhim et al., 2009).

6.4 DEVELOPMENT AND APPLICATION OF NANOFILLERS USED AS NANOFIBRILS EMBEDDED IN PROTEIN-BASED MATRICES

6.4.1 Gelatin

The main biopolymers used in the elaboration of biodegradable elements are polysaccharides and proteins. Within the latter we must highlight gelatin, with a production currently exceeding 300,000 tons from various sources, such as pig skin, beef, and bones among others (Hanani et al., 2014). Its importance lies in its functionality as a gelling agent and a filmogenic material, and that it is biodegradable. However, it is affected by moisture, attributed to the hydrophilic behavior of gelatin (Khan and Sadiq, 2021). For this reason, Santos et al. (2014) evaluated the effect of the addition of nanocellulose from cotton linter in tilapia gelatin films (Table 6.2), noting that also

TABLE 6.2
Application of Cellulose Nanofibrils Used as Nanofillers in Protein-Based Films

Matrix	CNF Source	Effect	Reference
Tilapia gelatin films	Cotton linter	Strengthening of mechanical and barrier properties	Santos et al., 2014
Wheat gluten films	Commercial	Improvement in mechanical properties	Rafieian et al., 2014
Wheat gluten films	Sunflower	Homogeneous morphology, reduced permeability to water vapor	Fortunati et al., 2016
Alginate films	Sisal fibers (*Agave sisalana*)	Reduced solubility in water	Deepa et al., 2016
Alginate films	Mulberry pulp (*Broussonetia papyrifera*)	The elastic modulus and the tensile strength of the alginate films were increased by 35% and 25%, respectively.	Wang et al., 2017
Whey Protein Isolate Films	Oat husk	Improves tensile strength, Young's modulus, and film solubility	Qazanfarzadeh and Kadivar, 2016
Films of soy protein and pine needle extract	Wood pulp	Improvement in mechanical properties, antioxidant capacity	Yu et al., 2018
Zein films	Commercial	Thermal stability	Shakeri and Radmanesh, 2014

CNF: Cellulose nanofibrils

in this matrix the nanocellulose is capable of improving the characteristics of the packaging without jeopardizing quality attributes, such as transparency. The authors point out that, thanks to their elaboration method using ultrasound, the dispersion of the filling was improved, to which they attribute the success of it as packaging.

One of the current trends is the development of packaging that not only protects the food but also provides extra protection against microorganisms; that is, active packaging. Li et al. (2019) developed packaging based on this biopolymer reinforced with cellulose nanofibrils grafted with dopamine, also adding silver nanoparticles. This mixture they designed achieved an efficient dissipation of the energy attributed to the multiple networks, which indicates improvements in some properties, such as tensile strength. This hybrid film achieved a high antimicrobial efficacy against various bacteria because of the addition of nanoparticles, and improved the poor characteristics of the biopolymer alone, becoming a novel strategy for the design of packaging materials.

6.4.2 Wheat Gluten

Gluten is a protein set that represents more than 80% of the proteins in wheat flour. It is attractive for its good film-forming properties, but it is necessary to modify the polymeric network through various mechanisms: The formation of intra and intermolecular covalent crosslinks among molecules is achieved by using heat treatments. Another mechanism is to mix gluten with a hydrophobic polymer, such as lipids or esters, and the third mechanism is to incorporate fillers (Guan et al., 2011; Tunc et al., 2007), such as CNF. Rafieian et al. (2014) optimized films made with this polymer reinforced with CNF to determine if the addition of this component has an impact on the mechanical and thermal attributes (Table 6.2), finding that the addition of CNF in adequate concentrations (11 g) does not cause agglomeration. The incorporation of CNF increased the mechanical properties, but there was no significant increase in the thermal properties.

Gluten is an extremely attractive product due to multiple factors. It is easy to obtain and process, and it has good barrier properties. However, like majority of biopolymers it is hygroscopic in nature (Mojumdar et al., 2011; Rafieian et al., 2014). Fortunati et al. (2016) used a lignocellulosic material from sunflower to produce cellulose nanofibrils and nanocrystals in order to reinforce wheat gluten–based films (Table 6.2). The authors found a homogeneous morphology, without visible nanofillers and also without the presence of holes, reducing the permeability of the packaging in addition to corroborating their results with improvements in thermal and mechanical properties.

6.4.3 Alginate

Alginate is a natural biopolymer, formed by linked β-d-mannuronic acid and α-1 -guluronic acid molecules (Norajit et al., 2010). The main source for it is the wall of seaweed. The alginate composites can been introduced in food packaging due to their multiple benefits, like non-toxicity and gel-forming properties, but as with others biopolymers, its limited mechanical properties and weak water resistance make it difficult to use, specifically in moist environments (Abdollahi et al., 2013). Deepa et al. (2016) prepared biodegradable alginate films using cellulose nanofibrils (CNF) and determined the effects of CNF content (Table 6.2). With the incorporation of CNF they managed to reduce the swelling and solubility of the films in water, the lowest results at 10% by weight of nanofiller. In another investigation, Wang et al. (2017) isolated nanofibers (CFs) and nanowhiskers (CNW) of cellulose from mulberry pulp and mixed them with alginate to make films (Table 6.2). They used different concentrations (2 to 6% by weight) of CF and CNW to verify the impact of the addition on the properties of the films, highlighting the thermal, mechanical, and barrier properties. The mechanical properties of the alginate films were increased with CNW. The elastic modulus and the tensile strength of the films increased 35% and 25%, respectively.

Han and Wang (2017) prepared antibacterial films using sodium alginate (SA), carboxymethylcellulose (CMC), and pyrogallic acid (PA) as a natural antibacterial agent. In this research, the microstructure and mechanical, antibacterial, and optical attributes, among others, of the mixed films were evaluated with the addition of various percentages of PA in the films. The results indicated that PA can interact with the film through hydrogen bonding. In addition, the inclusion of PA increased the moisture, consequently altering the permeability of films. Films with PA are effective against various pathogenic bacteria, generating a matrix with good impact to protect food products.

6.4.4 Whey Protein

Whey protein is made up of a set of proteins that are a by-product of the dairy industry. There are reports indicating that with this raw material can produce films with good barrier properties and that it is more flexible compared to films obtained with carbohydrates and lipids. But these films present the same problems as most of those made with biopolymers; it has poor mechanical attributes and its permeability is deficient due to a large amount of hydrophilic amino acids in its structure. This is the main reason the addition of fillers is needed in the structure of the biopolymer (Sothornvit et al., 2009; Zolfi et al., 2014). Accordingly, Qazanfarzadeh and Kadivar (2016) prepared films based on whey protein and strengthened with cellulose (nano) from oat husk (Table 6.2), managing to increase the tensile strength, Young's modulus, and the solubility by adding 5% of nanofiller. However, at a concentration of 7.5% the properties were affected, due to the agglomeration of the cellulose.

6.4.5 Soy Protein

Soy protein-based films have many attractive features, such as being economical and biodegradable, among others. The use of isolated soy protein (SPI) has been increasing, mainly because it has high protein content, which can be greater than 90%. In addition to that, it is currently a highly available material (Cao et al., 2007). Also, it has excellent film-forming properties and the advantage of an accessible price, since it is a highly produced product in much of the world (Gennadios et al., 1993). However, various strategies are required to improve and control the general attributes of these films to generate an option for use as possible protective food packaging. Yu et al. (2018), in their research, developed soy protein matrices reinforced with nanocellulose and added Cedrus pine needle extract (PNE), evaluating the physical attributes, barrier properties, and the antioxidant capacity of the resulting films (Table 6.2). According to this research, the addition reduced the moisture of the films by directly interfering with the hydrogen bonds between the proteins and the water. The nanofiller reduced the elongation at break and intensified the tensile strength. When the addition of PNE was higher (5–10%), the permeability was improved due to a decrease in hydrophilic parts in the matrix. In addition, films with PNE bioactive compounds, such as phenols, showed important antioxidant activity.

6.4.6 Zein

Zein is a prolamine that can be found in the endosperm of corn kernels. It has become a very attractive alternative due to the fact that it is approved as a generally recognized as safe (GRAS) ingredient by the US Food and Drug Administration (US-FDA) and can therefore be used for food processing and transformation, including as a coating (Luecha et al., 2010). Zein films are very fragile and are usually reinforced with cellulose or other ingredients, among them glycerol, oleic acid, and other fatty acids, to increase their elasticity, among other properties (Takagi and Asano, 2008). In their research, Shakeri and Radmanesh (2014) prepared films of zein along with CNF in varying the concentration, from 0 to 5% by weight (Table 6.2). The most interesting finding of this work is that cellulose, by presenting a crystalline structure, provided a higher thermal stability to the matrices, generating materials with better characteristics with respect to the food industry.

Vahedikia et al. (2019) examined zein films with the addition of cinnamon essential oil (CEO) and chitosan nanoparticles (CNP 2–4% w/w) to study the various attributes of the films, particularly the mechanical and antibacterial properties. The results showed that CEO-CNP increased the tensile strength and reduced the elongation in the matrix. In microbiological tests the films with addition of CEO alone and those with CEO-CNP considerably inhibited some gram bacteria, such as *S. aureus*. Therefore, these authors concluded that zein-based reinforced compounds may emerge as an alternative for the production of films for the packaging of certain foods.

6.5 CHITIN NANOFIBRILS

A special mention of chitin nanofibrils should be made here, because their use has also been tested in the development of packaging reinforcements. Although they are not as widely used as cellulose fibrils, they are a latent option to include in various polymeric matrices. Chitin is a polysaccharide, which is comprised of two linked acetamido-2-deoxy-d-glucopyranose molecules (1–4) that may be deacetylated to some extent. The importance of this compound lies in the fact that it is one of the most available materials and can be extracted from various sources, such as the cell wall of fungi or the exoskeleton of arthropods (e.g., shrimp), among others. As reported by the UN Food and Agriculture Organization (FAO), the total annual production of crustaceans and other marine derivatives results in approximately

50–70% residue, making this marine resource an important source of chitin at a very low cost (practically free). Chitin, after acid processing, has the ability to form nanofibrils, which can be used for applications in various areas such as food-oriented nanotechnology (Mincea et al., 2012).

Chitin nanofibrils, due to their crystalline structure, nanometric size, and natural origin, have proven to be an effective carrier for pharmaceuticals but also an excellent option to reinforce the fibers of polymeric nanocomposites for use in more advanced medicines and food packaging (Morganti et al., 2011). In addition, changes in society point to the increasing desirability of use of renewable resources, so the research and use of fibers from natural sources as components for reinforcement has increased significantly in recent years (Satyanarayana et al., 2009). These compounds are, in fact, more respectful of the environment and offer certain additional advantages, such as an adequate compatibility between the film and these fibers when compared to traditional materials (Matabola et al., 2009).

Shankar et al. (2015) elaborated on films of carrageenan reinforced with chitin nanofibrils (CNF). The results show that films made with carrageenan and CNF were more flexible and the nanofiller achieved adequate dispersion throughout the matrix. The tensile strength of the films was improved until the addition of 5% by weight, and it should be noted that some properties, such as elongation at break and permeability, were affected. Films containing carrageenan with CNF showed significant antibacterial activity against *L. monocytogenes.*

Smirnova et al. (2019) analyzed the properties of chitosan-based films to try to improve the deficiencies of these matrices by adding chitin nanofibrils as nanofillers. In this research, the addition of 5% by weight of the nanofibrils offered the best advantages in the analyzed attributes. Salaberria et al. (2015) studied the effect of chitin nanocrystals and nanofibers in the properties of thermoplastic starch films and found that the addition of chitin in concentrations of 5–20% was effective for the main purpose of improving the mechanical and antifungal properties, among others, over to those without nanofiller. In addition, the authors verified that the content and size of chitin are very relevant in the final characteristics of the films produced. These results prove the benefits of the addition of chitin in nanometric form applied in thermoplastic starch matrices to be incorporated in coatings and packaging.

6.6 NANORODS AND THEIR APPLICATION IN PACKAGING AS NANOFILLERS

Thanks to advances in nanotechnology, more and more materials have been designed for multiple purposes, and in the packaging industry this holds as nanorods emerge as an alternative to improve various characteristics of packaging, such as mechanical, barrier, antimicrobial properties, among others (Ejaz et al., 2018). In addition to the concern of keeping the characteristics of food intact, there is great interest in reducing the impact of waste in the environment, which is why the research focuses on developing nanocomposites in packaging(Jafarzadeh et al., 2017).

A variety of this type of nanoparticle has been developed to reinforce various matrices, usually up to 5% by weight, and this has helped make these new materials stronger, lighter, and less permeable. These improvements are due to the diversity of materials used and the variety of methods of synthesis, such as sol-gel, hydrothermal, chemical processes, among others, which generate different morphologies and sizes, all opening up new possibilities in functionality when compared with conventional materials (Abbas et al., 2019). Also, the intrinsic characteristics of the nanoparticles that are used as nanofillers are based on their size, composition, morphology, among others (Mahmud and Abdullah, 2007). Various metals or metallic oxides have attracted attention due to their stability and their safe condition for use in products intended for humans and for all fauna in general (Lin et al., 2009).

6.6.1 Zinc Oxide (ZnO)

Nanofillers are so effective because they generate high surface energy and have a large specific surface area, giving them a strong interfacial correlation among the matrix and the fillers (Kovačević et al., 2008). Among these, ZnO has generated interesting research. There is a wide variety of ZnO particles with different particle sizes. Normally, when acquired they are highly purified and are found in a range of 10 to 800 nm, and can be in powder form, suspension, emulsion, among others (Abbas et al., 2019).

Torabi and Nafchi (2013) proposed that applying zinc oxide as nanofiller in films (from 1 to 5% w/w) of sago starch and bovine gelatin could yield a matrix with UV protection and with potential applications in the food industry. To test this hypothesis, they evaluated the antibacterial, physicochemical, and mechanical properties, among others. They found that with 5% of nanofill, the oxygen permeability decreased up to 55%, generally improving the mechanical properties, and in addition achieving values close to 80% of inhibition activity against *E. coli*, suggesting that these films with added nanorods can be a possible active packaging.

Xing et al. (2021) have investigated the impact of incorporating ZnO in films made from bean starch, which is a widely consumed and distributed worldwide crop. But as described above, starch-based films offer poor attributes. The particles were synthesized by an electrochemical method and 5% was added to add to the filmogenic solution. The addition of the ZnO particles resulted in better thermal characteristics in the films, and the starch-ZnO relationship achieved a stronger interaction in films, as the nanorod acts as a physical crosslinking that reduces the thermal decomposition of starch. In addition, there was a reduction in the growth of *E. coli* and *S. aureus* proportional to ZnO added. The mechanism of this action is not yet clear but it can be related with the surface/volume ratio, as the generation of H_2O_2 is based on the area of ZnO, and the generation of more reactive oxygen species results in better antibacterial activity, another important feature in the development of films with a food protection application.

The market increasingly demands the use of environmentally friendly and safe materials for food packaging, and this has resulted in the development and coupling of polymers and nanostructures with potential to replace conventional packaging (Siracusa et al., 2008) and ZnO is a compound that has been recognized by the US-FDA as safe for this. The combination of nano-sized structures and biomaterials has generated a huge area of opportunity for research, leading to new applications of what we can call nanobiotechnology; specifically, combining nanorods with biomolecules for multiple purposes (Fritzsche and Taton, 2003). Rouhi et al. (2013) designed films of fish gelatin with ZnO to evaluate its mechanical and UV absorption. Finding that an addition of 5% ZnO resulted in enhancement of the mechanical properties and a 25% improvement in the tensile strength compared to the film without reinforcement, an effect generated by the interfacial interaction between the particles and the film. With the addition, they were able to reduce the transition from UV to 0, behavior related to the shape and structure of the nanofiller.

Ejaz et al. (2018) examined films made of bovine skin gelatin reinforced with ZnO and clove essential oil, observing that the addition of the nanofiller resulted in films with less flexibility but higher strength, in addition to improvements in oxygen permeability and protection against UV rays. The film was used to cover peeled shrimp inoculated with *L. monocytogenes* and *S. typhi*. The Zn^{2+} ions penetrated the cell wall of the microorganisms, destabilizing the microbial cell and causing its death, an effect attributed mainly to the activity of ZnO and the added essential oil.

6.6.2 Titanium Dioxide (TiO_2)

TiO_2 has generated special interest due to its unique and excellent properties in areas such as optics, electronics, and even biology and its potential applications in multiple materials. Currently, these particles can be synthesized by various methods, such as hydrothermal, chemical vapor deposition, or sol-gel method (Cao et al., 2011). The combination of biopolymers with nanocomposites has resulted in better properties than just the use of polymers, specifically and increase in barrier properties. It is considered an effective method for microbial control (Nassiri and Nafchi, 2013).

Shaili et al. (2015) have developed a novel packaging based on soluble soy polysaccharide extracted from the cell wall of the cotyledons, which they reinforced with different concentrations of TiO_2, up to 5% w/w. The incorporation of this nanorod made it possible to reduce the permeability to water vapor and to oxygen, because the incorporation of TiO_2 in the matrix creates a more complicated route for moisture or oxygen to pass through. The mechanical properties were also improved, an effect caused by the moisture concentration of the films and by the interfacial interaction that has been generated between the matrix and the filler. These films presented excellent activity against *E. coli* and *S. aureus*, behavior attributed to the oxidative attack that TiO_2 generates on the bacterial membranes, triggering lipid peroxidation.

Another example of research to generate options for packaging with superior characteristics is the work of Salarbashi et al. (2018), who developed films based on biodegradable soy polysaccharide reinforced with TiO_2 in different concentrations, finding that when the filler concentration did not exceed 3%, the surface of the matrix was morphologically uniform. However, the thermal analysis they carried out revealed that the storage modulus reached higher values, up to 3.62%, when they added 7% of filler. The nanocomposites made with the polysaccharide/TiO_2 showed excellent antimicrobial activity against *S. aureus*. In addition, they evaluated the anti- or pro-cancer activity in cell lines, an area of concern with use of particles such as TiO_2 in materials destined for consumption. The findings reveal that this new development in materials is a promising packaging system thanks to its characteristics as well as low health risk.

6.6.3 Silicon Dioxide (SiO_2)

The factor determining the safety of synthetic particles in products for humans is the use of amorphous SiO_2 particles. These particles offer certain advantages, such as their thermal stability and durability, among others. Unlike crystalline SiO_2 that has been reported as a compound causing carcinogenicity, amorphous SiO_2 is widely used in food processing thanks to its aforementioned advantages (Liu et al., 2019).

Bionanocomposites are a new generation of materials that present unique characteristics. They are formed from the combination of a biopolymer and a nano-scale inorganic material. The attractiveness of this mixture of components, as reviewed in this chapter, is the functional properties it generates, such as excellent mechanical attributes and less permeability to water, to name a few (Pavlath and Orts, 2009). The use of SiO_2 in natural biopolymer films has been investigated. Torabi and Nafchi (2013) evaluated the impact of SiO_2 in potato starch films, incorporating up to 5% w/w filler, evaluating physicochemical, barrier, and mechanical properties. Their findings indicate that by increasing SiO2, the characteristics of the films improve, mentioning that the nanoparticles could join with hydroxyl groups and through other bonds of the macromolecules, such as hydrogen bonds or Van der Wall's forces, make the interaction between nanorods and starch stronger. It was also observed that the permeability to water vapor decreased, attributed to the fact that the nanoparticles fill the pores of the macromolecule structures, decreasing the permeability to gas and water vapor.

Taking advantage of the fact that SiO_2 is a widely available nanorod that is biodegradable and profitable (Voon et al., 2012), Ghazihoseini et al. (2015) developed a film based on soluble soy polysaccharides reinforced with SiO_2 to evaluate its characteristics. They found that the incorporation of the reinforcement contributed to the reduction of permeability to water vapor and oxygen, generating a much more complex path for gaseous molecules to traverse. There was improved resistance to heat sealing and improved mechanical properties when fulfilling the purpose of filling voids, leading to greater structural rigidity, all proving the effectiveness of mixing this component with the polymeric matrix for use in packaging materials.

6.6.4 Magnesium Oxide (MgO)

MgO is a widely used compound that can be synthesized through inexpensive processes with common ingredients, such as magnesium salts among others. It has interesting characteristics, such as activity against various microorganisms, and is an excellent thermal insulator, in addition to its ability to block UV rays, among others (Mantilaka et al., 2014; Salehifar et al., 2016). It is these characteristics that have attracted the attention of research aimed at applying this compound as a nanofiller in films intended for the packaging industry. De Silva et al. (2017) developed chitosan-based films reinforced with MgO nanorods to try to improve their properties, finding that the mechanical properties were better optimized when they added 5% w/w, positively impacting tensile stress and Young's modulus, behavior attributed to the effect between the functional groups of chitosan with the nanofiller. In addition, the results indicate that the reinforced packaging generated better protection against UV rays and improved permeability to water vapor. A study carried out by Sanuja et al. (2014)

added clove essential oil to a film of chitosan reinforced with MgO and achieved the objective of improving the attributes. The permeability to water vapor was also decreased. With the general composition of this film, they generated an active packaging with effective antibacterial activity against *S. aureus*, mentioning that clove oil contains eugenol whose antimicrobial effect has been widely reported. Additionally, chitosan together with MgO affect the cell membrane, affecting the viability of these microorganisms.

6.7 CONCLUSIONS

The use of nanofibers is a growing trend in the development of packaging, because those of natural origin tend to have poor properties compared to packaging derived from plastic. A large amount of information has been generated regarding the characteristics that the packages acquire when they are reinforced, mainly with CNF (thanks to the fact that it is a very abundant and inexpensive material that normally takes advantage of waste). Most significant is the improvement in the barrier thermal and mechanical properties, taking into account the added concentration of filler. However, there are other materials that can improve the properties of packaging material, such as chitin nanofibrils, which reinforce polymeric matrices.

Likewise, the use of nanorods in packaging is an effective option, because these compounds are able to improve the matrices in their properties, and as an additional factor in most cases they are effective against various microorganisms, generating active packaging, a growing trend in food packaging.

The reinforcement of polymeric matrices with nanofillers opens the possibility of developing packaging that offers properties allowing it to better fulfill its function, which is to better contain the food and preserve its characteristics during storage. There is much opportunity for further research regarding the incorporation of these compounds, but very interesting bases have been laid that generate valuable information to build on for continued improvements in the packaging area and in the replacement of synthetic packaging with its massive pollutant impacts worldwide.

REFERENCES

Abbas, M., Buntinx, M., Deferme, W., and Peeters, R. 2019. (Bio) polymer/ZnO nanocomposites for packaging applications: A review of gas barrier and mechanical properties. *Nanomaterials* 9: 1494.

Abdollahi, M., Alboofetileh, M., Rezaei, M., and Behrooz, R. 2013. Comparing physico-mechanical and thermal properties of alginate nanocomposite films reinforced with organic and/or inorganic nanofillers. *Food Hydrocolloids* 32: 416–424.

Avérous, L., and Halley, P. J. 2009. Biocomposites based on plasticized starch. *Biofuels, Bioproducts and Biorefining* 3: 329–343.

Azeredo, H. M., Mattoso, L. H. C., Avena, Bustillos, R. J., Filho, G. C., Munford, M. L., Wood, D., and McHugh, T. H. 2010. Nanocellulose reinforced chitosan composite films as affected by nanofiller loading and plasticizer content. *Journal of Food Science* 75: N1–N7.

Balakrishnan, P., Gopi, S., MS, S., and Thomas, S. 2018. UV resistant transparent bionanocomposite films based on potato starch/cellulose for sustainable packaging. *Starch-Stärke* 70: 1700139.

Cao, C., Hu, C., Wang, X., Wang, S., Tian, Y., and Zhang, H. 2011. UV sensor based on TiO_2 nanorod arrays on FTO thin film. *Sensors and Actuators B: Chemical* 156: 114–119.

Cao, N., Fu, Y., and He, J. 2007. Preparation and physical properties of soy protein isolate and gelatin composite films. *Food Hydrocolloids* 21: 1153–1162.

Cenobio-Galindo, A. J., Pimentel-González, D. J., Del Razo-Rodríguez, O. E., Medina-Pérez, G., Carrillo-Inungaray, M. L., Reyes-Munguía, A., and Campos-Montiel, R. G. 2019. Antioxidant and antibacterial activities of a starch film with bioextracts microencapsulated from cactus fruits (*Opuntia oligacantha*). *Food Science and Biotechnology* 28: 1553–1561.

Chen, S., Wu, M., Wang, C., Yan, S., Lu, P., and Wang, S. 2020. Developed Chitosan/Oregano essential oil biocomposite packaging film enhanced by cellulose nanofibril. *Polymers* 12: 1780.

De Silva, R. T., Mantilaka, M. M. M. G. P. G., Ratnayake, S. P., Amaratunga, G. A. J., and de Silva, K. N. 2017. Nano-MgO reinforced chitosan nanocomposites for high performance packaging applications with improved mechanical, thermal and barrier properties. *Carbohydrate Polymers* 157: 739–747.

Deepa, B., Abraham, E., Pothan, L. A., Cordeiro, N., Faria, M., and Thomas, S. 2016. Biodegradable nanocomposite films based on sodium alginate and cellulose nanofibrils. *Materials* 9: 50.

Dinand, E., Chanzy, H., and Vignon, R. M. 1999. Suspensions of cellulose microfibrils from sugar beet pulp. *Food Hydrocolloids* 13: 275–283.

Dufresne, A. 2017. Cellulose nanomaterial reinforced polymer nanocomposites. *Current Opinion in Colloid & Interface Science* 29: 1–8.

Ejaz, M., Arfat, Y. A., Mulla, M., and Ahmed, J. 2018. Zinc oxide nanorods/clove essential oil incorporated type b gelatin composite films and its applicability for shrimp packaging. *Food Packaging and Shelf Life* 15: 113–121.

Emons, A. M. C., and Mulder, B. M. 2000. How the deposition of cellulose microfibrils builds cell wall architecture. *Trends in Plant Science* 5: 35–40.

Eslami, R., Azizi, A., and Najafi, M. 2021. Effects of different gums on tensile properties of cellulose nanofibrils (CNFs)-reinforced starch. *Polymers and Polymer Composites* 29: 436–443.

Fortunati, E., Luzi, F., Jiménez, A., Gopakumar, D. A., Puglia, D., Thomas, S., and Torre, L. 2016. Revalorization of sunflower stalks as novel sources of cellulose nanofibrils and nanocrystals and their effect on wheat gluten bionanocomposite properties. *Carbohydrate Polymers* 149: 357–368.

Fritzsche, W., and Taton, T. A. 2003. Metal nanoparticles as labels for heterogeneous, chip-based DNA detection. *Nanotechnology* 14: 63.

Gan, I., and Chow, W. S. 2018. Antimicrobial poly (lactic acid)/cellulose bionanocomposite for food packaging application: A review. *Food Packaging and Shelf Life* 17: 150–161.

Gennadios, A., Weller, C., and Testin, R. F. 1993. Temperature effect on oxygen permeability of edible protein-based films. *Journal of Food Science* 58: 212–214.

Ghanbari, A., Tabarsa, T., Ashori, A., Shakeri, A., and Mashkour, M. 2018. Thermoplastic starch foamed composites reinforced with cellulose nanofibers: Thermal and mechanical properties. *Carbohydrate Polymers* 197: 305–311.

Ghazihoseini, S., Alipoormazandarani, N., and Nafchi, A. M. 2015. The effects of nano-SiO2 on mechanical, barrier, and moisture sorption isotherm models of novel soluble soybean polysaccharide films. *International Journal of Food Engineering* 11:833–840.

Glenn, G. M., Imam, S. H., and Orts, W. J. 2011. Starch-based foam composite materials: Processing and bioproducts. *MRS Bulletin* 36: 696–702.

Guan, L., Jiménez, M. G., Walowski, C., Boushehri, A., Prausnitz, J. M., and Radke, C. J. 2011. Permeability and partition coefficient of aqueous sodium chloride in soft contact lenses. *Journal of Applied Polymer Science* 122: 1457–1471.

Hamid, K. H. A., Wan Yahaya, W. A., Mohd Saupy, N. A. Z., Almajano, M. P., and Mohd Azman, N. A. 2019. Semi-refined carrageenan film incorporated with α-tocopherol: Application in food model. *Journal of Food Processing and Preservation* 43: 13937.

Han, Y., and Wang, L. 2017. Sodium alginate/carboxymethyl cellulose films containing pyrogallic acid: Physical and antibacterial properties. *Journal of the Science of Food and Agriculture* 97: 1295–1301.

Hanani, Z. N., Roos, Y. H., and Kerry, J. P. 2014. Use and application of gelatin as potential biodegradable packaging materials for food products. *International Journal of Biological Macromolecules* 71: 94–102.

Hietala, M., Mathew, A. P., and Oksman, K. 2013. Bionanocomposites of thermoplastic starch and cellulose nanofibers manufactured using twin-screw extrusion. *European Polymer Journal* 49: 950–956.

Jafarzadeh, S., Alias, A., Ariffin, F., and Mahmud, S. 2017. Characterization of semolina protein film with incorporated zinc oxide nano rod intended for food packaging. *Polish Journal of Food and Nutrition Science* 67: 183–190.

Kargarzadeh, H., Johar, N., and Ahmad, I. 2017. Starch biocomposite film reinforced by multiscale rice husk fiber. *Composites Science and Technology* 151: 147–155.

Khalil, H. A., Davoudpour, Y., Islam, M. N., Mustapha, A., Sudesh, K., Dungani, R., and Jawaid, M. 2014. Production and modification of nanofibrillated cellulose using various mechanical processes: A review. *Carbohydrate Polymers* 99: 649–665.

Khan, M. R., and Sadiq, M. B. 2021. Importance of gelatin, nanoparticles and their interactions in the formulation of biodegradable composite films: A review. *Polymer Bulletin* 78: 4047–4073.

Kovačević, V., Vrsaljko, D., Lučić Blagojević, S., and Leskovac, M. 2008. Adhesion parameters at the interface in nanoparticulate filled polymer systems. *Polymer Engineering and Science* 48: 1994–2002.

Lee, H., Sundaram, J., and Mani, S. 2017. Production of cellulose nanofibrils and their application to food: A review. *Nanotechnology* 1: 1–33.

Li, K., Jin, S., Chen, H., and Li, J. 2019. Bioinspired interface engineering of gelatin/cellulose nanofibrils nanocomposites with high mechanical performance and antibacterial properties for active packaging. *Composites Part B: Engineering* 171: 222–234.

Lin, O. H., Akil, H. M., and Mahmud, S. 2009. Effect of particle morphology on the properties of polypropylene/nanometric zinc oxide (pp/nanozno) composites. *Advanced Composites Letters* 8: 77–83.

Liu, X., Chen, X., Ren, J., Chang, M., He, B., and Zhang, C. 2019. Effects of nano-ZnO and nano-SiO_2 particles on properties of PVA/xylan composite films. *International Journal of Biological Macromolecules* 132: 978–986.

Luecha, J., Sozer, N., and Kokini, J. L. 2010. Synthesis and properties of corn zein/montmorillonite nanocomposite films. *Journal of Materials Science* 45: 3529–3537.

Mahmud, S., and Abdullah, M. J. 2007. Tapered head facets of zinc oxide nanorods. *Solid State Science and Technology* 15: 108–115.

Mantilaka, M. M. M. G. P. G., Pitawala, H. M. T. G. A., Karunaratne, D. G. G. P., and Rajapakse, R. M. G. 2014. Nanocrystalline magnesium oxide from dolomite via poly (acrylate) stabilized magnesium hydroxide colloids. *Colloids and Surfaces A: Physicochemical and Engineering Aspects* 443: 201–208.

Matabola, K. P., De Vries, A. R., Moolman, F. S., and Luyt, A. S. 2009. Single polymer composites: A review. *Journal of Materials Science* 44: 6213–6222.

Mincea, M., Negrulescu, A., and Ostafe, V. 2012. Preparation, modification, and applications of chitin nanowhiskers: A review. *Reviews on Advanced Materials Science* 30: 225–242.

Mojumdar, S. C., Moresoli, C., Simon, L. C., and Legge, R. L. 2011. Edible wheat gluten (WG) protein films: Preparation, thermal, mechanical and spectral properties. *Journal of Thermal Analysis and Calorimetry* 104: 929–936.

Moradi, M., Tajik, H., Rohani, S. M. R., Oromiehie, A. R., Malekinejad, H., Aliakbarlu, J., and Hadian, M. 2012. Characterization of antioxidant chitosan film incorporated with Zataria multiflora Boiss essential oil and grape seed extract. *LWT* 46: 477–484.

Morganti, P., Morganti, G., and Morganti, A. 2011. Transforming nanostructured chitin from crustacean waste into beneficial health products: A must for our society. *Nanotechnology Science and Applications* 4: 123.

Mostafavi, F. S., and Zaeim, D. 2020. Agar-based edible films for food packaging applications-a review. *International Journal of Biological Macromolecules* 159: 1165–1176.

Nascimento, D. M., Dias, A. F., de Araújo Junior, C. P., de Freitas Rosa, M., Morais, J. P. S., and de Figueirêdo, M. C. B. 2016. A comprehensive approach for obtaining cellulose nanocrystal from coconut fiber. Part II: Environmental assessment of technological pathways. *Industrial Crops and Products* 93: 58–65.

Nassiri, R., and Nafchi, A. M. 2013. Antimicrobial and barrier properties of bovine gelatin films reinforced by nano TiO2. *Journal of Chemical Health Risks* 3: 12–28.

Nieto, M. B., and Akins, M. 2010. Hydrocolloids in bakery fillings. *Hydrocolloids in Food Processing* 43: 67.

Norajit, K., Kim, K. M., and Ryu, G. H. 2010. Comparative studies on the characterization and antioxidant properties of biodegradable alginate films containing ginseng extract. *Journal of Food Engineering* 98: 377–384.

Othman, S. H. 2014. Bio-nanocomposite materials for food packaging applications: Types of biopolymer and nano-sized filler. *Agriculture and Agricultural Science Procedia* 2: 296–303.303

Pavlath, A. E., and Orts, W. 2009. Edible films and coatings: Why, what, and how?. In Huber, K. C., and Embuscado, M. E. (eds.), *Edible Films and Coatings for Food Applications*, 1–23. Springer, New York, NY.

Prachayawarakorn, J., Chaiwatyothin, S., Mueangta, S., and Hanchana, A. 2013. Effect of jute and kapok fibers on properties of thermoplastic cassava starch composites. *Materials & Design* 47: 309–315.

Qazanfarzadeh, Z., and Kadivar, M. 2016. Properties of whey protein isolate nanocomposite films reinforced with nanocellulose isolated from oat husk. *International Journal of Biological Macromolecules* 91: 1134–1140.

Rachtanapun, P., Klunklin, W., Jantrawut, P., Jantanasakulwong, K., Phimolsiripol, Y., Seesuriyachan, P., and Ngo, T. M. P. 2021. Characterization of chitosan film incorporated with curcumin extract. *Polymers* 13: 963.

Rafieian, F., Shahedi, M., Keramat, J., and Simonsen, J. 2014. Mechanical: Thermal and barrier properties of nano-biocomposite based on gluten and carboxylated cellulose nanocrystals. *Industrial Crops and Products* 53: 282–288.

Rajendran, N., Puppala, S., Sneha Raj, M., Ruth Angeeleena, B., and Rajam, C. 2012. Seaweeds can be a new source for bioplastics. *Journal of Pharmacy Research* 5: 1476–1479.

Reddy, J. P., and Rhim, J. W. 2014. Characterization of bionanocomposite films prepared with agar and paper-mulberry pulp nanocellulose. *Carbohydrate Polymers* 110: 480–488.

Rhim, J. W., Hong, S. I., and Ha, C. S. 2009. Tensile, water barrier and antimicrobialproperties of PLA/nanoclay composite films. *LWT* 42: 612–617.

Rhim, J. W., Park, H. M., and Ha, C. S. 2013. Bio-nanocomposites for food packaging applications. *Progress in Polymer Science* 38: 1629–1652.

Rouhi, J., Mahmud, S., Naderi, N., Ooi, C. R., and Mahmood, M. R. 2013. Physical properties of fish gelatin-based bio-nanocomposite films incorporated with ZnO nanorods. *Nanoscale Research Letters* 8: 1–6.

Saba, N., Tahir, P. M., and Jawaid, M. 2014. A review on potentiality of nano filler/natural fiber filled polymer hybrid composites. *Polymers* 6: 2247–2273.

Salaberria, A. M., Diaz, R. H., Labidi, J., and Fernandes, S. C. 2015. Role of chitin nanocrystals and nanofibers on physical, mechanical and functional properties in thermoplastic starch films. *Food Hydrocolloids* 46: 93–102.

Salarbashi, D., Tafaghodi, M., and Bazzaz, B. S. F. 2018. Soluble soybean polysaccharide/TiO_2 bionanocomposite film for food application. *Carbohydrate Polymers* 186: 384–393.

Salehifar, N., Zarghami, Z., and Ramezani, M. 2016. A facile, novel and low-temperature synthesis of MgO nanorods via thermal decomposition using new starting reagent and its photocatalytic activity evaluation. *Materials Letters* 167: 226–229.

Santos, T. M., Men de Sá Filho, M. S., Caceres, C. A., Rosa, M. F., Morais, J. P. S., Pinto, A. M., and Azeredo, H. M. 2014. Fish gelatin films as affected by cellulose whiskers and sonication. *Food Hydrocolloids* 41: 113–118.

Sanuja, S., Agalya, A., and Umapathy, M. J. 2014. Studies on magnesium oxide reinforced chitosan bionanocomposite incorporated with clove oil for active food packaging application. *International Journal of Polymeric Materials and Polymeric Biomaterials* 63: 733–740.

Satyanarayana, K. G., Arizaga, G. G., and Wypych, F. 2009. Biodegradables composites based on lignocellulosic fibers—An overview. *Progress in Polymer Science* 34: 982–1021.

Sedayu, B. B., Cran, M. J., and Bigger, S. W. 2020. Reinforcement of refined and semi-refined carrageenan film with nanocellulose. *Polymers* 12: 1145.

Shaili, T., Abdorreza, M. N., and Fariborz, N. 2015. Functional, thermal, and antimicrobial properties of soluble soybean polysaccharide biocomposites reinforced by nano TiO_2. *Carbohydrate Polymers* 134: 726–731.

Shakeri, A., and Radmanesh, S. 2014. Preparation of cellulose nanofibrils by high-pressure homogenizer and zein composite films. *Advanced Materials Research* 829: 534–538.

Shankar, S., Reddy, J. P., Rhim, J. W., and Kim, H. Y. 2015. Preparation, characterization, and antimicrobial activity of chitin nanofibrils reinforced carrageenan nanocomposite films. *Carbohydrate Polymers* 117: 468–475.

Siracusa, V., Rocculi, P., Romani, S., and Dalla Rosa, M. 2008. Biodegradable polymers for food packaging: A review. *Trends in Food Science & Technology* 19: 634–643.

Smirnova, N. V., Kolbe, K. A., Dresvyanina, E. N., Grebennikov, S. F., Dobrovolskaya, I. P., Yudin, V. E., and Morganti, P. 2019. Effect of chitin nanofibrils on biocompatibility and bioactivity of the chitosan-based composite film matrix intended for tissue engineering. *Materials* 12: 1874.

Soltani, N., Bahrami, A., Pech-Canul, M. I., and González, L. A. 2015. Review on the physicochemical treatments of rice husk for production of advanced materials. *Chemical Engineering Journal* 264: 899–935.

Soni, R., Asoh, T. A., Hsu, Y. I., Shimamura, M., and Uyama, H. 2020. Effect of starch retrogradation on wet strength and durability of cellulose nanofiber reinforced starch film. *Polymer Degradation and Stability* 177: 109165.

Sothornvit, R., Rhim, J. W., and Hong, S. I. 2009. Effect of nano-clay type on the physical and antimicrobial properties of whey protein isolate/clay composite films. *Journal of Food Engineering* 91: 468–473.

Sun, J., Shen, J., Chen, S., Cooper, M. A., Fu, H., Wu, D., and Yang, Z. 2018. Nanofiller reinforced biodegradable PLA/PHA composites: Current status and future trends. *Polymers* 10(5): 505.

Syafri, E., Kasim, A., Abral, H., Sulungbudi, G. T., Sanjay, M. R., and Sari, N. H. 2018. Synthesis and characterization of cellulose nanofibers (CNF) ramie reinforced cassava starch hybrid composites. *International Journal of Biological Macromolecules* 120: 578–586.

Takagi, H., and Asano, A. 2008. Effects of processing conditions on flexural properties of cellulose nanofiber reinforced "green" composites. *Composites Part A: Applied Science and Manufacturing* 39: 685–689.

Torabi, Z., and Nafchi, A. M. 2013. The effects of SiO_2 nanoparticles on mechanical and physicochemical properties of potato starch films. *Journal of Chemical Health Risks* 3: 33–42.

Tunc, S., Angellier, H., Cahyana, Y., Chalier, P., Gontard, N., and Gastaldi, E. 2007. Functional properties of wheat gluten/montmorillonite nanocomposite films processed by casting. *Journal of Membrane Science* 289: 159–168.

Vahedikia, N., Garavand, F., Tajeddin, B., Cacciotti, I., Jafari, S. M., Omidi, T., and Zahedi, Z. 2019. Biodegradable zein film composites reinforced with chitosan nanoparticles and cinnamon essential oil: Physical, mechanical, structural and antimicrobial attributes. *Colloids and Surfaces B: Biointerfaces* 177: 25–32.

Voon, H. C., Bhat, R., Easa, A. M., Liong, M. T., and Karim, A. A. 2012. Effect of addition of halloysite nanoclay and SiO_2 nanoparticles on barrier and mechanical properties of bovine gelatin films. *Food and Bioprocess Technology* 5: 1766–1774.

Wang, L. F., Shankar, S., and Rhim, J. W. 2017. Properties of alginate-based films reinforced with cellulose fibers and cellulose nanowhiskers isolated from mulberry pulp. *Food HydrocollOids* 63: 201–208.

Xing, H., Zhang, T., Pan, F., and Liu, J. 2021. Application of electrochemically synthesized zinc oxide Nanorods/Fava bean starch biocomposite in food active packaging. *International Journal of Electrochemical Science* 16: 1–11.

Yu, Z., Sun, L., Wang, W., Zeng, W., Mustapha, A., and Lin, M. 2018. Soy protein-based films incorporated with cellulose nanocrystals and pine needle extract for active packaging. *Industrial Crops and Products* 112: 412–419.

Zhang, Q. X., Yu, Z. Z., Xie, X. L., Naito, K., and Kagawa, Y. 2007. Preparation and crystalline morphology of biodegradable starch/clay nanocomposites. *Polymer* 48: 7193–7200.

Zimmermann, T., Pöhler, E., and Geiger, T. 2004. Cellulose fibrils for polymer reinforcement. *Advanced Engineering Materials* 6: 754–761.

Zolfi, M., Khodaiyan, F., Mousavi, M., and Hashemi, M. 2014. The improvement of characteristics of biodegradable films made from kefiran–whey protein by nanoparticle incorporation. *Carbohydrate Polymers* 109: 118–125.

7 Oxide-Based Nanocomposites for Food Packaging Application

Amandeep Kaur Braich[1] *and Gurkirat Kaur*[2]
[1]Department of Food Science and Technology,
Punjab Agricultural University, Ludhiana, India
[2]Electron Microscopy and Nanoscience Lab,
Punjab Agricultural University, Punjab, India

CONTENTS

DOI: 10.1201/9781003207641-8

7.1 INTRODUCTION

Food packaging is one of the most important aspects of the modern food industry as only a suitable food package can protect food and maintain its nutritional quality along with convenient delivery to the consumers. The food packaging industry accounts for a 2% share of the gross national product, as modern life is incomplete and unthinkable without food packaging. The utmost purpose of a food package is to maintain food nutritional quality as well as food safety, from the farm to the hands of consumers (Pilevar et al., 2019). An appropriate food package should not only be used as a delivering agent for the food, it also has to provide a barrier between the external environment and the food material as foods generally are perishable in nature and easily deteriorate during storage and transportation. There are various kinds of food packaging available in the market but they have certain limitations, such as poor barrier properties and high investment costs. A suitable raw material used for preparation of food packaging material must be odorless, inert in nature, have high tensile strength and flexibility, a lower water vapor and gaseous transmission rate, be easy to fabricate, and should be low cost (Behboudi, 2017). Polyethylene and other plastic materials are widely used packaging material in the food industry, but their increasing use poses a threat to humans, which has shifted consumer preference as well as industrial interest toward other polymeric food packaging materials. Therefore, there is a need for excellent food packaging that can overcome the limitations of the prevailing food packaging materials. An appropriate food packaging material can be formulated by using nanotechnology with its enhanced barrier properties as well as improved weight, tensile strength, and antimicrobial action, along with ease in recycling.

Nanotechnology is said to be the innovative techniques of 21st century having many advantages over current techniques. Nanotechnology involves the manipulation of matter at a very small scale; that is, between 1 and 100 nanometers (Bajpai et al., 2018). Nanomaterials having a larger surface-to-volume ratio and enhanced barrier, mechanical, and thermal properties, which can make them a promising option for formulating food packaging materials (Pavlidou and Papaspyrides, 2008). The incorporation of nanomaterial in food packaging can lead to the development of an active as well as intelligent food packaging material. An active food packaging material made up of nanoparticles has antimicrobial properties that protect the food material from various kinds of microbial infestation during storage and distribution (Llorens et al., 2012). And an intelligent food packaging should provide complete information about the food, which can be achieved using nanotechnology. A nanocomposite is a mixture of polymer matrix with organic or inorganic nanofillers, having at least one with nanoscale geometries. With the fabrication of these nanocomposites, a more durable and environmentally friendly food packaging can be developed that limits the use of plastics-based materials in food packaging. Several organic as well as inorganic materials called nanofillers have been employed for making nanocomposites. Oxides are an inorganic nanomaterial used as a reinforcing agent in nanocomposites. Reinforcement with these nanofillers leads to enhanced barrier, mechanical, optical, and thermal properties, and to improved tensile strength of nanocomposites beyond those of the commonly used food packaging (Rubio et al., 2020). When these reinforced materials are blended with polymeric materials during nanocomposite formation, there are various kinds of interactions which tend to improve the properties of the polymeric materials. These reinforcing materials are in nanometric range, thus increasing the surface area of nanocomposites and improving their tensile strength and overall reactivity. Moreover, these nanomaterials make strong bonds and disperse in polymeric materials homogeneously, altering their mobility and reducing their interaction sites with water molecules, which further enhances the thermal, mechanical and barrier properties of composites (Bumbudsanpharoke et al., 2015).

There are various oxides, such as zinc oxide, aluminum oxide, copper oxide, titanium oxide, silica, boron oxide, magnesium oxide, and doped nano oxides, which have been used in formulating nanocomposites for food packaging purpose. Iron and graphene oxide possess strong antimicrobial action toward bacterial cells and make a good film with enhanced mechanical properties, but their use in food packaging is still under investigation (Konwar et al., 2016).These oxides have photocatalytic, antimicrobial, and ethylene scavenging properties which makes the nanocomposites a suitable food packaging material along with enhancing their mechanical and barrier properties (Tyler et al., 2019). Therefore, in this chapter we will discuss the various oxides used in nanocomposites, their properties, and their preparation methods along with the safety aspects.

7.2 NANO OXIDES IN FOOD PACKAGING

Figure 7.1 demonstrates the most commonly employed metal oxides in food packaging industry. In this section, the most commonly used nano oxides are discussed in detail.

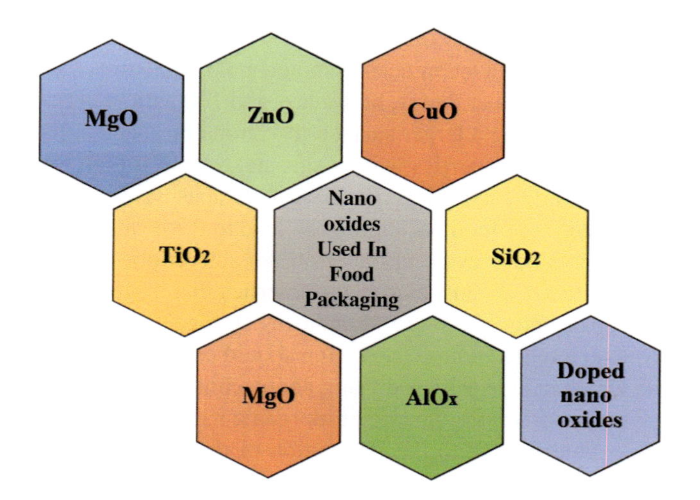

FIGURE 7.1 Different nano oxides used in food packaging.

7.2.1 Titanium Dioxide

Titanium dioxide (TiO_2) has been widely explored and used as a food additive among titanium-based inorganic materials because it is odorless and has a white coloration. Titanium dioxide has three different forms (anatase, rutile, and brookite) and each form has different properties due to differences in their band gap. Titanium dioxide is an inert material with many valuable properties, such as its photocatalytic nature and low in cost. Due to its photocatalytic nature, when TiO_2 is exposed to UV light, it starts generating reactive oxygen species which kill the microbial cells and other allergens by oxidation of their DNA and proteins (Carré et al., 2014). Also, titanium dioxide–based materials decrease the production of ethylene by oxidizes it into water and carbon dioxide, which is a very important factor in food packaging material, especially for fresh produce. The anatase form of titanium dioxide possesses the highest barrier property against light, and it gives a bright white appearance to the polymeric material (Oleyaei, 2016).

7.2.2 Zinc Oxide

Zinc oxide nanofillers are considered a novel class in formulating nanocomposites for improved shelf life of food products because of its remarkable antimicrobial action. Zinc oxide is an inorganic semiconductor with a white color and no odor, and with 3.3 eV band gap. It has been accepted as a generally recognized safe food additive by the US FDA (Espitia et al., 2012). Like titanium dioxide, ZnO occurs in three different forms: wurtzite, rock salt, and zinc blend. Among these groups, wurtzite is more thermodynamically stable in comparison to other forms. Zinc oxide–based nanocomposites have excellent photocatalytic properties, as zinc oxide nanoparticles have the ability to alter their spherical shape to form hexagonal nanosized rods. Also, due to its crystalline nature and surface functional groups, composites having zinc oxide possess a wide range of antimicrobial activity (Reyes-Torres et al., 2019).

7.2.3 Copper Oxide

Copper oxide use in food packaging material has been reported recently as these oxides have great potential in reducing the microbial load in food (Kuswandi and Moradi, 2019). Copper oxide–based material has antimicrobial properties against bacteria, fungi, and viruses. As copper is one of the necessary elements in maintaining whole metabolisms in microbes, the excess concentration of copper ions can lead to death of microbial cells. The inclusion of copper oxides in food packaging material results in excess of copper ions inside microbial cells. The excess copper ions raise the redox potential, which is harmful to microbes and leads to their death (Nan et al., 2008). The antimicrobial mechanisms of copper oxides against microbes are explained in detail in the antimicrobial properties of nanocomposites.

7.2.4 Silicon Dioxide

Silicon is one of the most profound elements on Earth having two different forms: silicate and silicon dioxide (SiO_2). Silicon dioxide, also known as silica, has been widely used in the food industry for formulating various kinds of nonstick coatings (Liu et al., 2018). Like all other metal oxides, nano silica is blended into polymeric materials for improving their heat sensitivity and mechanical strength. Also, silica acts to improve the catalyst activity in nanocomposites. Nano silica will attain different properties according to its reinforcement size, shape, and different morphology. Silica has great biocompatibility with polymeric material along with higher surface area and stability and lower toxicity and thermal conductivity, which makes silica a suitable material for incorporating into nanocomposites. However, due to the presence of silanol and siloxane groups on its surface, some surface modifications have to be made prior to introduction into some nanocomposites, as these make nano silica a hydrophilic particle, which further leads to formation of agglomerates due to altered dispersion phenomenon in whole matrices (Mallakpour and Naghdi, 2018). The most commonly used modified silicas in the food industry are 3-isocyanatopropyltriethoxysilane, tetrathoxysilane, and aminopropyltrietoxysilane. Nano silica as an addition to pure polymeric material is known to improve the mechanical and barrier properties of pure polymers, so that they can be used in food packaging.

7.2.5 Magnesium Oxide

Magnesium oxide (MgO), a colorless mineral, has been widely used in food industry due to its large-scale production. MgO has higher melting point owing to its higher thermal conductivity and stability and lower electrical conductivity in comparison to the other oxides, which makes MgO a suitable mineral for reinforcing into nanocomposites. Along with these properties, magnesium oxide has known for its higher plasticity and antimicrobial action against a wide range of microbes (Swaroop and Shukla, 2019). In some developed countries, like Europe and the United States, it has been employed as a replacement for other materials in the food packaging industry. The nanocomposites formulated with MgO have more flexibility, impermeability to water vapors and gases, recyclability, and thermal stability than the prevailing food packaging materials (Mirtalebi et al., 2019).

7.2.6 Aluminum Oxides

Aluminum oxides are used as a reinforcing material in polymeric matrixes in many forms, depending upon the compatibility of both materials. The most commonly used polymeric materials with different aluminum oxides are chitosan, kefiran, and polybutylene succinate (Moradi et al., 2019). Generally, aluminum oxide inclusion makes the final packaging material more transparent, flexible, and microwavable in nature. Aluminum (Al) is known for its barrier action against light and UV radiation, but coatings formulated with Al are very thick, non-transparent, non-microwavable, and less recyclable. With the addition of aluminum oxides in nanocomposites, the aroma retention and barrier action toward gases and water vapors of the material are improved (Hirvikorpi et al., 2011).

7.2.7 Doped Nano Oxides

Doping means the addition of some element or replacement of it with another in very small concentrations in order to alter the indigenous properties. Elements that are added are called dopants. Doped nano oxides have been formulated to enhance certain properties of nanocomposites. Various studies have concluded that the doping of some nano oxides with other elements improves the antimicrobial, barrier, and transparency of formulated nanocomposites. Also, the surface reactivity of nano oxide's lattice is increased without any alteration in structural form. In some of the studies, aluminum, manganese, and tantalum are used as doping material for replacing zinc in zinc oxide lattice. (Guo et al., 2015; Valerini et al., 2018).

7.3 PRODUCTION/PREPARATION OF OXIDE-BASED NANOCOMPOSITES-BASED FOOD PACKAGING

Preparation of oxide-based nanocomposites has been done by various fabrication techniques. But before discussing those techniques, production methods of oxide-based nanostructures are also important to discuss. Synthesis of these nanostructures is a challenging task because an ideal composition, size, and shape are to be reinforced into nanocomposites.

7.3.1 Synthesis Methods for Oxide-Based Nanostructures

Oxides can be integrated into polymeric matrices in various forms, such as nanoparticle, nanofilm, nanotubes, nanofiber, nanorods, or other nanostructures. The preparation methods for nanostructures are broadly divided into approaches; that is, top-down and bottom-up techniques. In top-down methods, a bulk material is broken down into smaller pieces by means of chemical, mechanical, or other form of energy. In bottom-up methods, small molecules or atoms are developed into nanometric range clusters. Generally, bottom-up approaches are employed in the food industry for preparation of different nanostructures. The assembly of molecular or atomic blocks is done by various techniques, such as vapor deposition by both physical and chemical methods, and by chemical or solution processes, depending upon the fabricated conditions (De la Fuente-Salcido 2018).

7.3.1.1 Physical Vapor Deposition (PVD)

Physical vapor deposition (PVD) procedures involve the vapor deposition of raw material over a substrate in a closed chamber. In this process, raw material is exposed to higher temperature conditions than its melting point so that evaporation of material can take place. These evaporated particles then settle over the substrate in the chamber. The procedure is followed in vacuum conditions so that a free path can be created for the movement and deposition of source material. The morphology and properties of the prepared nanostructure depend on the temperature and pressure conditions, and the time and kind of substrate used for deposition (Lahiri et al., 2017). PVD is completed using different procedures, such as thermal oxidation, sputtering deposition, electrochemical anodization, pulsed laser deposition, and molecular beam epitaxy (Prasanna et al., 2019). Thermal oxidation is one of the oldest and easiest procedures for preparation of nanostructures. This procedure can be carried out in atmospheric pressure conditions, without the use of any vacuum pump that would add extra costs. All the preparation conditions can be changed so that a desired thickness, shape, strength, adhesion, and structure can be obtained. Also, the electrical as well as optical properties can be altered in this method. The only limitation of this procedure is that it is carried out at a very high temperature, which limits the choice of polymer substrates to be used. To overcome this limitation, other methods are usually employed for preparing nanostructures on flexible polymeric substrates. In the electrochemical anodization method, as its name suggests, an anodization process is followed in a two-electrode system at ambient temperature conditions. The structural parameters can be altered in this method by controlling the anodization voltage, time, and electrolyte composition. However, the nanostructures fabricated with this technique are amorphous in nature, and can be further converted into a crystalline structure by means of post-growth thermal treatment. The other two methods, pulsed laser deposition and sputtering deposition, are generally employed for those oxides where deposition of source material is difficult over the substrate, or in case of semiconductors (Buzby et al., 2006).

7.3.1.2 Chemical Vapor Deposition (CVD)

Chemical vapor deposition (CVD) method prepares nanostructures through chemical reactions between two or more different precursors followed by deposition of source material over the substrate. This chemical reaction usually takes place near the surface of the substrate so that source material can make a solid deposit over the surface. The procedure is followed under a 200–1200°C temperature range. The morphological parameters of the fabricated nanostructure can be altered by modifying the temperature conditions and crystallinity, nucleation, and chemical composition of the precursors (Polarz et al., 2005). In case of doping materials, sometimes the crystalline nature of fabricated nanomaterial gets distorted as phase separation takes place due to a high concentration of doping material. In this regard, another method, atomic layer deposition (ALD) is an excellent approach for fabrication of nanostructures. In this method, an additional gaseous precursor is added over the surface of substrate along with an inert gas to eliminate all the byproducts and unreacted precursor. A very well-ordered structure with high uniformity can be fabricated by employing this method. AAO (anodic aluminum oxide) is the commonly used template in this method. One additional benefit of this method over the conventional CVD is that the desired length of nanostructure can be attained by increasing the number of deposition cycles (Zhang et al., 2015).

7.3.1.3 Chemical/Solution Processes

Chemical/solution methods are generally employed for the fabrication of oxide-based nanomaterials and nanocomposites. Sol-gel approach, microemulsion, hydrothermal, coprecipitation, and spin and dip coating methods are commonly used techniques among chemical processes (Richard et al., 2018). The choice of raw material to be used among chemical processes is higher than with the physical and chemical vapor deposition techniques.

Sol-gel approach is the most commonly employed method among the chemical processes. This approach can be employed for fabrication of all kinds of oxides, regardless of their composition, and for development of nanocomposites with both organic and inorganic polymeric materials. In this method, nanostructures are fabricated in the presence of some metal alkoxides as a precursor within a sol-gel route. There are five major steps in this process: hydrolysis of alkoxides, condensation, polymerization, drying, and calcination. The final products' morphological parameters, density, crystallinity, and alignment mainly depend upon the nucleation and seed layer conditions (Rana et al., 2017). However, this approach necessitates the use of alkoxides as precursors which brings up the cost of whole procedure.

Coprecipitation is another commonly used method for preparation of oxide-based nanostructures. This method involves the nucleation process, where numerous small particles are formed, followed by agglomeration of those particles. During the second process, agglomeration, all the supersaturation conditions, such as pH, temperature, viscosity, speed, and surface tension, alter the morphological parameters of the fabricated nanostructure. Both normal as well as reverse micelles were formed during this process (Ubani and Ibrahim, 2019). Coprecipitation is the simplest method for preparing nanostructures but it has two main drawbacks: low yield and high waste production. Some of the oxide-based nanostructures are also fabricated by modified coprecipitation methods, such as microwave and sonication assisted techniques (Mirtalebi et al., 2019).

Solvothermal/Hydrothermal is an approach involving the use of a solvent for preparing nanostructures. In this method, a chemical reaction of source material takes place with the addition of a solvent under high temperature and pressure conditions. Generally, the temperature of the whole chamber is maintained higher than the boiling point of solvent used (Ye et al., 2011). With this method, pure as well as doped nanostructures of desired shapes can be prepared by using different solvents. The morphological parameters of the final product can be altered by changing the chemical and other thermodynamical conditions.

7.3.1.4 Biosynthesis

Biosynthesis, as the name suggests, is an environmentally friendly approach for synthesis of oxide-based nanostructures. In this approach, some biological entity, such as fungi, bacteria, yeast, or some plant extract is used for development of nanoparticles.

The most commonly used biological sources for oxides are *S. oneidenis*, *Lactobacillus sp.*, *S. cerevisiae*, *F. oxysporum*, and yeast cells. (Jagadish et al., 2018). Biosynthesis is a simple and low-cost approach, overcoming drawbacks of other methods. Green chemistry principles are becoming more popular these days to prevent the use of hazardous chemical substances.

7.3.2 Processing of Oxide-Based Nanocomposites

When developing a promising nanocomposite, the interfacial interaction between the polymer matrix and the oxide nanostructure is one of the most important factors. The choice of the polymer matrix and its chemical nature, the aggregation behavior of the nanostructure, and its dispersion in polymer matrices all make the nanocomposite fabrication a very difficult task (Schadler, 2003). The commonly used methods for development of nanocomposites are in situ polymerization, sol-gel approach, and direct blending, as shown in Figure 7.2.

7.3.2.1 In Situ Polymerization

With the in-situ polymerization method, oxide-based nanostructures are dispersed in a monomeric solution, followed by polymerization of that dispersing solution. The standard methods for polymerization are followed (Ziola et al., 1992). The first step in this method is the formation of oil-in-water type mini emulsions using some organic monomer and a precursor. The polymerization of that dispersed solution results in the development of nanocomposite spheres. During polymerization, the hydrophilic material diffused toward the emulsion interfaces and nanostructures are glued onto the polymer surfaces via electrostatic forces (Wu et al., 2010). This approach has been used for developing nanocomposites that overcome the problem of aggregation of nanostructures' particles in polymeric matrices even after applying the external force.

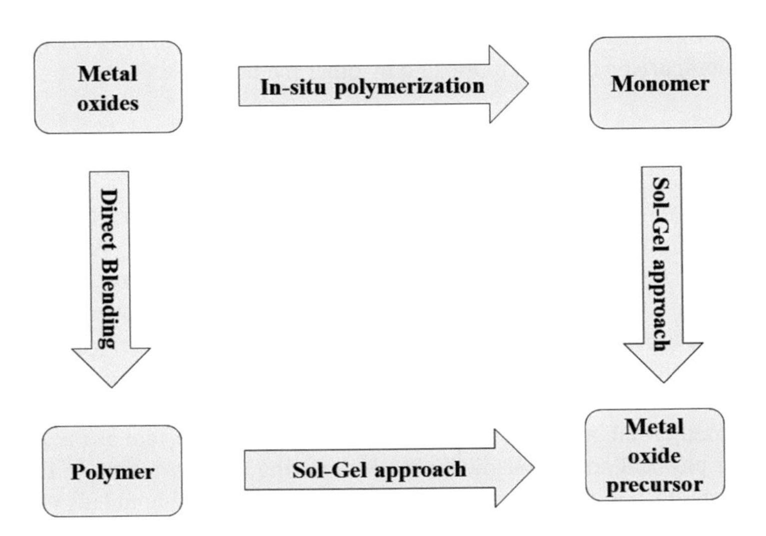

FIGURE 7.2 Different approaches for formulating nanocomposites.

7.3.2.2 Blending

Blending, or direct mixing, is one of the simplest techniques for fabricating oxide-based nanocomposites. This method involves the direct blending of oxide nanostructures with the polymers via two different approaches: melt (discrete phase) and solution blending (solution phase). The key factor for the effectiveness of this method is the dispersion grade of nanostructures within the polymer matrices. Melt blending is generally used as a large-scale, industry-level fabrication technique as a greater number of nanocomposites can be prepared by simple extrusion (Miyauchi et al., 2008). Another advantage of melt blending over solution blending is that it does not require any solvent for dispersion of nanostructures, which cuts out the cost factor. Also, extrusion is a very simple and ecofriendly technique with a low investment cost (Hong et al., 2006). On the other hand, solution blending involves the use of a solvent which facilitates the dispersion and de-agglomeration of nanostructures within a polymer matrix. Solution blending is completed in three steps; first, dispersing the nanostructure in a solvent by shear mixing, ultrasonicator, refluxing, or magnetic stirrer. After that, there is proper mixing of this solution with the polymeric matrix, followed by precipitation, or casting, for the recovery of the nanocomposite. This technique for developing nanocomposites is more widely employed at the lab scale.

7.3.2.3 Sol-Gel Method

Sol-gel method is the third and most commonly employed technique for fabrication of oxide-based nanocomposites. This method basically involves two main processes: hydrolysis and condensation. Hydrolysis leads to adhesion of the polymer material to the nanostructure by replacing the hydroxyl groups following condensation (Hench and West, 1990; Lu and Yang, 2009). The greatest advantage of using this approach is that it reduces the non-compatibility between the polymer (organic) and oxide nanostructure (inorganic) phases at the molecular level. As we have discussed earlier, nanocomposites formulated by this approach have more uniformity and higher purity in comparison to those fabricated by other methods.

7.4 PROPERTIES OF OXIDE-BASED NANOCOMPOSITES

The main aim of food packaging is to protect the food from the surrounding adverse environmental conditions during transportation as well as storage. When designing an appropriate food packaging material, a few properties of that material should be taken into consideration, as depicted in Figure 7.3. Moisture and microbial infestation are the primarily responsible for food deterioration during transportation and storage. To control these two factors, food packaging must have excellent barrier and antimicrobial action toward water vapors, gases, and pathogenic microbes. Mechanical, thermal, and optical properties are equally important as the packaging has to bear physical and mechanical abrasions during transportation. In the case of active packaging, a food packaging material's release kinetics should also be considered. In this section, we will discuss all these properties in detail.

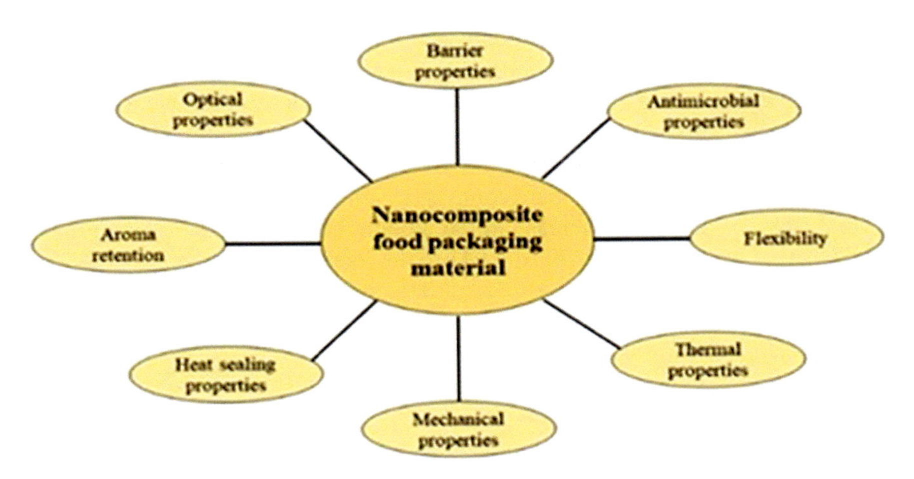

FIGURE 7.3 A graphical representation of prerequisites required for a suitable food packaging material.

7.4.1 Thermal Properties

Thermal properties of nanocomposites can be enhanced by the inclusion of oxide-based nanostructures, as most polymers possess poor thermal and chemical stability. Oxide-based nanostructures have excellent thermal and chemical properties, which makes them an appropriate candidate for introduction into polymeric matrixes. The thermal stability of the fabricated nanocomposites depends upon many factors: the interaction of polymeric matrix with the oxide nanostructures and the inherent thermal properties of both polymers and oxides. With the addition of oxide-based nanofillers, molecular-level interactions begin between the polymeric matrix and nanofillers, which causes the rearrangement of polymer molecules into an ordered state. This leads to alterations in the energy of the whole matrix and subsequently enhances the thermal properties of the nanocomposites. The resultant nanocomposites possess a higher heat distortion and melting temperature along with lower crystallization degree (Cheng et al., 2016). These factors can be determined by creating a thermal profile of the nanocomposite following standard procedures. Differential scanning calorimetry (DSC) and thermogravimetric analysis (TGA) are the two commonly used tools for determining thermal stability of nanocomposites. DSC measures the change in energy level, which further determines the melting, transition, and crystallization temperatures. TGA determines the change in weight that occurred in the oxidation or decomposition processes. These factors directly affect the thermal and chemical stability of the polymer matrixes (Billmeyer, 2007).

Roy and Rhim (2019) reported that thermal properties of carrageenan-based nanocomposite film were enhanced by integrating zinc-oxide nanoparticles into films. Their DSC and TGA results showed that the decomposition temperature of nanocomposite film was higher than the simple carrageenan films (271°C and 238°C, respectively) a clear indication that an addition of zinc oxide will cause the distortion temperature to start increasing. Similar results were observed in the case of polyaniline and cerium doped titanium dioxide–nanocomposites, where the glass transition temperature of the nanocomposites increased with the introduction of oxide nanostructures. This is due to the fact that there are strong interactions between the polyaniline and nanofillers, which lead to the ordered rearrangement of the material. The TGA analysis of these nanocomposites states that there was increase in the thermal stability due to an increase in the amount of char residue which provides the protection over the surface of film. Also, the increase in percentage addition of doped nano oxides is directly proportional to the increase in thermal stability of the nanocomposite.

7.4.2 Barrier Properties

The barrier properties of formulated nanocomposites are of utmost importance in food packaging material as shelf life and deterioration of food during transportation and storage is directly related to the moisture and gaseous composition of the surrounding environment. A suitable food packaging with excellent barrier action toward water vapors and gases can prevent the deterioration of the food inside, thus increasing its shelf life. By providing a gas barrier, nanocomposite packaging material can slow down processes like lipid oxidation, respiration, and enzymatic browning, which lead to the discoloration and softening of food products. Lipid oxidation deteriorates the quality of food products with a high fat content by causing oxidative rancidity and making them bitter in taste (Wihodo and Moraru, 2013). Furthermore, nanocomposites prevent the loss of the food's aroma and other volatile compounds as well as the pickup of foul odors from the surroundings. A packaging material with poor barrier properties leads to moisture and physiological weight loss, which further affects the texture and appearance of food products. Oxide-based nanocomposites are formulated with the aim of overcoming the poor barrier and mechanical properties of conventional packaging material. Packaging materials formulated with the inclusion of nanomaterials have enhanced barrier properties in comparison with conventional coatings and films with the pure polymer alone. Also, a very small amount of oxide nanostructure is needed for formulation of nanocomposites due to the larger surface-to-volume ratio of nanoparticles (Khan et al., 2012). Various factors affect the barrier action of nanocomposites against moisture and gases, such as percentage inclusion of nanofillers, intermolecular interactions between polymeric compounds and nanofillers, level of cross linkage, orientation angle of nanofillers, synthesis methods, thickness of packaging material, and so on. In case of nanocomposites, the gas permeation rate is quite low due to the impermeable nature of the nanomaterial, as gas must pass through a tough diffusion path as compared to polymeric matrix packaging material. With reinforcement by nanofillers, the inherent morphology of the polymeric compounds is altered, which hinders the passage of gases, thus maintaining the gaseous composition inside the packaging (Cheng et al., 2016).

Changes over the interfacial areas in the polymer itself start occurring with the inclusion of nanofillers. The gas and water molecules are not able to pass through the holes at the interfacial areas due to blockages created. But these mechanisms can only be achieved if there is an excellent computability between nanofillers

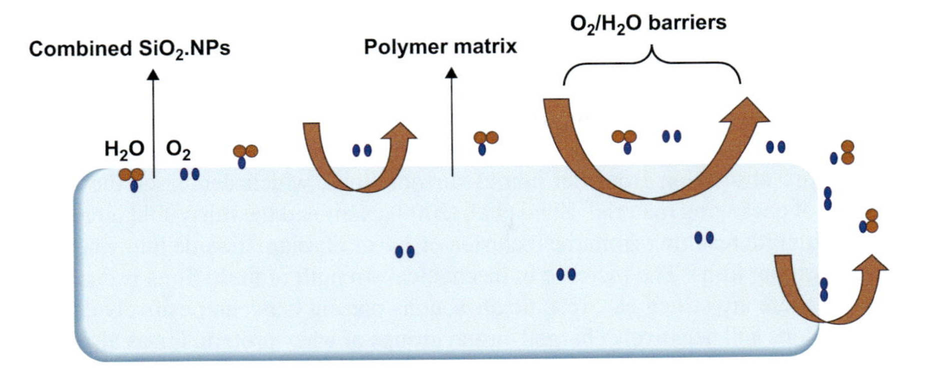

FIGURE 7.4 The barrier action of silicon dioxide-based nanocomposites against oxygen and water vapors.

and polymers. One study concluded that the addition of zinc-oxide nanostructures improves the barrier properties of chitosan and MMT clay matrix (Vaezi et al., 2019). Nanostructures in the polymer matrix arrange themselves in the perpendicular direction of the diffusion path, which creates a more torturous path for gases and water vapors. Also, it was noted that with a higher percentage inclusion of nanofillers, less impermeability of nanocomposites against moisture can be achieved, as nanofillers' particles aggregate and change their orientation angle. Bharathi et al. (2019) and Palomero et al. (2016) defined the optimum percentage inclusion of nanofillers among the nanocomposites as in the range of 2–8%. Silicon dioxide is proven to be one of the best choices for improving the barrier properties of packaging material. Nanocomposites reinforced with silica reduce the water and gaseous permeability up to 50%, along with ethylene scavenging action. The barrier action of silicon dioxide–based packaging material is demonstrated in Figure 7.4.

7.4.3 Mechanical Properties

Mechanical properties of packaging materials are an important aspect in the food industry as food products have to bear physical stress and abrasion during transportation. The mechanical strength of a material can be determined in terms of tensile strength, elasticity, and Young's modulus. The amount of stress a packaging material can bear without breaking determines its tensile strength and flexibility. Mechanical strength and stability of packaging films can be improved by the addition of nanomaterials, an improvement of what was a major drawback in conventional packaging material. The increase in mechanical strength of a composite by reinforcing nanofillers can be explained by two things. The first reason is the formation of hydrogen bonds and ionic interactions between nanofillers and polymers and within two nanoparticles (Sanchezgarcia et al., 2010). The second reason is the stress transfer at the interface of nanofillers and polymer matrix which causes a reinforcing effect (Abdollahi et al., 2013). However, mechanical properties of a formulated nanocomposite depend upon many factors, like the use of cross-linkages, formulating parameters, hydrophilicity or hydrophobicity of polymer matrix, concentration of nanofillers, surrounding environmental factors of transportation and storage, and, generally, moisture and humidity (Bharathi et al., 2019). The percentage inclusion of nanofillers should be optimum; higher percentage additions would not increase the mechanical strength to a higher extent as particles start aggregating during the formulation of nanocomposites. Likewise, the hydrophilic nature of polymers leads to moisture absorption from the humid surroundings, which decreases the tensile strength of packaging material. Zhuo et al. (2009) examined the thirty-fold increase in tensile strength resulting from the inclusion of 1% of titanium dioxide into whey protein packaging films. The increase in mechanical strength of these films is due to the fact that there are strong electrostatic attractions present between positively charged titanium ions and negatively charged amino groups of whey protein. It was also noted that higher percentage addition of titanium dioxide (more than 1%) decreases the elastic module and tensile strength, due to agglomeration of nanoparticles in the polymer matrix. Thus, reinforcement of nano oxides at an optimum concentration is an effective tool for improving mechanical properties of food packaging material.

7.4.4 Antimicrobial Properties

The antimicrobial property of food packaging material is of utmost importance as microbial infestation in food is a health issue. Conventional food packaging material might possess good barrier and mechanical properties but they do not have any kind of antimicrobial action. Thus, the reinforcement of nano oxides in nanocomposites is a beneficial step in preventing microbial infestation and food deterioration. Most of the metal oxides, such as zinc oxide, copper oxide, titanium dioxide, and boron oxide, have excellent antimicrobial action against *S. aureus*, *L. monocytogenes*, and *E. coli*. The variations in the stoichiometry and oxygen atom defects are largely responsible for the excellent antimicrobial action of metal oxides (Jafarzadeh et al., 2020). A comparative study conducted by Azam et al. (2011) examined the antimicrobial properties of copper-oxide, zinc-oxide, and ferrous-oxide nanoparticles. The results revealed that zinc oxide showed the highest antimicrobial action toward both gram positive and gram negative bacteria.

The antimicrobial action of nano oxides can be explained by two theories. During the application of food packaging material with nano oxides, the ions from metal oxides dissolve the germifuga and destroy the energy metabolism of microbial cells and consequently retard cell division of the microbes (Montazer et al., 2016). Another theory, called contact reaction theory, states that when positively charged antimicrobial nanocomponents encounter negatively charged microbial cells, they absorb each other's charge. This leads to break through of the cell membrane of microbes, denaturing the protein, and ultimately killing the microorganism (Seil and Webster, 2012). There are, however, various factors which affect the antimicrobial properties of nanocomposites, such as the size, shape, and percent addition of the nano oxides introduced, and the kind of microbes.

The antimicrobial action of copper oxide–based nanocomposites was determined by conducting various studies (Muthulakshmi et al., 2019). As was discussed earlier, copper is one of the important metals for regulating metabolism in microbes. A higher concentration of copper has a detrimental effect on the growth mechanisms

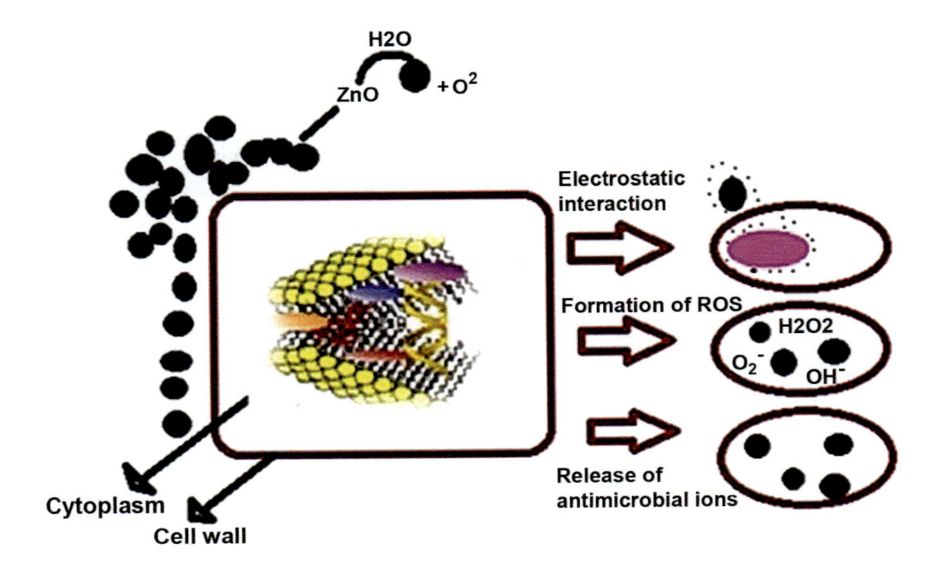

FIGURE 7.5 Antimicrobial action of zinc oxide nanoparticles.

of microbes. Basically, copper oxide retards the growth of microbes by stopping the enzymatic activity, ending the exchange of ions, generating the reactive oxygen species, denaturing proteins, and disrupting the cell wall. Similarly, zinc oxide has a broad-spectrum antimicrobial activity. Zinc-oxide nanoparticles, when incorporated in nanocomposites, retard the growth of microbes by the formation of reactive oxygen species, electrostatic interactions, and the release of antimicrobial ions from the nanoparticle's surface, as depicted in Figure 7.5. When released from the surface of nanoparticles, Zinc ions attach to the inside of the microbial cell wall and act as inhibitors, thus, stopping the synthesis of nucleic acids. Along with this, these ions activate some autolytic enzymes inside the microbial cells, which leads to lesions and ultimately the death of the microbes (Lemire et al., 2013).

7.4.5 Photocatalytic Properties

Metal oxides are semiconductors by nature that act as photocatalytic materials and have a special electronic structure (Smith et al., 2010). Semiconductor nanostructures have higher energy conduction bands with lower energy valence bands, unlike metals. Also, a band gap is present between the bands, which makes nano oxides a suitable material for introducing into nanocomposites and improving their photocatalytic properties. A series of highly active electrons (e^-) and positively charged holes (h^+) are produced when these nanofillers are irradiated by light with greater energy than their inherent band gap. Consequently, two highly active electron-hole pairs are generated, which further react with the water molecules and dissolved oxygen to generate high-activity hydroxyl radicals (OH) and superoxide radicles (O_2^-) (He, 2011). These two highly active radicals then react with organic compounds and, consequently, degrade ethylene and other pathogenic cells. The photocatalytic

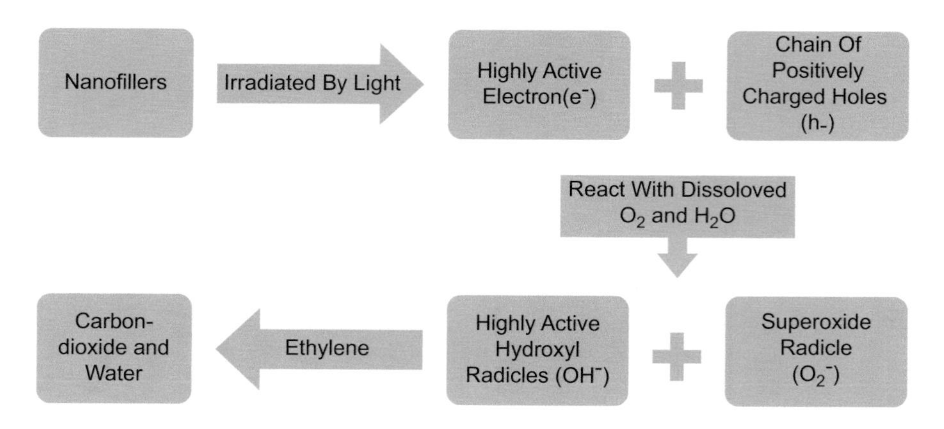

FIGURE 7.6 Photocatalytic action of nano oxides.

action of nano oxides is demonstrated in the Figure 7.6. Photocatalytic properties are an important topic when considering fresh fruits and vegetables as they directly affect ethylene production and degradation, which is a main cause of spoilage and softening of fresh produce during transportation and storage. The larger surface area of nano oxides gives them high redox potential, owing to quantum size effect. The higher redox potential facilitates the separation of photogenerated carriers, which further increases the photocatalytic effect of the nanocomposite.

Nano–titanium dioxide is the most used photocatalytic nanomaterial. It can prolong the shelf life of stored food products by hastening the oxidative decomposition of ethylene and other gases generated during storage (Mihindukulasuriya and Lim, 2014). Nanocomposites with nano–titanium dioxide and nano-silver particles could delay the production of free radicals, possess ethylene scavenging activity, regulate air components of the micro-environment, and retard microbial growth (Fang et al., 2016). Photocatalytic effects can be determined by the degradation extend of methylene blue when exposed to UV light. Xie and Hung (2018) reported that with the addition of titanium dioxides in polymeric packaging films, the dye degradation was noted at higher percentage than with the polymer alone. Along with photocatalytic effect, titanium dioxide (TiO_2) also increased the bactericidal properties of the packaging material.

7.4.6 Optical Properties

Optical properties of food packaging material are of interest when the packed food products are light sensitive. The optical properties of a nanocomposite depend upon various factors, such as the inherent optical properties of both the nanofillers and polymeric compounds and the intermolecular interactions and bonds developed between both components. The surrounding environmental conditions, especially UV light, also have a significant effect on the optical properties. Recent studies have concluded that optical properties of packaging material can be enhanced by inclusion of nano oxides (Goswami et al., 2018; Mojaki et al., 2018). Most of the materials possessing excellent optical properties are difficult to synthesis from a single crystal.

These materials need proper protection from environmental stress, which can be easily provided by polymer matrix in nanocomposites. Metal oxides such as zinc oxide and titanium dioxide possess good optical properties when added, in nano scale, in packaging material. Titanium dioxide is known to be a high-index nanomaterial for enhancing optical properties. Zinc-oxide nanoparticles, when incorporated in composite films, improved the optical properties of nanocomposite. The photoluminescence spectrum and UV visible analysis of nanocomposites showed a shift in blue region. That shift was generated by high energy and the charge between nano oxide and polymeric chains.

7.4.7 Dielectric Properties

Most of the metal oxides such as titanium dioxide, zinc oxide, copper oxide, and magnesium oxide are n-type semiconductors. There are some defects or oxygen vacancies present in their structure which are responsible for the permittivity and electric properties of the oxides (Huang et al., 2017). Dielectric properties of these oxides can be enhanced by doping a nano oxide with some other metal. Inclusion of another element increases the band gap of nano oxides, which further affects the electron mobility, binding energy, degree of crystallinity, and dielectric properties. Dielectric properties are one of the less exploited areas of nanocomposites. Some research has been conducted which concluded that zinc-oxide nanoparticles enhanced the dielectric properties of food packaging nanocomposites. Dielectric properties of a nanocomposite depend upon various factors, such as the size, number of defects/vacancies, and density of nanoparticles introduced, as well as on the operating frequency and thickness of the packaging film (Kaur et al., 2021). The dielectric properties of nanocomposites can be explained by a phenomenon called percolation. When this occurs, there is a gradual inclusion of nano oxides into the polymeric matrices, where the nano-scale size of particles reduces the space between the particles for same volume fraction, which increases the dielectric values. The inclusion of nano oxides in polymer matrix increases the permittivity due to their large interfacial area in comparison to the micro particles. The concentration of nano-oxide inclusion required to cause a shift in dielectric value is known as percolation value (Zois et al., 2003). Also, the nano oxide's surface is composed of many defects/vacancies, which cause the unequal distribution of charge, leading to polarization. The different kinds of space polarizations are directly proportional to the percent inclusion of nano oxides. Rahman et al. (2018) formulated chitosan-based nanocomposites with the addition of different concentrations of zinc nano oxides and concluded that with an increase in the percent inclusion, the dielectric values and conductivity of the composite films increased as well.

7.4.8 Heat-Sealing Properties

The heat-sealing properties of a food packaging material are an important consideration as packaging material has to withstand heat while sealing. However, heat-sealing properties are a less evaluated area among nanocomposites. Heat-sealing properties generally depend upon the mechanical properties of a packaging material.

A packaging material with good tensile strength, less percentage elongation at break, and higher Young's modulus has good heat-seal strength. Along with mechanical properties, the moisture uptake and absorption of a material as well as strong hydrogen bonding between the nanofiller and composite material also affects the heal-seal strength (Abdorreza et al., 2011). Food packaging material reinforced with nano oxides has improved heat-seal strength as, with the inclusion of nano oxides, tensile strength, and other parameters of mechanical strength are improved. Also, there is a strong intermolecular interaction between the nanofiller and polymer matrix which develops strong hydrogen bonding, consequently increasing heat-sealing strength. It has also been observed that the moisture content of the packaging material is decreased with the inclusion of nano oxide, which decreases the flexibility of material, affecting heat-sealing properties (Teymourpour et al., 2015). The heat-sealing strength of a commonly used food packaging material like polyethylene is said to be around 600 N/m. The nanocomposites formulated with nano oxides have seal strength in the same range.

7.4.9 Water Solubility and Moisture Absorption Properties

Water solubility and moisture uptake are two parameters which are interrelated with both the mechanical strength and barrier properties of a food packaging material. Most of the packaging materials have higher water solubility and moisture uptake when stored in humid environmental conditions, which makes them not suitable for food packaging. To overcome this problem, nanocomposites proved to be a good option. A few studies have been conducted, observing that with the inclusion of nano oxides the water solubility and moisture uptake decreased. In nanocomposite packaging materials, nano oxides have very strong intermolecular interactions with the polymer matrix, which reduces the availability of hydroxyl groups to interact with water vapors, consequently decreasing the moisture absorption. Nano oxides tend to form hydrogen bonds with the different groups of polymer matrix, which reduces the active sites for diffusion and absorption of water vapors into the nanocomposites; thus, moisture resistance of the packaging film is increased (Fei et al., 2013). This results in restricted movement of water molecules as well lower-weight polymer molecules, which decreases the water solubility of nanocomposite packaging films. It was also reported that with an increase in percent inclusion of nano oxides there was a significant decrease in the solubility index and moisture absorption of nanocomposites (Goudarzi et al. 2017; Oleyaei, 2016). Along with this phenomenon, nano oxides possess hydrophobicity, which increases the barrier action of nanocomposites.

7.4.10 Printing Properties

The food packaging material should not only have excellent barrier and mechanical strength but it must have good printing properties. The printing properties are determined in terms of residual solvent value while printing, regardless of the nature of the solvent. Packaging materials with lower residual solvent levels are considered safe for foods. The most commonly used polar and non-polar solvents are acetone, toluene, ethanol, ethyl benzene, and acetate. Solubility of any solvent is directly proportional to the amorphous region of the matrix. With the reinforcement of nanofillers, the amorphous fraction of the matrix is decreased as the nanofillers increase the crystalline fraction due to crystallization of the polymer matrix. These nanofillers acts as nucleating agents for crystallization and, thus, lower the level of solvent solubility (Pukanszky et al., 1994). During printing, the organic solvents are diffused into the packaging material. Composite films become swollen where they came in contact with the solvents, leading to the diffusion of printing ink particles into the packaging material. Most of the time these ink particles are desorbed when the packaging material is exposed to the heat of the drying process after printing. But the solvent particles which penetrate deep into the packaging material cannot be desorbed, leaving residual solvents behind. Li et al. (2014) investigated the effect of nano silicon dioxide on polypropylene packaging material and revealed that with inclusion of nano oxide, the solubility level of polar solvent is decreased as there was more incompatibility between those solvents and the nano-composed polypropylene. The solubility index of the formulated nanocomposite was decreased with the increase in the percent addition of nano oxide, which reduces the diffusion of the solvent ink particles into the film, even in the amorphous fraction. Consequently, there was less residual solvent present in the packaging material. Therefore, nano oxides are proven to be a promising tool for reducing the level of residual solvents, improving the printing properties of food packaging material.

7.4.11 Release Kinetics of Active Packaging

Active packaging of a food supports the overall quality and shelf-life enhancement of that particular food. Generally, the active elements used in active packaging are antimicrobial agents, ethylene absorbers, antioxidants, and sachets for oxygen and water vapor absorption as well as for other undesirable odors present in the surrounding environment (Biji et al., 2015). Among these elements, the addition of antimicrobial agents in the packaging material is the most common. Various studies have been conducted for investigating the effect of including nano oxides as active compounds in packaging material, concluding that these packaging films have great potential in increasing shelf life of food products (Heydari-Majd et al., 2019; Rokbani et al., 2019). However, the release of active compounds depends upon various factors, such as the type of release medium, time, concentration of nano oxides, and the presence of other additives in nanocomposites. The kinetic release of active compounds is determined by a model, called Peleg model. The percent inclusion of nanofiller has an important role in the fast or slow release of active compounds; a higher percent inclusion of nanofiller results in a higher release of the active compound, and vice versa. This is due to the fact that a lower amount of nanofiller is fully dispersed in the polymeric matrix, with strong intermolecular interactions, and, thus, the release of active compound is quite difficult. At higher concentration of nanofillers, nanostructures start aggregation and form larger particles which consequently reduce their interactions and bonding with polymeric matrix, leading to the faster release of active compounds. In another study, the kinetic release of thyme essential oil in a nanocomposite fabricated with nano silica and potato starch was investigated (Cui et al., 2021). It was observed that the release rate of active compounds was abruptly high in the initial hours and then increases gradually in the remaining time. This is due to the fact that some of the oil was not incorporated and diffused properly into the polymer matrix and it comes out faster in the early stage. The second factor that affects the release is the releasing medium. The medium is slowly dissolved into the nanocomposite structure and degrades the wall structure, releasing the active compounds. Also, due to the concentration gradient difference, active compounds start moving into the releasing media from the nanocomposite structure.

7.5 APPLICATIONS OF OXIDE-BASED NANOCOMPOSITES IN FOOD PACKAGING

There are a number of oxide-based nanocomposites formulated and investigated for maintaining the shelf life of food products. As we have already discussed in the above sections, nano oxides introduced into packaging material improves their properties while maintaining the nutritional quality of food products. The applications in food packaging material are listed in tabular form (Table 7.1).

TABLE 7.1
Applications of Oxide-Based Nanocomposites in Food Packaging

Nano oxide	Polymer Matrix	Particle Size	Findings	References
Aluminum oxide (Al_2O_3)	Polylactic acid, chitosan, and sodium alginate	25 nm	Improved barrier properties and hydrophobicity of films	Hirvikorpi et al. (2011)
	Biodegradable kefiran	20 nm	Improves tensile strength; lowers water vapor permeability, solubility and, moisture absorption	Moradi et al. (2019)
	Polyethylene and polypropylene	25 nm	Improved barrier properties	Struller et al. (2014)
	Gamma nanoplatelets	Nano sized	Improved mechanical and thermal properties	Wagih et al. (2019)
Boron oxide (B_2O_3)	Polyacrylic acid	28–54 nm	Higher thermal stability, strong antimicrobial action against *E. coli* and *S. aureus*	Beyli et al. (2018)
Copper oxide (CuO)	Polypropylene	10 nm	Excellent antimicrobial action against *E. coli*	Delgado et al. (2011)
	LDPE	50 nm	Significantly decreases coliform bacterial load on cheese	Beigmohammadi et al. (2016)
Magnesium oxide (MgO)	Polylactic acid (PLA)	<60 nm	Improved tensile strength and thermal stability; excellent barrier and antibacterial action	Swaroop and Shukla (2018)

(Continued)

TABLE 7.1 *(Continued)*
Applications of Oxide-Based Nanocomposites in Food Packaging

Nano oxide	Polymer Matrix	Particle Size	Findings	References
	Chitosan	42 nm	Increases tensile strength and elastic modulus by 86% and 38%; improved thermal stability and UV barrier action	De Silva et al. (2017)
Silicon oxide	Chitosan	15 nm	Increases mechanical and barrier properties of films; maintains all post-harvest parameters and overall quality	Tian et al. (2019)
	LDPE	<250 nm	Improved gas barrier and antimicrobial properties; maintains sensory score of stored shrimps	Luo et al. (2015)
	Carrageenan and konjac-glucomannan	20 nm	Decreases water vapor and gaseous permeability rate; improves tensile strength; extends the storage period for mushrooms	Zhang et al. (2019)
	Biodegradable polylactic acid	60 nm	Improved barrier and optical properties; highly transparent films developed	Bang and Kim (2012)
	Agar and sodium alginate	25–35 nm	Improves tensile strength, barrier action, and thermal stability; decreases water solubility	Hou et al. (2019)
	Polypropylene	50 nm	Increases degree of crystallinity, thermal stability; decreases water and gaseous permeability	Li et al. (2014)
Titanium dioxide	Pectin	3–5 nm	Antimicrobial action against *E. coli*	Nesic et al. (2018)
	Wheat gluten and nanocellulose	20 nm	Increases all the mechanical parameters of paper; strong antimicrobial action against *S. cerevisiae,* both G - and G+ bacteria	El-Wakil et al. (2015)
	Potato starch	21 nm	Blocked 90% of UV light; improves mechanical, thermal, and water related properties	Oleyaei et al. (2016)

(Continued)

TABLE 7.1 ***(Continued)***
Applications of Oxide-Based Nanocomposites in Food Packaging

Nano oxide	Polymer Matrix	Particle Size	Findings	References
	Polyvinyl alcohol and chitosan	15–30 nm	Improved mechanical and barrier properties, excellent antibacterial action	Lian et al. (2016)
	PLA and oleic acid modified TiO_2	4.5 nm	Significant reduction in water and gaseous permeability, better blockage properties toward UV light	Baek et al. (2018)
	Cellulose acetate, polycaprolactone and polylactic acid	21 nm	Increase in photocatalytic and antimicrobial action	Xie and Hung (2018)
Zinc oxide	Chitosan and CMC	6 nm	Improved mechanical, barrier, and antimicrobial properties; maintains quality parameters of coated cheese	Youssef et al. (2018)
	Allyl isothiocyanate	Nano sized	Effectively reduced microbial load in liquid egg albumin stored in glass jars	Jin and Gurtler (2011)
	Chitosan and polyethylene	35–45 nm	Decreases water solubility by 80%, excellent antimicrobial against food pathogens	Al-Naamami et al. (2016)
	Pullulan films	Nano sized	Increased antimicrobial properties and shelf life of meat and poultry products	Morsy et al. (2014)
	LDPE	30 nm	Excellent antibacterial action against *B. subtilis* and *E. areogenes*, negligible migration of zinc ions	Esmailzadeh et al. (2016)
Doped nano oxides				
TiO_2 with Ag	LDPE	Nano sized	Strong antimicrobial action against microbes, prevents growth of green mold, maintains all other post-harvest parameters	Wang et al. (2010)

(Continued)

TABLE 7.1 *(Continued)*
Applications of Oxide-Based Nanocomposites in Food Packaging

Nano oxide	Polymer Matrix	Particle Size	Findings	References
TiO_2 with Ag and Cu	Polyvinylchloride	10–15 nm	Improved tensile strength and thermal stability, strong antimicrobial properties	Krehula et al. (2017)
Zn with Ag	LDPE	70 nm	Significant increase in antimicrobial properties	Emamifar et al. (2011)
Zn with Al	PLA	7–9 nm	Improved UV blocking and optical properties, strong antibacterial action toward *E. coli*	Valerini et al. (2018)
TiO_2, SiO_2 with Ag	LDPE	40–60 nm	Regulated gaseous concentration inside package, maintains nutritional quality of mushrooms up to 14 days	Donglu et al. (2016)

7.6 TOXICITY ISSUES AND SAFETY REGULATIONS OF NANO OXIDES IN FOOD PRODUCTS

Nanomaterials have emerged as an innovative and effective tool to preserve fresh produce, prolong its shelf life, and prevent microbial spoilage. But the safety and toxicological aspects surrounding migration of these nanofillers into the food through the packaging material cannot be ignored. There are a few criteria formulated, such as exposure, migration, leaching, persistency, sustainability, and toxicological assessment, for various nanofillers used in food products. On the basis of these criteria, a nanofiller can be graded as safe for the human consumption.

7.6.1 Migration of Nano Oxides

Migration is a non-intended transfer of nano oxides from the packaging material into the packed food products. These migrated particles not only affect the organoleptic properties of the food but they could cause serious health hazards. The leaching of nanofillers into the food product is completed in two stages. In the first stage, the nanofillers that are not fully dispersed in the polymer matrix or present over the surface layers can get transferred to the food product. The second migration occurs when the nanofillers passed through the inner layers of food product having some voids and are leached into the food product (Huang et al., 2011). The migration of nanofillers depends upon various parameters, such as the barrier properties, particulate size, water solubility, viscosity, and additives or cross linkers and on the surrounding environmental conditions like moisture, temperature, pH, or physical stress. Bradley et al. (2011) point out that nanofillers have a large surface-to-volume ratio and are chemically active particles, so they might interact with the polymeric matrix, reducing the rate of migration in some cases. In the case of active packaging, active compounds are intentionally added in the nanocomposites and delivered in the food products in a controlled manner. Therefore, these active compounds must be food grade and safe for human consumption. Studies have been conducted where it was observed that migration of aluminum and silicon dioxide is quite low as compared to other nano oxides. Also, the migration rate increases with a rise in temperature and longer time period (Echegoyen and Nerin, 2013). Another study investigated the migration rate of silver and zinc oxide from low density polyethylene (LDPE) nanocomposites, and found that a very small number of nanoparticles migrated into orange juice after 112 days of storage (Emamifar et al., 2011). Studies on the migration rate of nano oxides proved conclusively that these nanoparticles are fully dispersed and settled in polymer matrix; thus, there are very few chances of migration. But in the case of zinc oxide or other materials which are ionized, these might migrate from the active packaging system that is needed for the antimicrobial action against microbes.

7.6.2 Safety Aspects of Nano Oxides

Testing has found that nano-oxide particles are a greater threat than organic nanoparticles. The inorganic nano oxides are found in various organs of the body, causing chronic disorders by increasing the rate of oxidation reactions and inflammation. As already discussed above, the migration rate depends upon various factors, and in the same way the deposition and toxicity caused by nanofillers in the human body also depend upon certain factors. The nanoparticle size and shape, the absorption rate of the human body, and toxicological aspects of that particular nano oxide and its migration rate all determine the potential health hazards associated with exposure to nanofillers (Cushen et al., 2012). The absorption and migration rates of smaller nanoparticles are far higher than those of larger nanoparticles. Smaller particles are easily distributed among the various organs due to their size, and thus are more harmful to human health. Few toxicity tests have been conducted to investigate the cytotoxicity of nano oxides when exposed to human and rodent cells. The most commonly used procedures to investigate the cytotoxicity of nanofillers are cell count, live or dead assay, tryan blue dye exclusion, and protein content. Micronucleus, comet, and ames assay are widely used tests to investigate geno toxicity of nanofillers (Maisanaba et al., 2015). Allergies and heavy metal release are the main adverse effects of nano scale material exposure. The possible risks of migration or consumption of nanofillers is still under investigation. Few studies have been conducted to measure the potential risk of nanofillers in foods and human consumption of that food. One study found that SiO_2 nanoparticles can induce allergen-specific Th2-type allergic immune responses in female mice exposed to nanoparticles (Yoshida et al., 2011). Metal oxide–based nanomaterials, such as ZnO and CuO, can cause heavy-metal release, and their long-term accumulation may cause adverse effects. The release of heavy metal ions from nanomaterials can lead to an increase of intercellular reactive oxygen species (ROS), consequently leading to lipid peroxidation and DNA damage. When eaten by humans, titanium dioxide nanoparticles present in food accumulate in the lungs, kidneys, and liver and cause damage to immune and circulatory systems (Tarhan, 2020).

7.6.3 Regulatory Bodies Associated with Inclusion of Nanofillers in Food Material

7.6.3.1 US Food and Drug Administration

The US Food and Drug Administration developed a Nanotechnology Task Force, a regulatory body for formulating various parameters and methods for nanotechnological developed products. This task force deals with all manner of safety and toxicity issues regarding nanotechnology, and is responsible for reporting all the data to the FDA. The FDA further provides the analysis methods and recommendations regarding biological interactions of nanofillers, and other regulatory policies. Titanium and silicon dioxide appear in the section of food contact elements considered safe when employed within the FDA's given range of safety. Also, zinc oxide has been graded as a generally recognized as safe substance (GRAS).

7.6.3.2 European Union

According to the European Commission (EC), nanofillers are aggregated materials having half of particles of a size in the range of 1 nm to 100 nm, whether formulated in a laboratory or present in nature as such. In the case of materials that might pose a threat to the environment or human health, the threshold of 50% is decreased to a range of 1 to 50%. The EC allows and recommends the use of only those nanofillers which are listed in their Annexure No.1 and have size range near 100 nm, or are smaller than that. A few nano oxides are considered authorized materials for use in nanocomposites for the purpose enhancing barrier action and other properties. Silicon dioxide and zinc oxide, in both their amorphous and bulk forms, are authorized by the EC. The use of unauthorized materials is only allowed if they have to be used as functional barriers in the case of food products. The unauthorized materials must be used in amounts lower than the detection limit and should be noncarcinogenic.

7.7 LIMITATIONS AND FUTURE TRENDS

Although nanotechnology-based food packaging materials provide many benefits and overcome the drawbacks of conventional packaging material, there are a few limitations that need to be discussed as well. The major drawbacks regarding nanocomposites are commercialization and consumer acceptance of nano-based packaging material. Even though there are many studies regarding nano oxides' properties, still there is a lack of complete exfoliation of the nano oxides, specific purification, desired compatibility between nano oxides and polymer matrix, and appropriate selection of raw materials and surface modifications. There are many challenges that need to be overcome before there can be large scale commercial applications of nanocomposite food packaging material. Also, the cost benefit analysis should be taken into consideration since some of the formulating techniques require large investments of money and technical staff. There are very few commercialized industries which deal with nano-based packaging material. When it comes to consumer acceptance of nano food, most consumers are still not fully aware of the beneficial effects of nano oxides' inclusion in packaging material. A survey conducted by Siegrist et al. (2007) revealed that consumers were hesitant to buy food products packed in nanomaterials, especially the active packaging system, because of the release of active compounds into the food. This hesitance can be overcome if appropriate and detailed knowledge is provided to them about the migration and toxicity and about regulatory bodies overseeing nano foods. Another major drawback is the unintentional leaching and migration of nano fillers into the food material, which negatively affect consumer acceptance of nano-based packaging material. Therefore, more focused research should be pursued in this particular area. Regarding other strategies or methods to improvise nano oxide–based packaging material, some researchers suggest that there should be an additional layer fabricated over the packaging material to decrease the leaching or migration of nano oxides (Vaha-Nissi et al., 2015). And in case of active packaging systems, there should be precise and controlled release of active ions into the food products, and that this must be within the safe range for human consumption. Another strategy to increase the consumer acceptance is to make consumers aware of the benefits of nano-based packaging by introducing proper labelling systems; and their effects over the food material passed by regulatory bodies. There are very few manufacturing companies and brands formulating this packaging material, so a government body should be involved to develop this industry.

7.8 CONCLUDING REMARKS

Nano oxides–based packaging materials are becoming one of the innovative and effective tools for maintaining the nutritional quality and extending the shelf life of different food products. The introduction of nano oxides in very small amounts can improve the barrier and mechanical strength of packaging material and overcome the other limitations of conventional packaging material. Regarding the formulating techniques, there are many techniques available but more focused research should be done in the area of green synthesis of nano oxide–based packaging materials. But there are a few issues which must be dealt before the commercialization of these packaging materials. The safety issues and toxicity studies related to nano-scale materials should also be considered in future research, as it might be possible that nano oxides can migrate into food products through unintentional leaching.

REFERENCES

Abdollahi, M., Alboofetileh, M., Rezaei, M., and Behrooz, R. 2013. Comparing physico-mechanical and thermal properties of alginate nanocomposite films reinforced with organic and/or inorganic nanofillers. *Food Hydrocolloids* 32: 416–24.

Abdorreza, M. N., Cheng, L. H., and Karim, A. A. 2011. Effects of plasticizers on thermal properties and heat sealability of sago starch films. *Food Hydrocolloids* 25: 56–60.

Al-Naamani, L., Dobretsov, S., and Dutta, J. 2016. Chitosan-zinc oxide nanoparticle composite coating for active food packaging applications. *Innovative Food Science and Emerging Technologies* 38: 231–37.

Azam, A., Ahmed, A. S., Oves, M., Khan, M., Habib, S. S., and Memic, A. 2011. Antimicrobial activity of metal oxide nanoparticles against Gram-positive and Gram-negative bacteria: A comparative study. *International Journal of Nanomedicine* 7: 6003–9.

Baek, N., Young, T. K., Marcy, J. E., Duncan, S. E., and O'Keefe, S. F. O. 2018. Physical properties of nanocomposite polylactic acid films prepared with oleic acid modified titanium dioxide. *Food Packaging and Shelf Life* 17: 30–38.

Bajpai, V. K., Kamle, M., Shukla, S., and Mahato, D. K. 2018. Prospects of using nanotechnology for food preservation, safety and security. *Journal of Food and Drug Analysis* 26: 1201–14.

Bang, G., and Kim, S. W. 2012. Biodegradable poly (lactic acid)- based hybrid coating materials for food packaging films with gas barrier properties. *Journal of Industrial and Engineering Chemistry* 18: 1063–68.

Beigmohammadi, F., Peighambardoust, S. H., and Hesari, J. et al. 2016. Antibacterial properties of LDPE nanocomposite films in packaging of UF cheese. *LWT-Food Science and Technology* 65: 106–11.

Beyli, P. T., Dogan, M., Gündüz, Z., Alkan, M., and Turhan, Y. 2018. Synthesis, characterization and their antimicrobial activities of boron oxide/Poly (acrylic acid) nanocomposites: Thermal and antimicrobial properties. *Advanced Material Science* 18: 28–36.

Behboudi, A., Jafarzadeh, Y., and Yegani, R., 2017. Polyvinyl chloride/polycarbonate blend ultrafiltration membranes for water treatment. *Journal of membrane science*, *534*:18–24.

Bharathi, S. K. V., Rohini, B., Moses, J. A., and Anandharamakrishnan, C. 2019. Nanocomposite for food packaging. In Anandharamakrishnan, C. and Parthasarathi, S. (Eds.), *Food Nanotechnology: Principles and Applications*, 275–305. Boca Raton: CRC Press.

Biji, K. B., Ravishankar, C. N., Mohan, C. O., and Gopal, S. T. K. 2015. Smart packaging systems for food applications: A review. *Journal of Food Science and Technology* 52: 6125–35.

Billmeyer, F. W. J. R. 2007. *Text Book of Polymer Science* (3rd ed.). USA: John Wiley.

Bradley, E. L., Castle, L., and Chaudhry, Q. 2011. Applications of nanomaterials in food packaging with a consideration of opportunities for developing countries. *Trends in Food Science and Technology* 22: 604–10.

Bumbudsanpharoke, N., Choi, J., and Ko, S. 2015. Applications of nanomaterials in food packaging. *Journal of Nanoscience and Nanotechnology* 15: 6357–72.

Buzby, S., Franklin, S., and Ismat Shah, S. 2006. Synthesis of metal-oxide nanoparticles: Gas-solid transformations. In Rodriguez, J. A., and Fernandez-Garcia, M. (Eds.), *Synthesis, Properties and Applications of Oxide Nanomaterials*, 119–134. John Wiley & Sons, Inc.

Carré, G., Hamon, E., Ennahar, S., Estner, M., Lett, M.-C., and Horvatovich, P. 2014. TiO_2 photocatalysis damages lipids and proteins in Escherichia coli. *Applied and Environmental Microbiology* 80: 2573–81.

Cheng, S., Zhang, Y., Cha, R., Yang, J., and Jiang, X. 2016. Water-soluble nanocrystalline cellulose films with highly transparent and oxygen barrier properties. *Nanoscale* 8: 973–8.

Cui, Y., Cheng, M., Han, M., Zhang, R., and Wang, X. 2021. Characterization and release kinetics study of potato starch nanocomposite films containing mesoporous nano-silica incorporated with thyme essential oil. *International Journal of Biological Macromolecules* 184: 566–73.

Cushen, M., Kerry, J., and Morris, M. et al. 2012. Nano-technologies in the food industry-recent developments, risks and regulation. *Trends in Food Science and Technology* 24: 30–46.

De Silva, R. T., Mantilaka, M. M. M. G. P. G., Ratnayake, S. P., Amaratunga, G. A. J., and Nalin de Silva, K. M. 2017. Nano-MgO reinforced chitosan nanocomposites for high performance packaging applications with improved mechanical, thermal and barrier properties. *Carbohydrate Polymers* 157: 222–28.

Delgado, K., Quijada, R., Palma, R., and Palza, H. 2011. Polypropylene with embedded copper metal or copper oxide nanoparticles as a novel plastic antimicrobial agent. *Letters in Applied Microbiology* 53: 50–54.

Donglu, F., Wenjian, y, and Kimatu, B. M. et al. 2016. Effect pf nanocomposite-based packaging on storage stability of mushrooms (*Flammulina velutipes*). *Innovative Food Science and Emerging Technologies* 33: 489–97.

Echegoyen, Y., and Nerin, C. 2013. Nanoparticle release from nano-silver antimicrobial food containers. *Food and Chemical Toxicology* 62: 16–22.

El-Wakil, N. A., hassan, E. A., Abou-Zeid, R. E., and Durfresne, A. 2015. Development of wheat gluten/nanocellulose/titanium dioxide nanocomposites for active food packaging. *Carbohydrate Polymers* 124: 337–46.

Emamifar, A., Kadivar, M., Shahedi, M., and Soleimanian-Zad, S. 2011. Effect of nanocomposite packaging containing Ag and ZnO on inactivation of *Lactobacillus plantrum* in orange juice. *Food Control* 22: 408–13.

Esmailzadeh, H., Sangpour, P., Shahraz, F., Hejazi, J., and Khaksar, R. 2016. Effect of nanocomposite packaging containing ZnO on growth of *Bacillus subtilis* and *Enterobacter areogenes. Materials Science and Engineering:C* 58: 1058–63.

Espitia, P. J. P., Soares, N. d. N. F. F., Coimbra, J. S., de Andrale, d. r., N. J., Criuz, R. S., and Mzedeiros, E. A. A. 2012. Zinc oxide nanoparticles: Synthesis, antimicrobial activity and food packaging applications. *Food and Bioprocess Technology* 5: 1447–64.

Fang, D., Yang, W., and Kimatu, B. M. et al. 2016. Effect of nanocomposite-based packaging on storage stability of mushrooms (*Flammulina velutipes*). *Innovations in Food Science and Emerging Technology* 33: 489–97.

Fei, P., Shi, Y., Zhou, M., Cai, J., Tang, S., and Xiong, H. 2013. Effects of nano-TiO_2 on the properties and structures of starch/poly (e-caprolactone) composites. *Journal of Applied Polymer Science* 130: 4129–36.

Goswami, A., Bajpai, A. K., Bajpai, J., and Sinha, B. K. 2018. Designing vanadium pentoxide-carboxymethyl cellulose/polyvinyl alcohol-based bionanocomposite films and study of their structure, topography, mechanical, electrical and optical behavior. *Polymer Bulletin* 75: 781–807.

Goudarzi, V., Shahabi-Ghahfarrokhi, I., and Babsei-Ghazvini, A. 2017. Preparation of eco-friendly UV-protective food packaging material by starch/TiO2 bio-nanocomposite: Characterization. *International Journal of Biological Macromolecules* 95: 306–13.

Guo, B.-L., Han, P., and Guo, L.-C. et al. 2015. The anti-bacterial activity of Ta-doped ZnO nanoparticles. *Nanoscale Research Letters* 10: 336.

He, H. Y. 2011. Recent study on nano TiO_2 photocatalyst: A review in modification, synthesis technique, and operation parameters. *Micro and Nanosystems* 3: 14–25.

Hench, L. L., and West, J. K. 1990. The sol-gel process. *Chemical Reviews* 90: 33–72.

Heydari-Majd, M., Ghanbarzadeh, B., Shahidi-Noghabi, M., Najafi, M. A., and Hosseini, M. 2019. A new active nanocomposite film based on PLA/ZnO nanoparticle/essential oils for the preservation of refrigerated Otolithes ruber fillets. *Food Packaging and Shelf Life* 19: 94–103.

Hirvikorpi, T., Vähä-nissi, M., Nikkola, J., Harlin, A., and Karppinen, M. 2011. Thin Al_2O_3 barrier coatings onto temperature-sensitive packaging materials by atomic layer deposition. *Surface Coating Technology* 205: 5088–92.

Hirvikorpi, T., Vaha-Nissi, M., and Harlin, A. et al. 2011. Enhanced water vapor barrier properties for biopolymer films by polyelectrolyte multilayer and atomic layer deposited Al_2O_3 double-coating. *Applied Surface Science* 257: 9451–54.

Hong, J. S., Kim, Y. K., Ahn, K. H., Lee, S. J., and Kim, C. 2006. Interfacial tension reduction in PBT/PE/clay nanocomposite. *Rheological Acta* 46: 469–78.

Hou, X., Xue, Z., and Xia, Y. et al. 2019. Effect of SiO_2 nanoparticle on the physical and chemical properties of eco-friendly agar/sodium alginate nanocomposite film. *International Journal of Biological Macromolecules* 125: 1289–98.

Huang, D., Liu, Z., Li, Y., and Liu, Y. 2017. Colossal permittivity and dielectric relaxation of (Li, In) co-doped ZnO ceramics. *Journal of Alloys and Compounds* 698: 200–6.

Huang, Y., Chen, S., Bing, X., Gao, C., Wang, t, and Yuan, B. 2011. Nanosilver migrated into food-simulating solutions from commercially available food fresh containers. *Packaging Technology and Science* 24: 291–97.

Jafarzadeh, S., Salehabadi, A., and Jafari, S. M. 2020. Metal nanoparticles agents in food packaging. *Handbook of Food Nanotechnology*, 379–414.

Jagadish, K., Shiralgi, Y., and Chandrashekar, B. N. 2018. Ecofriendly synthesis of Metal/Metal oxide nanoparticles and their application in food packaging and food preservation. In Grumezescu, A. M., and Holban, A. M. (Eds.) *Handbook of Food Bioengineering*, 197–216.

Jin, T., and Gurtler, J. B. 2011. Inactivation of salmonella in liquid egg albumen by antimicrobial bottle coatings infused with allyl isothiocyanate, nisin and zinc oxide nanoparticles. *Journal of Applied Microbiology* 110: 704–12.

Kaur, D., Bharti, A., Sharma, T., and Madhu, C. 2021. Dielectric properties of ZnO-based nanocomposites and their potential applications. *International Journal of Optics* https://doi.org/10.1155/2021/9950202

Khan, A.S., Yu, S. and Liu, H., 2012. Deformation induced anisotropic responses of Ti–6Al–4V alloy Part II: A strain rate and temperature dependent anisotropic yield criterion. *International Journal of Plasticity* 38:14–26.

Konwar, A., Kalita, S., Kotoky, J., and Chowdhury, D. 2016. Chitosan-iron oxide coated graphene oxide nanocomposite hydrogel: A robust and soft antimicrobial biofilm. *ACS Applied Materials and Interfaces* 8: 20625–34.

Krehula, L. K., Papic, A., Krehula, S., Gilja, V., Foglar, L., and Hrnjak-urgic, Z. 2017. Properties of UV protective films of poly(vinly-chloride)/TiO_2 nanocomposites for food packaging. *Polymer Bulletin* 74: 1387–1404.

Kuswandi, B., and Moradi, M. 2019. Improvement of food packaging based on functional nanomaterial. *Nanotechnology: Applications in Energy, Drug and Food*, 309–44. Cham: Springer International Publishing.

Lahiri, R., Ghosh, A., Dwivedi, S. M. M. D., Chakrabartty, S., Chinnamuthu, P., and Mondal, A. 2017. Performance of erbium-doped TiO2 thin film grown by physical vapor deposition technique. *Applied Physics A* 123: 573.

Lemire, J. A., Harrison, J. J., and Turner, R. J. 2013. Antimicrobial activity of metals: Mechanisms, molecular targets and applications. *Nature Reviews Microbiology* 11: 371–84.

Li, D., Zhang, J., Xu, W., and Fu, Y. 2014. Effect of SiO_2/EVA on the mechanical properties, permeability and residual solvent of polypropylene packaging films. *Polymer Composites* 37: 101–7.

Lian, Z., Zhang, Y., and Zhao, Y. 2016. Nano-TiO_2 particles and high hydrostatic pressure treatment for improving functionality of polyvinyl alcohol and chitosan composite films and nnao-TiO_2 migration from film matrix in food stimulants. *Innovative Food Science and Emerging Technologies* 33: 145–53.

Liu, X., Wang, L., Qiao, Y., Sun, X., Ma, S., and Cheng, X. 2018. Adhesion of liquid food to packaging surfaces: Mechanisms, test methods, influencing factors and anti-adhesion methods. *Journal of Food Engineering* 228: 102–17.

Llorens, A., Lloret, E., Picouet, P. A., Trbojevich, R., and Fernandez, A. 2012. Metallic-based micro and nanocomposites in food contact materials and active food packaging. *Trends in Food Science and Technology* 24: 19–29.

Lu, C., and Yang, B. 2009. High refractive index organic-inorganic nanocomposites: Design, synthesis and application. *Journal of Materials Chemistry* 19: 2884.

Luo, Z., Xu, Y., and Ye, Q. 2015. Effect of nano-SiO2-LDPE packaging on biochemical, sensory and microbiological quality of pacific white shrimp *Penaeus vannamei* during chilled storage. *Fisheries Science* 81: 983–93.

Maisanaba, S., Prieto, A. I., Pichardo, S., Jorda´-Beneyto, M., Aucejo, S., and Jos, A. 2015. Cytotoxicity and mutagenicity assessment of organomodified clays potentially used in food packaging. *Toxicology in Vitro* 29: 1222–30.

Mallakpour, S., and Naghdi, M. 2018. Polymer/SiO_2 nanocomposites: Production and applications. *Progress in Materials Science* 97: 409–47.

Mihindukulasuriya, S. D. F., and Lim, L. 2014. Nanotechnology development in food packaging: A review. *Trends in Food Science and Technology* 40: 149–67.

Mirtalebi, S. S., Almasi, H., and Alizadeh Khaledabad, M. 2019. Physical, morphological, antimicrobial and release properties of novel MgO-bacterial cellulose nanohybrids prepared by in-situ and ex-situ methods. *International Journal of Biological Macromolecules* 128: 848–57.

Miyauchi, M., Li, Y., and Shimizu, H. 2008. Enhanced degradation in nanocomposites of TiO_2 and biodegradable polymer. *Environmental Science and Technology* 42: 455.

Mojaki, S. C., Mishra, S. B., Mishra, A. K., and Mofokeng, J. P. 2018. Influence of polysiloxane as nanofiller on the surface, optical and thermal properties of guar gum grafted polyaniline matrix. *International Journal of Biological Macromolecules* 114: 441–52.

Montazer, M., Keshvari, A., and Kahali, P. 2016. Tragacanth gum/nano silver hydrogel on cotton fabric: In-situ synthesis and antibacterial properties. *Carbohydrate Polymers* 154: 257–66.

Moradi, Z., Esmaiili, M., and Almasi, H. 2019. Development and characterization of kefiran-Al_2O_3 nanocomposite films: Morphological, physical and mechanical properties. *International Journal of Biological Macromolecules* 122: 603–9.

Morsy, M. K., Khalaf, H. H., Sharoba, A. M., El-Tanahi, H. H., and Cutter, C. N. 2014. Incorporation of essential oils and nanoparticles in pullulan films to control foodborne pathogens on meat and poultry products. *Journal of Food Science* 79: 675–82.

Muthulakshmi, L., Varada Rajalu, A., Kaliaraj, G. S., Siengchin, S., Parameswaranpillai, J., and Saraswathi, R. 2019. Preparation of cellulose/copper nanoparticles bio nanocomposite films using a bioflocculant polymer as reducing agent for antibacterial and anticorrosion applications. *Composites Part B: Engineering* 175: 107177.

Nan, L., Yang, W., and Liu, Y. et al. 2008. Antibacterial mechanism of copper-bearing antibacterial stainless steel against *E. coli. Journal of Material Science and Technology* 24: 197–201.

Nesic, A., Gordic, M., Davidovic, S., Radovanovic, ′Z., Nedeljkovic, J., Smirnova, I., and Gurikov, P. 2018. Pectin-based nanocomposite aerogels for potential insulated food packaging application. *Carbohydrate Polymers* 195: 128–135.

Oleyaei, S. A., et al. 2016. Modification of physiochemical and thermal properties of starch films by incorporation of TiO_2 nanoparticles. *International Journal of Biological Macromolecules* 89: 256–64.

Palomero, F., Morata, A., Suárez-Lepe, J., Calderón, F., and Benito, S. 2016. Antimicrobial packaging of beverages. In *Antimicrobial Food Packaging*, 281–96. Elsevier, Spain.

Pavlidou, S., and Papaspyrides, C. D. 2008. A review on polymer-layered silicate nanocomposites. *Progress in Polymer Science* 33: 1119–98.

Pilevar, Z., Bahrami, A., Beikzadeh, S., Hosseini, H., and Jafari, S.M., 2019. Migration of styrene monomer from polystyrene packaging materials into foods: Characterization and safety evaluation. *Trends in Food Science & Technology* 91: 248–261.

Polarz, P., Roy, A., and Merz, M. et al. 2005. Chemical vapor synthesis of size-selected zinc oxide nanoparticles. *Small* 1 540–52.

Prasanna, S. R. V. S., Balaji, K., Pandey, S., and Rana, S. 2019. Metal oxide based nanomaterials and their polymer nanocomposites. *Nanomaterials and Polymer Nanocomposites* 123–144.

Pukanszky, B., Belina, K., Ockenbauer, R. A., and Maurer, F. H. J. 1994. Effect of nucleation, filler anisotropy and orientation on the properties of PP composites. *Composites* 25: 205–14.

Rahman, M.J., Ambigaipalan, P., and Shahidi, F., 2018. Biological activities of camelina and sophia seeds phenolics: Inhibition of LDL oxidation, DNA damage, and pancreatic lipase and α-glucosidase activities. *Journal of Food Science* 83(1): 237–245.

Rana, A. U. H. S., Chang, S.-B., Chae, H. U., and Kim, H.-S. 2017. Structural, optical, electrical and morphological properties of different concentration sol-gel ZnO seeds and consanguineous ZnO nanostructured growth dependence on seeds. *Journal of Alloys and Compounds* 729: 571–82.

Reyes-Torres, M. A., Mendoza-Mendoza, E., and Miranda-Herna´ndez, A. M. et al. 2019. Synthesis of CuO and ZnO nanoparticles by a novel green route: Antimicrobial activity, cytotoxic effects and their synergism with ampicillin. *Ceramic International* https://doi.org/10.1016/J.CERAMINT.2019.08.171

Richard, D., Romero, M., and Faccio, R. 2018. Experimental and theoretical study on the structural, electrical and optical properties of tantalum-doped ZnO nanoparticles prepared via sol-gel acetate route. *Ceramic International* 44: 703–11.

Rokbani, H., Daigle, F., and Ajji, A. 2019. Long- and short-term antibacterial properties of low-density polyethylene-based films coated with zinc oxide nanoparticles for potential use in food packaging. *Journal of Plastic Film Sheeting* 0(0): 1–18.

Roy, S., and Rhim, J. W. 2019. Carrageenan-based antimicrobial bionanocomposite films incorporated with ZnO nanoparticles stabilized by melanin. *Food Hydrocolloids* 90: 500–7.

Rubio, F.T.V., Haminiuk, C.W.I., Martelli-Tosi, M., da Silva, M.P., Makimori, G.Y.F., and Favaro-Trindade, C.S., 2020. Utilization of grape pomaces and brewery waste Saccharomyces cerevisiae for the production of bio-based microencapsulated pigments. *Food Research International* 136: 109470.

Sanchezgarcia, M. D., Lagaron, J. M., and Hoa, S. V. 2010. Effect of addition of carbon nanofibers and carbon nanotubes on properties of thermoplastic biopolymers. *Composites Science and Technology* 70: 1095–105.

Schadler, L. S. 2003. Polumer-based and polymer-filled nanocomposites. *Nanocomposite Science and Technology*. Wiley Online Library, Weinheim.

Seil, J. T., and Webster, T. J. 2012. Antimicrobial applications of nanotechnology: Methods and literature. *International Journal of Nanomedicines* 7: 2767–81.

Siegrist, M., Cousin, M. E., Kastenholz, H., and Wiek, A. 2007. Public acceptance of nanotechnology foods and food packaging: The influence of affect and trust. *Appetite* 49: 459–66.

Smith, W., Mao, S., and Lu, G. et al. 2010. The effect of Ag nanoparticle loading on the photocatalytic activity of TiO2 nanorod arrays. *Journal of Physical Chemistry Letters* 485: 171–75.

Struller, C. F., Kelly, P. J., and Copeland, N. J. 2014. Aluminum oxide barrier coatings on polymer films for food packaging applications. *Surface and Coatings Technology* 241: 130–37.

Swaroop, C., and Shukla, M. 2018. Development of blown polylactic acid-MgO nanocomposite films for food packaging. *Composites A: Applied Science and Manufacturing* 124: 105482.

Tarhan, O. 2020. Safety and regulatory issues of nanomaterials in foods. *Handbook of Food Nanotechnology*, pp. 655–703. Elsevier, Turkey.

Teymourpour, S., Nfachi, A. M., and Nahidi, F. 2015. Functional, thermal and antimicrobial properties of soluble soybean polysaccharide biocomposites reinforced by nano TiO_2. *Carbohydrates Polymers* 134: 726–31.

Tian, F., Chen, W., and Wu, C. et al. 2019. Preservation of Ginkgo biloba seeds by coating with chitosan/nano-TiO2 and chitosan/nano-SiO_2 films. *International Journal of Biological Macromolecules* 126: 917–25.

Tyler, T., Boeckmann, R.J., Smith, H.J., and Huo, Y.J., 2019. *Social Justice in a Diverse Society*. Taylor & Francis Group, New York.

Ubani, C. A., and Ibrahim, M. A. 2019. Complementary processing methods for ZnO nanoparticles. *Materials Today: Proceedings* 7: 646–54.

Vaezi, K., Asadpour, G., and Sharifi, H. 2019. Effect of ZnO nanoparticles on the mechanical, barrier and optical properties of thermoplastic cationic starch/montmorillonite biodegradable films. *International Journal of Biological Macromolecules* 124: 519–29.

Vaha-Nissi, M., Pitkanen, M., Salo, E., Slevanen-Rahijarivi, J., Putkonen, M., and Harlin, A. 2015. Atomic layer deposited thin barrier films for packaging. *Cellulose Chemistry and Technology* 49: 575–85.

Valerini, D., Tammaro, L., and Di Benedetto, F. et al. 2018. Aluminum-doped zinc oxide coatings on polylactic acid films for anti-microbial food packaging. *Thin Solid Films* 645: 187–92.

Wagih, A., Abu-Oqail, A., and Fathy, A. 2019. Effect of GNPs content on thermal and mechanical properties of a novel hybrid cu-Al2O3/GNPs coated Ag nanocomposite. *Ceramics International* 45: 1115–24.

Wang, K., Jin, P., Shang, H., and Li, H. et al. 2010. A combination of hot air treatment and nano-packaging reduces fruit decay and maintains quality in postharvest Chinese bayberries. *Journal of the Science of Food and Agriculture* 90: 2427–32.

Wihodo, M., and Moraru, C. I. 2013. Physical and chemical methods used to enhance the structure and mechanical properties of protein films: A review. *Journal of Food Engineering* 114: 292–302.

Wu, Y., Zhang, Y., Xu, J., Chen, M., and Wu, L. 2010. One-step preparation of PS/TiO_2 nanocomposite particles via miniemulsion polymerization. *Journal of Colloidal and Interface Science* 343: 18–24.

Xie, J., and Hung, Y. C. 2018. Novel chitosan films with laponite immobilized ag nanoparticles for active food packaging. *Carbohydrate Polymers* 199: 210–18.

Xie, J., and Hung, Y. C. 2018. UV-A activated TiO2 embedded biodegradable polymer film for antimicrobial food packaging application. *LWT* 96: 307–14.

Ye, J., Liu, W., and Cai, J., et al. 2011. Nanoporous anatase TiO2 mesocrystals: Additive-free synthesis, remarkable crystalline-phase stability, and improved lithium insertion behavior. *Journal of American Chemical Society* 133: 933–40.

Yoshida, T., Yoshioka, Y., and Fujimura, M. et al. 2011. Promotion of allergic immune responses by intranasally-administrated nanosilica particles in mice. *Nanoscale Research Letters* 6: 1–6.

Youssef, A. M., El-Sayed, S. M., El-Sayed, H. S., Salama, H. H., Assem, F. M., and El-Salam, M. H. A. 2018. Novel bionanocomposite materials used for packaging skimmed milk acid coagulated cheese (Karish). *International Journal of Biological Macromolecules* 115: 1002–11.

Zhang, R., Wang, X., Li, L., Cheng, M., and Zhang, L. 2019. Optimization of konjac glucomannan/carrageenan/nano-SiO2 coatings for extending the shelf-life of *Agaricus bisporus*. *International Journal of Biological Macromolecules* 122: 857–65.

Zhang, Y., Liu, M., Ren, W., and Ye, Z.-G. 2015. Well-ordered ZnO nanotube arrays and networks grown by atomic layer deposition. *Applied Surface Science* 340: 120–25.

Zhuo, J. J., Wang, S. Y., and Gunasekaran, S. 2009. Preparation and characterization of whey protein film incorporated with TiO_2 nanoparticles. *Journal of Food Science* 74: 50–56.

Ziola, R. F., et al. 1992. Matrix-mediated synthesis of nanocrystalline γ-Fe_2O_3: A new optically transparent magnetic material. *Science* 257: 219.

Zois, H., Apekis, L., and Mamunya, Y. P. 2003. Dielectric permittivity and morphology of polymer composites filled with dispersed iron. *Journal of Applied Polymer Science* 88: 3013–20.

8 Nanostructure-Based Bioactive Packaging

Priyanka Suthar[1], Kajal Dorge[2], Nazmin Ansari[2], Shafiya Rafiq[3], and Mohammad Javed Ansari[4]

[1]Department of Food Science and Technology, Dr. Y. S. Parmar University of Horticulture and Forestry, Nauni, Himachal Pradesh, India

[2]Department of Food Technology and Nutrition, Lovely Professional University, Punjab, India

[3]Amity Institute of Biotechnology, Amity University Rajasthan, Jaipur, India

[4]Department of Botany, Hindu College Moradabad (Mahatma Jyotiba Phule Rohilkhand University), Bareilly, UP, India

CONTENTS

DOI: 10.1201/9781003207641-9

8.1 INTRODUCTION

Food packaging is considered a unit material that provides a barrier to various environmental factors, like UV light, water, water vapor, pressure, heat, and gases, in order to offer protection to food from microbial or chemical contaminants, prolong its shelf life, and ensure safe and secured food transport. Traditional packaging material in the food industry includes various petroleum-based polymers. These polymers are much cheaper and possess better barrier properties then biopolymers films. Therefore, biopolymers are not always helpful in food packaging industry (Zhang et al., 2021). However, newer technology has been studied extensively and the application of nanotechnology is one of these. The application of nanotechnology in the production of different novel packaging material has had remarkable growth in the past decade and it is expected to have a role in the food packaging industry in the future. This exponential growth in packaging is due to better knowledge of nanotechnology, the result of the combined efforts of academic and industrial research on the development of new nanotechnology-based products with enhanced functional, technological, and physical properties along with advances in material science, analytical techniques, and processing technology. Simultaneously, increased knowledge about sustainable, greener, and renewable concepts pushed the development of nanostructure-based packaging for the food industry. The acceptance of this system at social and commercial levels clears its future innovation pathway and helps in the understanding the broad application of nanotechnology in food packaging. The major concerns for food manufacturers are the production of waste and high energy consumption; however, nanotechnology properly addresses these issues. The concept of nanotechnology was introduced by Richard Feynman in 1959, and the term "nanotechnology" was first used by Norio Taniguchi in 1974. Nanotechnology involves the fabrication, characterization, and manipulation of nano—that is, with a range of less than 100 nm—molecules. The application of these nano-sized molecules in polymers have been extended to design, manufacturing, processing, and application into polymer materials prepared from nano particles (Sharma et al., 2017).

The potential of nanotechnology has gained the attention of researchers from various fields, like biological science, physics, chemistry, and engineering. It is predicted that nanotechnology will create a major thrust for the development of novel packaging material, owing to its unique and improved physicochemical properties when material is brought to nanoscale. Due to the extremely small size of nano particles, it holds a huge surface-to-volume ratio and, thus, surface activity. Nanomaterials also improved the electrical conductivity, mechanical strength, and thermal stability, and thus can be used in food packaging in order to enhance both barrier and mechanical properties. Nanotechnology also plays a role in developing intelligent packaging systems.

8.2 TRADITIONAL PACKAGING

In the later years of primitive or ancient society, the concept of packaging was already developed (Mustafa et al., 2012). From early times, a small need for packaging was already noticed, for either storage or transportation. Later, due to many social developments and human bartering of goods to far places, the need for transportation of a large quantity of goods came into demand. These societies then found "natural" packaging, like leaves and shells, from nature and containers were created from hollowed bamboo, tree bark, animal organs, and woven grasses. Later, as ores and chemical compounds were discovered, pottery and metal containers were used as packaging. In traditional packaging, packaging material is most commonly obtained from leaves from various trees, such as palm, palas, banana, pandan, and the leaves of casuarinas and bamboo. Coconut leaves were used for wrapping rice (called as ketupat), banana leaf was used to wrap rice and confectionary fats, Palas leaf wrapped sticky rice, padan or pine leaf was used for dodol (sweet from palm sugar), rubber leaf was used to pack fermented rice or fermented tapioca, casuarina leaves were used to wrap temp or fermented soya bean, and bamboo was used for holding water.

Before the development and advancement of packaging technology, the concept of attraction was not important due to socio-economic conditions. Today packaging is whole new concept involving product features and directly impacting of consumer purchases, and it also reflects consumer psychology. The main functions of packaging are to provide protection to goods, be easy to use and transport, and to convey the product information. The main objective is to prevent early spoilage of food. The method of traditional packaging of edible goods often involved direct wrapping in bamboo leaves. It is important to understand the basics of traditional packaging as it helps in distinguishing the separate techniques, styles, and types of packaging. Modern packaging material includes synthetic material that has efficiently replaced the natural traditional packaging system. This is probably due to the changes in lifestyle and limited availability of natural resources causing depletion of traditional packaging systems. It is evident that today's giant food processing industries, and even small food vendors, greatly depend on plastic-based packaging material, even though these possess hazardous chemical compounds. Following are a few packaging materials that have been used since ancient times.

8.2.1 Leaves

Already mentioned, banana and other leaves were a commonly used material for wrapping food, like cheese or other confectionary. Cornhusk was used to wrap block brown sugar, and all kind of foods were cooked in the leaves. The pan leaf was used to wrap spices in India. Leaves are cost effective and readily available for packaging purposes.

8.2.2 Vegetable Fibers

The fibers obtained from vegetables are used to produce yarn, string, or a cord which is further converted into a packaging material. The materials derived are flexible, and to some extent they are resistant to tearing. However, these materials are permeable to both water vapor and water. The main advantage of this packaging is that it is lightweight for transportation and handling. The other important advantage of natural material is its biodegradable property when left untreated.

8.2.3 Bamboo and Rattan

These materials were used in making baskets. Bamboo pots and bamboo stem were also used extensively.

8.2.4 Coconut Palm

Green coconut palms were used to prepare bags and baskets to carry meat and vegetables in different parts of the world. Palyra palm leaves are woven into boxes in which cooked foods are transported.

8.2.5 Treated Skins

Leather has been in the market from many centuries as a non-breakable container. Water and wines were frequently stored and transported in leather containers.

8.2.6 Earthenware

These materials were widely used all over the world to carry solids and liquid foods, like beer, curd, honey, etc. Corks, leaves, plastic sheeting, wax, and wooden lids were used to seal the earthen pots. It provides gas, moisture, and lightproof containers, if sealing is done properly. There are two main types of earthen pots: glazed and unglazed. The unglazed pots are porous and suitable for cooling food, such as curd, whereas glazed earthenware is better for storing wine and oil as these pots are moisture proof and airtight after sealing. They are also lightproof and if clean properly they restrict entry of microbes, rodents, and insects. Earthenware should not have lead content if an acidic food is stored in such containers.

8.3 BIOACTIVE PACKAGING

Bioactive packaging is a modern packaging technology that has an added bioactive compound or functional ingredient within a packaging unit material which can interact with food products through controlled release of substances into the food. This concept sometimes resembles the characteristic of active packaging but the objective of each is pursued differently. Bioactive packaging has a positive, direct impact on consumer health through a healthier packaged food while active packaging is only concerned with extending shelf life and preserving and enhancing the state of packaged food. Presently, industrial demand is for the development of technology that ensures that the stability of bioactive compounds remains strong. The main objective of bioactive packaging is to preserve bioactive components utilizing biomaterials, either through micro- and nanoencapsulation methods or by incorporating functional components into the package walls. Encapsulation technologies for food components are still in the research stage, and a fully optimized commercial procedure is still a long way off. The information needed to build novel formulations, on the other hand, may be found in the pharmaceutical and biomedical fields, where the creation of matrixes for controlled release of bioactive components (drugs) is already a reality and an active research area that is constantly improving. The bioactive packaging idea encompasses a number of methods that may be used to preserve or stabilize functional components, and can be classified into the following categories:

1. Bioactive molecules or nano-components in biodegradable and/or feasible packaging systems are integrated and released in a regulated manner.
2. Micro- and nanoencapsulation of such bioactive components in packaging or within the food.
3. Enzymatically active packaging that improves health by transforming particular food-borne components.

8.3.1 Edible Films and Coatings

Edible films and coatings are novel developed technology in the food packaging sector and are applied as a layer to food items to help preserve, distribute, and commercialize the food by protecting it from chemical, microbial, and physical deterioration. These films can help enhance the quality. Along with that, it helps in handling, and to maintain the integrity of the packed item by including antioxidants, antimicrobials, nutraceuticals, flavoring compounds, and other ingredients. Biopolymers and mixes, which are potential alternatives to traditional plastic packaging, are among the ingredients utilized in film production. Certain qualities of raw materials, such as inertness, nontoxicity, and barrier properties to microorganisms, must be addressed while designing packaging in order to ensure the food's preservation and protect it from deterioration and external conditions. Different sections of agricultural commodities like fruits and plants, as well as some agricultural waste found in nature, have been shown to be a potential resource to be used as raw material for the purpose of renewable materials as fillers for nanocomposite industries. Biopolymers such as polysaccharides, proteins, and lipids can be used to create edible films and coatings due to their biodegradable properties. Edible coating is usually applied when it is at a liquid state by using various technologies, like immersion or spraying, whereas film is obtained from a molding process and followed by drying to make it a solid sheet before being applied on the food surface in the shape of bags, packages, or capsules. The utilization of films and coatings provides the benefit of combining polymer matrix with the various active compounds, for betterment of the functional properties of a packaging unit. They can also be taken alongside foods. Carbohydrates and proteins as films provide barrier properties against oxygen at low to moderate relative humidity, and they have high mechanical characteristics; but, due to their water-loving nature, they are poor water vapor barriers.

As a result, researchers are continually experimenting with new combinations of various bioactives with polymer matrix to improve the mechanical, thermal, and barrier properties. Plasticizers added to the filmogenic solution may enhance the elongation at the break and reduce the modulus of elasticity and elongation stress in obtained film. Plasticizers work by changing the mechanical characteristics of the film, which results in changes to its barrier qualities. The main application of plasticizers in biopolymer production is to improve the free volume of the system. It is also helpful in improving flexibility and extensibility of developed films by minimizing intermolecular forces and by the enlargement of the polymer chains, resulting in fewer cracks in the film, which eventually enhances its strength.

8.3.2 Antioxidant Properties in BioPolymer Packages

Among the numerous forms of active packaging, the ones with the antioxidant impact are among the most significant for the business, particularly in the food industry. One of the most common degrading processes in food is oxidation. Antioxidant active packaging, in particular, aims at preventing or reducing the oxidation of specific food components, like lipids and proteins, which results in the loss of physical qualities including sensory characteristics in food products. Oxygen degrades the quality of various ranges of food items. Vegetable oils high in polyunsaturated fatty acids tend to oxidize. The use of antioxidant films and coatings with these products provides a chance to tackle these challenges. One of the primary goals of developing new packaging material is the inclusion of antioxidant compounds into the polymeric matrix of films.

Tomato paste was used in the production of mortadella and it confirmed the increased value in the lycopene content in the samples (Arroyo et al., 2019). The results suggested that significant reduction of lipid oxidation was reported. In another study by Barbosa-Pereira et al. (2014), a film was created with rosemary extract along with two synthetic antioxidants, propyl gallate and BHT, into the packaged meat samples. It was reported that active film contact with meat sample had approximately 60% reduced thiobarbituric acid relative substances (TBRA) even after nine days of storage when compared with controlled sample.

8.3.3 Bio-Based and Biodegradable Food Packaging Materials

Polymers derived from renewable resources, according to the definition of European Bioplastics, can be divided into three groups, based on the origin of the raw ingredients and the method of production.

1. Polymers derived from biomass like polysaccharides from maize, wheat starch, and rice; hemicellulose from barley and potato; gums from guar. Protein polymers such as alginate, pectin and carrageenam, chitin and chitosan. Along with these, animal origin proteins were whey, casein, and collagen; and plant origin proteins are soy, gluten, and zein.
2. Monomer-based polymers from renewable source are bio-polyfins, bio-polyethylene terephthalate, and polylactic acid.
3. Microorganism-generated polymers include polyhydroxyalkanoates, like polyhydroxyvalerate; polyhydroxybutyrate; and polyhydroxybutyrate-co-valerate copolymers.

Petroleum-based biodegradable monomers, like polybutylene adipate (its copolymers) with polyethylene terephthalate, polybutylene succinate (its copolymers) with polybutylene adipate, polyglycolic acid, polypropylene carbonate, and polycaprolactone could be added to this as a fourth category. It's important to remember that biodegradability is determined by the polymer chain's ultimate chemical composition, but by not the origin of raw material used in it. As a result, the biodegradable polymer might be made using monomers derived from renewable or petrochemical sources (Siracusa and Rosa, 2018).

8.4 NANO BIOACTIVE COMPOUNDS IN PACKAGING TECHNOLOGY

The invention of nanotechnology and enhanced knowledge of its food application led to the development of active or smart packaging. This novel technology has potential to bring about a revolution in the food and agricultural sector (Imran et al., 2010). The novel packaging technology in food is the most promising owing to the nanotechnology and its wide applications in the food industry. Many companies are already using this technology for developing films, carbon nanotubes, and waxy nano coatings to enhance the shelf life, antimicrobial property, barrier property, and safety of consumed food and beverages. Nano materials can also be used as a vehicle to provide vitamins, flavors, antibiotics, enzymes, and anti-browning agent to improve overall quality of foods. Among all, bioactive packaging is widely experimented and it possesses great potential in the field of food safety and preservation, or shelf-life prolongation. Examples of bioactive packaging are O_2 scavenging systems, ethanol generation, or moisture absorption systems (Imran et al., 2010).

The new theory of a bio-switch nano-system follows the approached that when load of microorganism reaches to certain level, amylase starts to secrete and degrade the encapsulated antimicrobials. Researchers are now developing sensors that track the altered properties based on internal or external conditions. The constructed sensors have CNTs (carbon nanotubes) which are cost-effective, fast, and can be recycled, which are advantageous over convectional technological methods like spectroscopic, enzymatic, chromatographic, which are time consuming and much more expensive in comparison with nanotechnology. Packaging system that include nanomaterial is intelligent or smart, which in packaging means they respond to the changes in environmental conditions, make consumers aware, or repair itself (self-healing) toward adulteration or contamination. Micro- and nanoencapsulated repairing materials are used in self-healing packaging units. A small amount of healing agent present in encapsulated form in the coatings is usually released either by a triggering mechanism or crack propagation. The safety and security of food may enhanced by developing "pathogen-repulsive" surfaces or packaging material which changes the color in the presence of microbial toxins. It was reported that there is an increased number of nanotechnology developments in either intelligent or active packaging type (Imran et al., 2010).

8.4.1 Nanocomposites in Food Packaging

Nanocomposites are considered a new and effective option to conventional technology for upgrading the properties of polymer. Nanocomposites provide good barrier properties, good heat resistance, and improved mechanical strength compared to neat polymers. For example, the use of nano-sized montmorillonite clay improves the mechanical and thermal property of nylon. When nanocomposites are used in food packaging, it provides better thermal resistance during food processing, transportation, and storage. Particle fillers used are nanoclays, montmorillonite (MMT), carbon nanotubes, graphene nanosheets, and kaolinite. The montmorillonite clays possess nanometer scale platelets of magnesium aluminum silicate. The thickness is kept at 1 nm and 100 to 500 nm diameter for platelets ratio. Finally, the clay structure is formed by 100 layered platelets stacked into particles of 8 to 10 μm in diameter. The carbon-based graphene nanoplates (GNPs) have excellent heat resistance and are good barrier nanocomposites with promising food packaging applications. Carbon nanotubes (CNTs), another carbon-based nanofiller, were studied due to its intrinsic mechanical and electrical properties. But the utilization of carbon nanotubes as nanocomposite is limited due to its processing and high cost. Biopolymers have been now a new attraction for researchers to replace the plastic packaging material due high interest in sustainable development. Example of biopolymers include starch, cellulose, polysaccharides, proteins, microbial products (polyhdroxybutyrate), and polymer synthesized chemically from naturally derived monomers (polylactic acid) (Arora and Padua, 2010).

Polymer nanocomposites (PNCs) are developed by dispersing an inert nanoscale filler into the polymeric matrix. Filler materials have clay and silicate nanoplatelets, silica nanoparticles, carbon nanotubes, starch nanocrystals, chitin or chitosan nanoparticles, graphene, nanowhiskers or nanofibers, g1qraphene, and others. Enhancing barrier properties is the most important application of PNCs, but PNCS also provide the most strong, more flame resistance, and improved thermal resistance than control polymers without nanocomposite (Duncan, 2011).

8.4.2 Nanotechnology in Active and Antimicrobial Packaging

In packaged foods, a certain amount of headspace gas is always present. Also, the permeation of oxygen through plastic material is an important concern. The presence of molecular O_2 is usually reduced to various intermediate species, and forms OH radicals, H_2O_2, and water. Except water, the intermediate products are highly reactive in nature and, hence, it is important to remove oxygen during packaging, or to control the residual oxygen level by using several scavengers in order to limit the food spoilage rate. Oxygen scavengers work on oxidation phenomena. The oxidation of fats and oils leads to rancidification and develops the off-odors and off-flavors, color changes, nutritive losses, and slowed metabolism process of food. The use of oxygen scavengers is economical and very effective in modifying the atmosphere. For preserving ethylene-sensitive crops, ethylene scavengers are used (Prasad and Kochhar, 2014). Polyolefin are a commonly used polymer in food packaging, with excellent moisture barrier applications. However, these polymers don't have good barrier property toward oxygen. High density polyethylene films modified with iron containing kaolinite were developed to create oxygen scavenging packaging films. It was reported that oxygen scavenging activity reached 43 ml O_2/g at 100% relative humidity and 37 ml O_2/g at relative humidity of 50% with the modified film (Mihindukulasuriya and Lim, 2014).

Antimicrobial systems in packaging are a recently advanced development that is designed to restrict the growth of spoilage-causing microorganism in packaged food, along with the physical protection and preservation functions of packaging material. Among active packaging (AP), interest in antimicrobial packaging has attracted most of the research in the past decade. Nanotechnology offers a new platform for development of antimicrobial packaging systems, or to improve the surface activity of active substances. The high surface-to-volume ratio and enhanced surface activity of nano-sized antimicrobial agents may restrict targeted microbes more effectively then the micro or macro scale sized antimicrobial agents. Nanoparticles get attached to bacterial cells easily and interact directly with them causing changes and damage in microbial function by creating gaps and pits and permeability, suppressing enzymatic activity, and cell death of the microbes. The antimicrobial activity of many nanoparticles with metal, like silver, copper, gold; and metal oxides, like titanium dioxide, zinc oxide, or magnesium oxide, was reported. The natural substances like chitosan/chitin also possess antimicrobial activity at nanoscale (Yildirim and Röcker, 2018).

Silver nanoparticles are found to be effective against a number of bacteria like *E.coli, Staphylococcus* (*aureus* and *epidermidis*), *Enterococcus faecalis, Pseudomonas* (*aeruginosa, flurescens, putida,* and *olevorans*), *Vibrio cholera, Bacillus* (*cereus, subtils,* and *anthracis*), *Shigella flexneri, Salmonella enterica, Proteus mirabilis, Typhimurium, Micrococous luteus, Listeria monocytogenes,* and *Klebsiella pneumonia.* Titanium dioxide nanoparticles are photocatalyzed and, thus, active only in the presence of UV light. TiO_2 nanoparticles are reported as effective against many food spoilage bacteria, like *Vibrio parahaemolyticus, Salmonella choleraesuis* subspp., and *Listeria monocytogens* under UV illumination but not in dark (Duncan, 2011).

8.4.3 Nanoencapsulation of Bioactive Compounds

Nanoencapsulation is a process whereby a bioactive ingredient is used as a core matrix which is trapped inside a wall matrix which resists the degradation by different kinds of enzymatic actions as well as surrounding factors. Using an appropriate wall matrix or carrier provides protection for sensitive bioactive compounds from heat and enzymes. These wall materials help in maintaining the nutritional properties of compounds and also in reducing the unwanted taste from some compounds. Liposomes are most appropriate for encapsulation and for the bioactive compound delivery for a longer time duration compared to other nano-carriers. The size of nano-capsules varies from 1 to 100 nm. Nanoencapsulation is also practiced in packaging of fruits and meat to enhance its shelf life. Along with that it also possesses good nutritional characteristics. Many nano-carriers are used for encapsulation, such as casein, zein, chitosan, gelatin, arabinogalactan, polyethylene glycol, pilyaniline, poly-D lactide-co glycolide, poly L-lysine, and L-lactide-co glycolide. Most common encapsulation types reported are polymeric nano-carriers, lipid-based nano-carriers, and hybrid nano-carriers (Noore et al., 2021). Nanoencapsulation of bioactive compounds is done by using different technologies, like electrospinning, elecrospraying, high-pressure homogenization, nano-spray dryer, micro-/nano fluidics, ultrasonication, polymerization, supercritical-based technology, and coacervation or ionic gelation (Noore et al., 2021). There are many studies reported with the utilization of

TABLE 8.1
Characteristics of Different Types of Nanostructure-Based Bioactive Packaging Materials

Nanostructure	Key Features	Examples or Types	References
Nanoparticles (NPs)	Green synthesis of NPs is possible using plant extracts	Silver NPs, gold NPs, palladium and platinum NPs, copper NPs, zinc oxide NPs and Titanium oxide NPs	Jadoun et al., 2021
Nanocomposites	The mixture of many constituents helps in exploiting synergistic effects with improved optical, electronic, catalytic, and magnetic properties.	Inorganic/organic polymer nanocomposite, polymer/layered silicate nanocomposite, Polymer/ceramics nanocomposites, magnetic nanocomposite, nanoclay	Hassan et al., 2021; Sriplai and Pinitsoontorn, 2021
Nanoemulsions	Encapsulation is possible with nanoemulsions to protect and for the control release of different bioctives	Single emulsion (o/w) or (w/o) and double emulsions (w/o/w) or (o/w/o). essential oil based emulsions	Barradas and de Holanda e Silva, 2021; Naseema et al. 2021
Nano-encapsulation	Controlled release system for bioactives molecules	Coating materials can be natural or modified polysaccharides, proteins, gums, synthetic polymers and lipids. Core material can be flavor compounds or bioactive compounds	Boostani and Jafari, 2021; Saifullah et al., 2019

nanoencapsulation of phenolic compounds into the active packaging. Makwana et al. (2014) encapsulated cinnamaldehyde by using nanoliposmes. The results showed excellent antimicrobial activity against *E. coli* and *B. cereus*. In another study, chitosan-based nanoparticles were used in an encapsulation system for tea polyphenols to provide antioxidant activity with 80% even after the sixth week of storage (Liu et al., 2015). The antioxidant films, when encapsulated polyphenols were added, showed higher DPPH scavenging efficiency. The characteristics of different types of nanostructure-based bioactive packaging materials is shown in Table 8.1

8.4.4 Nanoemulsions in Packaging

Emulsions are characterized as a dispersion of two immiscible liquid phases in which the dispersed phase consists of spherical droplets and the continuous phase consists of liquids. Emulsions are, basically, categorized into two types, microemulsions and nanoemulsions, based on their stability. The microemulsion droplets are thermodynamically stable with >100 nm size. These properties can be altered by slight variations in the environmental conditions, like temperature and composition.

Nanoemulsions require two immiscible liquids along with emulsifiers. The core shell structure is present in o/w and w/o nanoemulsion. In o/w nanoemulsion systems, surface active molecules are present on amphiphilic shell, whereas lipophilic core possesses non-polar molecules. The oil phase in nanoemulsions are made from triglycerols, monoacylglycerols, diaclygycerols, and free fatty acids. Other non-polar compounds, like essential oils, mineral oils, wax, lipid substitutes, and oil-soluble vitamins, can also be used as oil phase. Viscosity, refractive index, phase behavior, density, and interfacial tension greatly affect the stability, formation, and functional properties of nanoemulsions (Aswathanarayan and Vittal, 2019). In a recent study by Otoni et al. (2014), the authors developed an edible film for food packaging based on pectin, papaya puree, and cinnamaldehyde emulsions. The cinnamaldehyde droplets were reduced to nanoscale by using homogenization process. The results showed the improved antimicrobial property of edible film against food-borne pathogens.

8.4.5 Use of Nano-Technique in Intelligent Packaging

Intelligent packing materials are different from active packing as they don't release any components into the food matrix. Intelligent packaging helps in improving the HACCP (Hazard Analysis and Critical Control Point). Smart or intelligent packaging gives information about the packed food materials to consumers, such as on temperature, pH, and oxygen concentration, with the help of several detectors which sense and record if any changes have taken place. Changes in temperature mostly lead to the alteration in water activity which ensures the shelf life of food stuff. Hence, the monitoring of temperature becomes critical during the whole food chain. The changes in temperature can be recorded by using time-temperature indicators (TTI). Silver nanoplates are used to show the thermal history by changes in color. An iridescent technology is used in nanocrystalline cellulose films to check the humidity level (Babu, 2021). Intelligent packaging provides the information outside of the package and directly measures the quality of food stuff present inside the package. Intelligent package should help consumers by ensuring food safety and providing information and it may warn about possible problems (Huff, 2008).

8.5 CHALLENGES AND FUTURE SCOPE

The impressive growth and demand for nanotechnology regarding food packaging can be considered a promising development which has the potential to overcome several challenges present in food packaging industries. Nanotechnology is one of the most efficient, diverse, and sustainable technologies which improves the food packaging system. Several advantages of nanotechnology in food packaging have already been reported in many studies. Nanotechnology has played important roles in food safety, quality, security, and shelf-life extension. The many advantageous properties of nanotechnology support both producers and consumers. However, there is lack of significant studies on the direct impact on human health of nanoparticles. The migration assessment of nanoparticles from the food packaging unit to food is another major concern. Utilization of intelligent packaging and active packaging still has great potential in the actual market, yet it presents several disadvantages, like additional cost and consumer acceptance.

REFERENCES

Arora, A., and Padua, G. W. 2010. Nanocomposites in food packaging. *Journal of Food Science* 75(1): R43–R49.

Arroyo, B. J., Santos, A. P., de Melo, E. D. A., Campos, A., Lins, L., and Boyano-Orozco, L. C. 2019. Bioactive compounds and their potential use as ingredients for food and its application in food packaging. *Bioactive Compounds* 2019 (pp. 143–156). Elsevier, Amsterdam.

Aswathanarayan, J. B., and Vittal, R. R. 2019. Nanoemulsions and their potential applications in food industry. *Frontiers in Sustainable Food Systems* 3: 95.

Babu, P. J. 2021. Nanotechnology mediated intelligent and improved food packaging. *International Nano Letters*: 1–14.

Barbosa-Pereira, L., Aurrekoetxea, G. P., Angulo, I., Paseiro-Losada, P., and Cruz, J. M. 2014. Development of new active packaging films coated with natural phenolic compounds to improve the oxidative stability of beef. *Meat Science* 97(2): 249–254.

Barradas, T. N., and de Holanda e Silva, K. G. 2021. Nanoemulsions of essential oils to improve solubility, stability and permeability: A review. *Environmental Chemistry Letters* 19(2): 1153–1171.

Boostani, S., and Jafari, S. M. 2021. A comprehensive review on the controlled release of encapsulated food ingredients; Fundamental concepts to design and applications. *Trends in Food Science & Technology* 109: 303–321.

Duncan, T. V. 2011. Applications of nanotechnology in food packaging and food safety: Barrier materials, antimicrobials and sensors. *Journal of Colloid and Interface Science* 363(1): 1–24.

Hassan, T., Salam, A., Khan, A., Khan, S. U., Khanzada, H., Wasim, M., Khan, M. Q., and Kim, I. S. 2021. Functional nanocomposites and their potential applications: A review. *Journal of Polymer Research* 28(2): 1–22.

Huff, K. 2008. Active and intelligent packaging: Innovations for the future. *Department of Food Science & Technology*, 1–13. Virginia Polytechnic Institute and State University, Blacksburg.

Imran, M., Revol-Junelles, A. M., Martyn, A., Tehrany, E. A., Jacquot, M., Linder, M., and Desobry, S. 2010. Active food packaging evolution: Transformation from micro-to nanotechnology. *Critical Reviews in Food Science and Nutrition* 50(9): 799–821.

Jadoun, S., Arif, R., Jangid, N. K., and Meena, R. K. 2021. Green synthesis of nanoparticles using plant extracts: A review. *Environmental Chemistry Letters* 19(1): 355–374.

Liu, F., Antoniou, J., Li, Y., Yi, J., Yokoyama, W., Ma, J., and Zhong, F. 2015. Preparation of gelatin films incorporated with tea polyphenol nanoparticles for enhancing controlled-release antioxidant properties. *Journal of Agricultural and Food Chemistry* 63(15): 3987–3995.

Makwana, S., Choudhary, R., Dogra, N., Kohli, P., and Haddock, J. 2014. Nanoencapsulation and immobilization of cinnamaldehyde for developing antimicrobial food packaging material. *LWT-Food Science and Technology* 57(2): 470–476.

Mihindukulasuriya, S. D. F., and Lim, L. T. 2014. Nanotechnology development in food packaging: A review. *Trends in Food Science & Technology* 40(2): 149–167.

Mustafa, M., Nagalingam, S., Tye, J., Shafii, A. H., and Dolah, J. 2012, October. Looking back to the past: Revival of traditional food packaging. In *2nd Regional Conference on Local Knowledge (KEARIFAN TEMPATAN). Jerejak Island* (pp. 1–17).

Naseema, A., Kovooru, L., Behera, A. K., Kumar, K. P., and Srivastava, P. 2021. A critical review of synthesis procedures, applications and future potential of nanoemulsions. *Advances in Colloid and Interface Science* 287: 102318.

Noore, S., Rastogi, N. K., O'Donnell, C., and Tiwari, B. 2021. Novel bioactive extraction and nano-encapsulation. *Encyclopedia* 1(3): 632–664.

Otoni, C. G., de Moura, M. R., Aouada, F. A., Camilloto, G. P., Cruz, R. S., Lorevice, M. V., de, F. F., Soares, N., and Mattoso, L. H. 2014. Antimicrobial and physical-mechanical properties of pectin/papaya puree/cinnamaldehyde nanoemulsion edible composite films. *Food Hydrocolloids* 41: 188–194.

Prasad, P., and Kochhar, A. 2014. Active packaging in food industry: A review. *Journal of Environmental Science, Toxicology and Food Technology* 8(5): 1–7.

Saifullah, M., Shishir, M. R. I., Ferdowsi, R., Rahman, M. R. T., and Van Vuong, Q. 2019. Micro and nano encapsulation, retention and controlled release of flavor and aroma compounds: A critical review. *Trends in Food Science & Technology* 86: 230–251.

Sharma, C., Dhiman, R., Rokana, N., and Panwar, H. 2017. Nanotechnology: An untapped Resource for food packaging. *Frontiers in Microbiology* 8: 1735.

Siracusa, V., and Rosa, M. D. 2018. Sustainable packaging. In *Sustainable Food Systems from Agriculture to Industry*. 275–307. Academic Press, Cambridge, MA, USA.

Sriplai, N., and Pinitsoontorn, S. 2021. Bacterial cellulose-based magnetic nanocomposites: A review. *Carbohydrate Polymers* 254: 117228.

Yildirim, S., and Röcker, B. 2018. Active packaging. In *Nanomaterials for Food Packaging*: 173–202. Elesiver, Amsterdam, Netherlands.

Zhang, W., Jiang, H., Rhim, J. W., Cao, J., and Jiang, W. 2021. Tea polyphenols (TP): A promising natural additive for the manufacture of multifunctional active food packaging films. *Critical Reviews in Food Science and Nutrition*: 1–14.

Part II

Role of Nanotechnology in Shelf-Life Extension

9 Shelf Life of Foods through Nanosensors Application

Aysha Sameen[1], Amna Sahar[2], Sipper Khan[3], Tayyaba Tariq[1], and Bushra Ishfaq[4]
[1]National Institute of Food Science and Technology, University of Agriculture, Faisalabad, Pakistan
[2]Department of Food Engineering, University of Agriculture, Faisalabad, Pakistan
[3]University of Hohenheim, Institute of Agricultural Engineering, Tropics and Subtropics Group, Stuttgart, Germany
[4]Food Technology Section, Post-Harvest Research Centre Ayub Agricultural Research Institute, Faisalabad, Pakistan

CONTENTS

DOI: 10.1201/9781003207641-11

9.1 INTRODUCTION

Human imagination and dreams often give rise to new science and technology. Nanotechnology was also born out of such dreams as the 21st-century frontier. Nanotechnology deals with nanomaterials that have at least one dimension ranging from 1 to 100 nm that enables the development of novel applications. Although humans were always in contact with nanoparticles (Np) throughout human history, the industrial revolution has dramatically increased this rate. The theory of nanoparticles, a "nanometer," was first proposed by a 1925 Nobel Prize Laureate in chemistry, Richard Zsigmondy. He was the first to do the microscopic study for measurement of a particles' size, like gold colloids, and coined the term nanometer to illustrate the size of particles (Zhou, 2013).

Specific properties are exhibited by certain nanoparticles depending upon their structures, such as size (1–100 nm), color, shape (cylindrical or spherical), pore size, structure (amorphous or crystalline), surface area and volume ratio, sensitivity to environmental factors (sunlight, air, heat, and moisture), surface properties including reactivity, surface charge, and surface charge density (Salavati-Niasari et al., 2008). Nanoparticles are categorized into organic and inorganic materials depending on design and development at the nanoscale, with appropriate biological, chemical, and physical properties. The physical and chemical properties of a particle are changed after reducing the particle size below a threshold, as compared to the same substance with larger particle size. It has potential applications in the field of material structure, medical diagnostics and therapeutics, energy production, and molecular computing from the last decade (Patel et al., 2018).

9.2 NANOSENSORS

Nano-foods are the foods prepared by application of nanotechnology techniques in the food industry, including processing, production, packaging, analysis, and storage. Nanomaterials enhance the health benefits of the foods through improved food security, quality, and nutrient delivery techniques. Several industries, researchers, and organizations are designing direct applications of nanotechnology by the development of innovative products, methods, and techniques (Dasgupta et al., 2015).

Nanotechnology applications can be categorized into two groups, including nanosensing and nanostructured ingredients. The baseline steps of product development ranging from food processing to food packaging are included in group nanostructured ingredients. Nanostructured ingredients can be utilized as antimicrobial agents, anticracking agents, food additives, filters to enhance mechanical strength, transporters

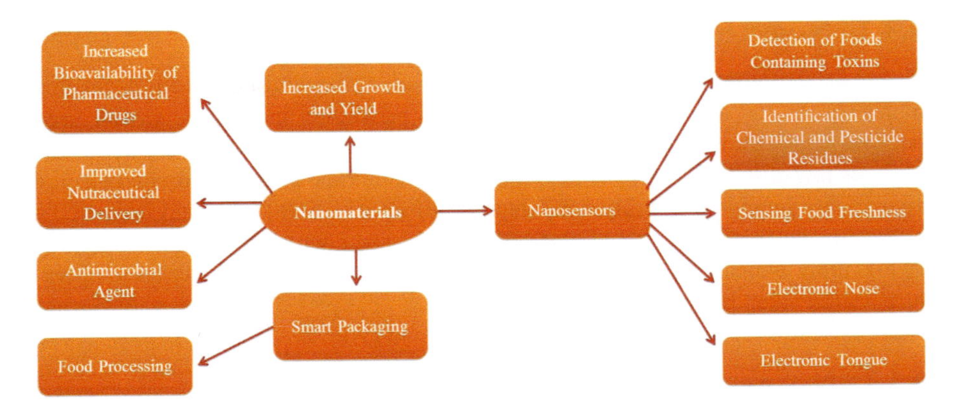

FIGURE 9.1 Applications of nanotechnology in the food industry.

for nutrients delivery, and stabilizers of packaging material, etc., whereas better safety evaluation and food quality can be achieved by using food nanosensing (Ezhilarasi et al., 2013). Different applications of nanotechnology in the food industry are shown in Figure 9.1.

9.3 CLASSIFICATION OF NANOMATERIALS

Nanomaterials are being widely studied and applied in the food industry. They exhibit different properties as compared to the large particles with the same chemical composition prepared by different methods. Two forms of nano foods are present, depending on their application; nano insides (food additives) and nano outsides (food packaging) (Ranjan et al., 2014). The following classification is based on the composition of nanomaterials.

9.3.1 Nanoparticles

The physicochemical and biological properties of nanoparticles make them suitable for pharmaceutical applications. The pharmaceutical industry uses nanoparticles as delivery systems to synthesize hydrophobic drugs. Particle size affects the bioavailability of encapsulated bioactive ingredients. Food ingredients will be more readily available to the body if the particles containing them are smaller in size. Smaller particles are easily digested and penetrate the muscle layer (Erfanian et al., 2015; Heo et al., 2016; Sessa et al., 2014; Shaikh et al., 2009; Tsai et al., 2011). Lipid-based nanoencapsulation technology is considered the most feasible application of nanoparticles (McClements, 2015). High-pressure homogenization and microfluidization are some of the widely used nanotechnologies to produce lipid-based nanoparticles. On the other hand, it is difficult to produce some protein and lipid-based nanoparticles economically on a commercial basis although they are good at their functional performance (Fathi et al., 2012).

9.3.2 Nanoclays

Nanomaterial-based food packaging has replaced conventional food packaging materials for several reasons. Nanoclay's food packaging material is widely used because of its motorized and thermal properties. One thing that makes it more feasible is its low cost. For example, nanocomposite membrane loaded with vinyl alcohol appreciably improved stability with water permeance of 6500 GPU and improved a selectivity value of 46 (Jose et al., 2014). Woven carbon fiber loaded with 3% nanoclay significantly increased interlaminar fracture strength, and glass transition temperature was elevated by 6°C (Gabr et al., 2015).

9.3.3 Nanoemulsions

Nanoemulsions are used to distribute nutrients and nutraceutical products through food. It is a colloidal dispersion comprised of liquids that are immiscible to each other, normally stabilized with the help of a suitable emulsifying agent. The emulsifying agent stands in the interface of these two immiscible layers (oil and aqueous layer). Spherical nanoparticles of one of the liquids are distributed in a continuous phase of the other liquid. The non-polar end of the emulsifier is directed toward the oil phase, while its polar end is directed toward the aqueous phase. The emulsion system is known as "oil-in-water" if lipid droplets are suspended in a continuous aqueous phase or "water-in-oil" if the aqueous phase is involved in the lipid phase (Joye and McClements, 2014; McClements and Xiao, 2012).

9.3.4 Nanolaminates

Nanotechnology has developed various techniques to construct unique laminate coatings which can be used in the food industry to increase the value of food items. Chemical or physical bounding of nanometer dimensions is done with two or more layers of nanolaminate material. Now, polysaccharides, proteins, and lipids are used to construct desirable nanolaminates (Ravichandran, 2010). They can offer extra advantages over traditional packaging substances and feature many essential programs in the food industry. Nanolaminates as edible coatings are integrated on a broad diversity of foods, which include fruits, vegetables, meats, bakery products, and chocolates; such coatings may act as a protective cover for gas, lipids, and moisture. Moreover, they can be used as carriers or delivery system of practical additives, such as colors, flavors, antioxidants, and antimicrobial components, and they provide improve the textural properties of many food items (Ranjan et al., 2014).

9.3.5 Nanocapsules

Nanocapsules are small-sized vehicles used to enhance drug delivery, absorption of food, and bioactive compounds. The diverse application of natural polymers or lipid nanocapsules is done in food and the most commonly used is casein micelle (CM) (Chen et al., 2006). Nanocapsules can be retained in the intestinal epithelium of the gastrointestinal tract due to their bioadhesive property. Sensory properties of food products can be enhanced by the incorporation of nanocapsules in sprinkled or mucosal surfaces, solid foods, or liquid substrates without affecting their sensory qualities (Qureshi et al., 2012).

9.3.6 Nanotubes

Nanotubes are numerous globular proteins that function as enzyme immobilization techniques to retard microbial growth in food commodities. The protein source can be used to synthesize nanotubes; for example, a milk protein named alpha-lactalbumin. The pasteurization of nanostructures can be carried out at 72°C for 40 seconds to increase the capacity of an encapsulating agent (Singh, 2016).

9.3.7 Nanofibers

An electrospinning technology is used to fabricate nanofibers where a spinneret with a small capillary opening is allowed to pass a solution in which a strong electric-powered area is implemented to provide thin, stable polymer strands of nanofibers. These nanofibers are suitable for the structural milieu for synthetic foods and the growth of certain bacterial cultures due to their diameters (10 to 1000 nm). Nanofibers have limited application in the food industry as they are not composed of meal-grade substances (Shukla, 2012).

9.4 NANOSENSORS

Sensors are defined as the devices used to detect, quantify, or locate matter or energy by giving a signal for the measurement or detection of a chemical or physical property to which the device responds. Most sensors contain two components: a transductor and a receptor. (Lee and Rahman, 2014). Nanosensors are nanotechnology-enabled sensors grouped based on the following properties: whether the sensitivity of the sensor is on the nanoscale, the size of the sensor, or whether the spatial interaction distance between the object and sensor is given in nanometers. Broadly, it is a device that eventually converts a physical stimulus, including luminescent, thermal, electric, or magnetic effect, into an electrical signal. There are some crucial parameters needing to be considered, including resolution, selectivity, sensitivity, calibration characteristics, and stability, during their functional operation in any field of the response (Khanna, 2008).

Nanoscale technologies are regarded as the next generation of biosensors for three major reasons. First, the limits of detection (LOD), assay specificity, and sensitivity are improved. Second, sample throughput is increased, and third, the cost or complexity of the assay is reduced (by the detection through "in situ" performance). Nanosensors present a small area of sensing surface and, therefore, they need less analyte to trigger a response (Driskell and Tripp, 2009). A bacterium is at least an order of magnitude bigger in size compared to nanomaterial and, therefore, a noteworthy number of nanomaterials are available to bind each bacterial cell, and as a result amplification of signals will be carried out (Gilmartin and O'Kennedy, 2012). Smaller sensing areas mostly allow the fabrication of higher-density arrays, and this property boosts the number of analytes (e.g., biomarkers or pathogens) that are easily detected through a single test without increasing the requirements of the sample. Nanosensors are also used to eliminate or reduce the steps of sample processing, thus helping to reduce the cost and assay complexity. Moreover, it is critical to note that many nanosensor devices that are self-described as "label free" often need an element for recognition, like an antibody for oligomers for selective sensor detection. It is obligatory to set a balance between selectivity and complexity so that the designed nanosensor contains affordable cost and fulfills its intended purpose (Kuswandi et al., 2017).

9.5 CLASSIFICATION OF NANOSENSORS

Nanosensors can be classified according to their structure, energy source, and applications (Figure 9.2).

9.5.1 Classification Based on Structure

Nanosensors are categorized into three types based on structure: optical nanosensors (measure amplitude, energy, polarization, decay phase, and delay time), electromagnetic nanosensors (measure mass transduction and electrochemical mechanisms), and mechanical nanosensors (devices that measure chemical and physical characteristics in a nanoscale region) (Agrawal and Prajapati, 2012).

9.5.2 According to Energy Source

Based on energy source, nanosensors are classified into two groups, including active nanosensors that require an energy source for working, such as a thermistor; and passive nanosensors which do not need any energy source, such as a piezoelectric and thermocouple sensor (Lim and Ramakrishna, 2006).

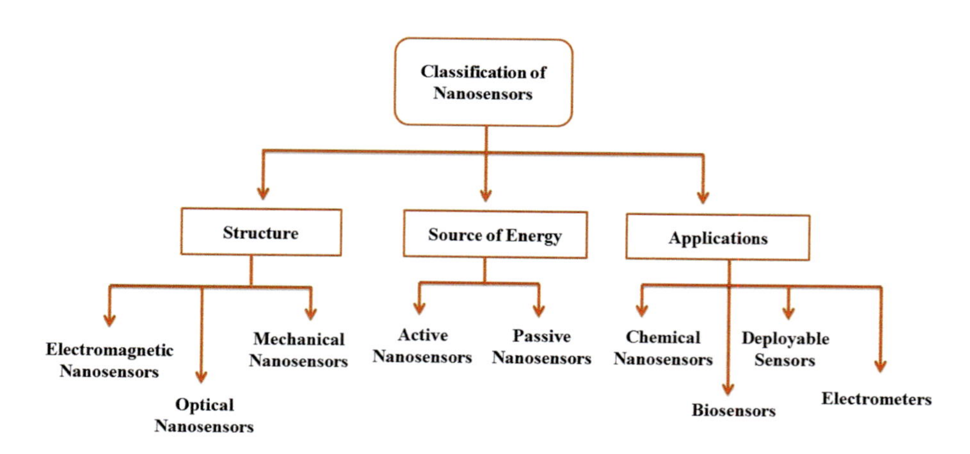

FIGURE 9.2 Classification of nanosensors based on structure, energy source, and applications.

9.5.3 Classification Based on Applications

According to application, there are four types of nanosensors, including chemical nanosensors (made up of electronics and capacitive cantilevers, and are capable of detecting single biological and chemical molecules), deployable sensors (utilized in national security and military applications), electrometers (comprised of torsional mechanical resonators), and biosensors (made of polymers to detect DNA, cancer, etc.) (Saini et al., 2017).

9.6 DETECTION OF FOODS CONTAINING TOXINS

Food contamination is caused by a variety of factors, including naturally occurring toxicants caused by bacteria and the overuse of chemicals, veterinary medications, and pesticides used during food production. Chemical and contaminant detection in food items is a requirement for avoiding negative public health consequences. As a result, in the food industry, process analysis, and quality control of food products are critical. Chemical and biological contaminants in food must be detected because, unlike physical contaminants, they cannot be seen directly. For diverse food applications, biological and chemical contaminants must be detected, in addition to sugar, alcohols, amino acids, flavors, and sweeteners (Kuswandi et al., 2017).

Chemical and biological pollutants are detected by chemical sensors and bionanosensors, respectively. Pathogens and toxicants are biological contaminants, whereas pesticides, veterinary medications, and food additives are chemical contaminants. The major sources of biological contaminants are natural toxicants, such as aflatoxins and ochratoxins, as well as foodborne pathogenic bacteria, such as *Salmonella*, *Campylobacter*, *E. coli*, *Listeria monocytogenes*, *Bacillus cereus*, *S. aureus*, and other bacteria that are produced during the production, distribution, and preservation of food (Murphy et al., 2006).

One of the ways to detect bacterial toxins in a food product is Enzyme-linked Immunosorbent Assay (ELISA). It can only be utilized for detecting and not for observing the functional activity of toxicants. DNA probes are another approach for detecting bacteria's toxigenic action. Although this method has been demonstrated to be highly accurate and sensitive, a positive result does not mean that harmful genes have been produced in bacterium strains. Nanomaterials are manufactured in various sensors to improve food safety and quality to overcome these restrictions. The development of cost-effective and efficient nanosensors that can replace traditional methods for quality measurement in the food sector, such as chromatography and spectroscopic techniques, is in high demand. Because just a few nanosensors have been discovered to be beneficial for food processing and quality control, ongoing efforts are needed to fabricate low-cost, highly sensitive nanosensors and nanobiosensors that can be used in real conditions. Unwanted chemical substances can be found as byproducts of food synthesis and distribution as well as in polluted environments and other natural sources. High-performance liquid chromatography (HPLC), gas chromatography (GC)/MS, ELISA, liquid chromatography/mass spectroscopy (MS), and capillary electrophoresis are all commonly used techniques for detecting chemical pollutants. These procedures require the use of expensive equipment and

TABLE 9.1
Nanomaterial-Based Detection of Food-Contaminating Toxins

Type of Nanoparticle	Method/Technique	Analyte	Reference
Gold	Electrochemical: enzyme-linked immunosorbent assay, immunochromatographic and cyclic voltammetry	Brevetoxins and botulinum neurotoxin type B	Chiao et al. (2004); Zhou et al. (2009)
Quartz, iron oxide nanopipettes	Enzyme-linked immunosorbent assay, ion nanogating, and immunoassay	Mycotoxin: HT-2 and zearalenone	Actis et al. (2010); Mak et al. (2010)
Superparamagnetic, iron oxide, and gold	Enzyme-linked immunosorbent assay and immunoassay	Aflatoxin M1 and aflatoxins B1	Mak et al. (2010); Radoi et al. (2008); Sharma et al. (2010); Wang et al. (2011); Xiulan et al. (2005); Zhang et al. (2013)
Gold–graphene oxide–ionic liquid, zinc oxide, and cerium dioxide	Electrochemical: cyclic impedance and voltammetry	Ochratoxin-A	Ansari et al. (2010); Kaushik et al. (2009); Norouzi et al. (2012)

highly trained workers. Despite their high accuracy and sensitivity, these approaches have limited practical application. In this scenario, nanosensors and nanobiosensors could be a convenient solution for providing accuracy and sensitivity (Kuswandi et al., 2017). Different nanomaterials used for the detection of food-contaminating toxins are explained in Table 9.1.

9.7 DETECTION OF FOODS WITH CHEMICAL AND PESTICIDE RESIDUES

The most essential features of healthy food are safety and quality, and these have universally become an indispensable feature of human life attracting broad public attention. Consumer interest in food products free from adulteration and hazardous chemicals is increasing day by day due to growing accessibility and awareness among consumers. Chemicals are added either intentionally or unintentionally into foods, either from natural sources, environmental pollution, or through food processing, distribution, and storage. The excessive usage of chemicals ranging from agrochemicals (pesticides and veterinary drugs) to chemical and industrial hazards such as heavy metals, melamine, dioxins, and acrylamide has deleterious effects on human and animals health (Kuswandi et al., 2017).

Conventionally, various techniques are used on site for chemical and pesticide residue analysis, such as high-pressure liquid chromatography, gas chromatography, mass spectrometry, atomic absorption spectrometry, inductively coupled plasma mass spectrometry, and inductively coupled plasma atomic emission spectrometry. These devices are very complex to use and have the additional disadvantages of requiring a skilled person and intricate sample preparation, as well as being high cost and time-consuming (Umapathi et al., 2021; Zhao et al., 2018). For those reasons, there is a need for the development of a cheap and stable nanosensor system for rapid detection of chemical and pesticide residue as a tool to ensure the safety and health of consumers. In the literature, there are currently different sensing strategies adopted by scientists and engineers to meet the goals of food safety, such as colorimetric sensors, electrochemical sensors, and optical sensors.

Colorimetric sensing strategies are commonly adopted for the on-site detection of specific chemicals. It can be done with the naked eye owing to its easiness, is inexpensive, and does not require trained persons. Most importantly it is a non-labeled detection method with good sensitivity and selectivity, and with non-specialized equipment requirements (Chawla et al., 2018; Liu et al., 2019; Patel et al., 2019; Singh et al., 2020; Sulaiman et al., 2020).

For the detection of chemicals such as melamine, metal nanoparticles are applied. Melamine is a derivative of a trimer of cyanamide and triazine with high nitrogen content and used in food products and animal feed. Gold nanoparticles were used by Wu et al. (2015) for their detection based on energetic transition. Likewise, Kumar et al. (2016) developed a nanosensor from gold nanoparticles stabilized with sodium citrate. An aggregation of particles gave a visible color change for analysis of a milk sample. Colorimetric methods using nanomaterial gold nanoparticles stabilized with polyethylene glycol are used for melamine detection in processed raw milk (Chen et al., 2019), and iron detection is done by bathocuproine disulfonic acid (BCS)-intercalated MgAl-LDH nanosheets in beverages (Hamid and Fat'hi, 2018). Dibutyl phthalate traces in milk and red wine are detected by metal-organic frameworks (CuMOF) (Zhu et al., 2018). Fluorancese methods are used for the detection of benzoic acid and total paraben content in beverages. Detection is by quantum dots (mercaptosuccinic acid capped cadmium telluride) (Prapainop et al., 2019), and ascorbic acid in fruit juices is detected by using ratiometric fluorescent MnO_2 nanosheet (Lyu et al., 2019).

For the first time, simultaneous detection of bisphenol A and sudan I in tomato paste, ketchup sauce, and chili powder was done by using an electrochemical sensor based on molybdenum tungsten disulfide nanoparticles modified screen-printed electrode ($MoWS_2$/SPE) with a detection limit of 0.01 μM (Ghazanfari et al., 2021). The harmful food dyes (fast green and metanil yellow) were simultaneously detected in fruit juice by the development of a glassy carbon electrode (GCE), of which the surface was adapted with calixarene and gold nanoparticles. This nanosensor was tested by electrochemical impedance spectroscopy, cyclic voltammetry, and differential pulse voltammetry (Shah, 2020).

However, there is a high residual level of antibiotics in animal-derived food owing to the exploitation of antibiotics in the breeding industry. The most frequently used antibiotic is tetracycline and its derivatives, and its detection was achieved by carbon dots due to the fluorescence quenching effect of tetracycline with a limit of detection of 7.5 nM (Yang et al., 2014). A similar assay was applied for the detection of doxycycline in raw milk (Song et al., 2017). The detection of hydrogen peroxide was done by an electrochemical sensor (amperometric) using carbon nanotubes and nano iron oxide for accessing the milk quality (Thandavan et al., 2015). The monitoring of boric acid by localized surface plasmon resonance sensor was done using gold nanoparticles (Morsin et al., 2012). Formaldehyde in fish samples was done by an electrochemical based on chitosan deposition and gold nanoparticles on a glassy carbon electrode with a detection of a limit of 0.1 ppm (Noor Aini et al., 2016). Recently, Tripathy et al. (2019) have established a smartphone-based by electrospun halochromic nanofibers miniaturized milk adulteration detector, paper-like material made up of nylon nano-sized fibers.

The most frequently observed heavy metals in food are Cu^{2+}, Hg^{2+}, and Fe^{3+}, which are present in water or soil. These are not biodegradable by microorganisms (Fan et al., 2020) and enter the food chain when absorbed by plants and livestock, and pose a serious health effect (Gori et al., 2019). Carbon quantum dots, rare earth metal ions, quantum dots, and ratiometric sensors have been prepared by combing organic fluorescent dyes for detecting Cu^{2+} in fruit and vegetable samples (Rao et al., 2016). Moreover, the single-walled carbon nanohorns decorated screen-printed electrode displayed good recoveries of Cd^{2+} and Pb^{2+} ions in samples of milk and honey elaborates how the manufactured electrochemical sensor have excellent practicability in applications food safety (Yao et al., 2019). Carbon quantum dots detect Cu^{2+} in fruit juices (Zhao et al., 2018) while (He et al., 2020) produced *N*-doped carbon dots from citric acid are used for the detection of Hg^{2+} in seafood. For food safety, the toxic metal Hg^{2+} in black tea was detected by chemosensor rhodamine complex with upconversion-luminescent nanoparticles ($NaYF_4$: Yb, Er) (Annavaram et al., 2019).

In the past, pesticides were used only to protect crops from insects, pests, fungi, and weeds before or after harvesting. Nowadays, there is a trend to increase the yield of agricultural products by extensive utilization of pesticides (Narenderan et al., 2020; Singh et al., 2020) that not only accumulate in the food and environment in the form of residues but also is a serious threat for non-targeted species, such as humans and animals, and cause serious health disorders (Tümay et al., 2021). Currently, in food products, the following colorimetric sensing approaches are used based on the recognition of color alteration signals allied with pesticide residues:

i. nanoparticle aggregation
ii. lateral flow immunoassays (LFAs)
iii. label-free hydrogel particles
iv. integrated digital readers
v. portable and paper-based field devices

All these approaches are based on qualitative estimation of analytes, except integrated digital readers that not only quantify but also capture the digital images for study via webcam, smartphone camera, or digital camera (Umapathi et al., 2021).

i. **Nanoparticle aggregation:** The nanoparticle aggregations generally recognized worldwide by scientists are gold (AuNPs) and silver (AgNPs) nanoparticles. In potato, wheat samples, and water, the residues of paraquat and thiram fungicide were identified by CN-DTC-AgNPs colorimetric probes (Rohit and Kailasa, 2014), the citrate-capped silver nanoparticle (Cit-AgNP) aggregation were used for triazophos concentration in apple and water samples. Recently, Hu et al. (2019) developed a colorimetric chemosensor for monitoring dimethoate pesticides based on the inhibition of peroxidise mimicking the activity of AuNPs with the limit of detection of 4.7 µgL-1.

ii. **Lateral flow immunoassays (LFAs):** The LFAs, which confirm the absence or presence of targeted analyte, works on the principle of the law of mass action and self-diagnostic. Advancement of this technology from dipstick (AuNPs) (Lisa et al., 2009) to digital platform (2D) platinum nanoparticle-anchored nickel hydroxide (Pt-Ni$(OH)_2$) nanosheets which were augmented with two way LFAs (Cheng et al., 2019) and AuNP-based immune chromatographic assay (Li et al., 2018). The detection of 6-benzyl amino purine and chlorpyrifos residues in bean sprouts, grain, and lettuce leaves, respectively, by immune chromatographic assay within 10 min (Kim et al., 2011b). The LFAs technology is expected to be commercialized soon in the future.

iii. **Label-free hydrogel particles:** Several methods of colorimetric sensing based on hydrogel technology have been developed. These gels having the characteristics of stimulating a polymer network reaction, consequently developing easily and rapidly a periphery of a definite reaction system in a solid state. A portable hydrogel test kit with MnO_2 nanosheets by incorporating a 3,3',5,5'-tetramethylbenzidine sensing probe was developed by Jin et al. for the detection of oxalates which was integrated by a smartphone. Another portable label-free inverse long-range macroporous hydrogel particles approach was used for methanephosphonic acid (MPA) pesticide residue detection by Huang et al. (2018).

iv. **Integrated digital readers:** The integrated digital readers offer both the qualitative and quantitative determination of the pesticide residues. The organophosphate pesticides, such as parathion, parathion methyl, fenitrothion, carbaryl, chlorpyrifos, and malathion, are rapidly analyzed by the development of an ambient light sensors dipstick and proposed an ideal approach for detection at home or in the field (Fu et al., 2019).

v. **Portable and paper-based field devices:** The paper-based colorimetric sensing recognitions are based on inhibition of the activity of enzymes such as butyryl-cholinesterase or acetylcholinesterase by the targeted pesticide analyte and serve as an excellent indicator for these contaminations (Badawy and El-Aswad, 2014; Guo et al., 2013; Kaur and Singh, 2020). Mostly recognized for the recognition of organophosphate and carbamate residues in fruits and vegetables (Sun et al., 2017), these sensors gathered attention owing to their ease of use, cheapness, portability, and ease of disposal.

Zhao et al. (2021) developed an automatic and portable dual-readout detector for quick carbamate pesticides detection, made up of 3D-printed microfluidic nanosensors and based on the agglomeration effect of AuNPs and the cross-responsive mechanism. In optical sensors, mostly quantum dots, carbon dots, nano porphyrin, gold, and silver nanoparticles are applied for pesticide residues detection (Chen et al., 2021). Recently, chlorantraniliprole and parathion detection in apple and tomato was carried out by an electrochemical sensor; a novel hybrid ferrocene-thiophene modified by a carbon nanotube. The limit of detection for chlorantraniliprole and parathion in the linear range of 0.01–7.00 µmol/L and 0.02–6.50 µmol/L were found as 8.1 nmol/L and 5.3 nmol/L, respectively (Tümay et al., 2021). Regarding the application of optical sensors, including colorimetric, fluorescent, and surface-enhanced, Raman scattering–based systems using various materials, like metallic nanoparticles, silicon dioxide, magnetic nanoparticles, carbon quantum dots, nanohorns, metal oxide nanoparticles, and quantum dots, have been successfully used to detect different food contaminants, such as metallic species and pesticides in foodstuff (Yadav et al., 2019).

9.8 NANOSENSORS FOR DETECTION OF FOOD FRESHNESS

Freshness indicators have the ability to provide direct product quality information resulting from chemical changes or microbial growth within food items. To monitor the freshness of the food, product nanosensors are gaining popularity among the population as it is not possible to monitor the freshness of the individual package through laboratory-based testing (Jiang et al., 2015; Realini and Marcos, 2014).

These freshness indicators are correlated to the quality of the product and were usually applied in the form of labels on the container. Classically, these indicators emphasize the recognition of the primary kind of changes in freshness, like pH and gas composition, and communicate a response, generally, in the form of color which can be easily detected and allied to the freshness of the food commodities (Mohammadalinejhad et al., 2020; Roy and Rhim, 2021; Taherkhani et al., 2020). These changes can be related to the metabolic activities of microorganisms and the formation of volatile nitrogen and sulfur comprising compounds, organic acids, glucose, CO_2, amines, and ethanol. There are two approaches to detecting these indications; through direct (biosensors) and indirect methods (color indicators) (Sohail et al., 2018). Recently, electrospun nanofibers have been employed owing to their excellent properties as food freshness indicators in different food products (Forghani et al., 2021).

9.8.1 Quality Assessment Due to Improper Storage

The quality of food products deteriorates due to improper storage, which may be due to fluctuations in storage temperature and oxygen exposure. This fluctuation leads to a deleterious effect on the food products. Traditionally used time and temperature indicators are expensive and lack flexibility in programming (Kumar et al., 2017). With the advancements in the field of nanotechnology, these limitations are overcome by nanoparticle-based time and temperature indicators. These indicators serve as external markers on packaged products and are explicated either by smartphone or the naked eye. The nanocomposites of plasmonic nanomaterials exhibited sharp color changes when coming in contact with the different environmental conditions, which are usually preferred for the development of these time and temperature sensors (Mohammadpour and Naghib, 2021).

In this context, the colorimetric sensing of volatile amines in meat products is done by fresh tag indicator and label sensors. Fresh tag indicators are preferable for monitoring fishery products and the label sensor for real-time checking of freshness of broiler chickens (Kuswandi et al., 2014). The presence of trimethylamine indicates the loss of freshness in meat and is normally produced by the metabolism of animal proteins. For trimethylamine detection in fish samples, nanocomposite sensors of tin dioxide-zinc oxide were effectively employed by Zhang and Zhang (2008). With the advancements in this field, there was the development of visual sensors (change in color) to monitor the freshness of fish based on the presence of hypoxanthine; that is, the end product of purine metabolism that is formed during the decay of animal meat (Chen et al., 2017). The chitosan-gold nanoparticles (Wang et al., 2018), and gelatin-gold nanoparticles (Wang et al., 2015) serve as a frozen detector in those products in which quality is affected by freezing temperatures. Similarly, Albelda et al. (2017), for meat freshness identification, developed a graphene-titanium dioxide-based nanocomposite. In meat of microbiological origin, SensorQ indicators are employed for recording the development of biogenic amines (Chun et al., 2014).

In fruits, the first intelligent sensor used RipeSense that changes color according to ripeness. As the fruit ripens, the aroma is released to initiate the reaction and changes color from red to orange and finally to yellow depending on the maturity level. Radiofrequency identification (RFI) tags are also employed for the quality assessment of food commodities based on freshness in meat, vegetables, and milk. These are integrated with other sensors, like gas and temperature sensors to monitor the freshness of pork (Sen et al., 2013), or temperature and humidity sensors for codfish (Smits et al., 2012), and O_2, and CO_2 sensors in vegetables (Eom et al., 2012). For monitoring the freshness, marketing, and distribution of packaged milk (Potyrailo et al., 2012), as well as in a modified atmosphere (Martínez-Olmos et al., 2013), this RF technology has also been recognized. Recently, self-healing nanofiber mats were introduced into the cold food supply chain as time and temperature indicators (Choi et al., 2020). Oxygen exposure of the packed food items triggers the spoilage of food due to the oxidation of antioxidants and, ultimately, degradation of freshness and quality. These exposures in packaged food items might be due to improper packaging or poor sealing or to damaging of packaging materials (Kumar et al., 2017). Recently, an oxygen sensor was developed for detecting concentrations as low as 0.5% based on methylene blue, glycerol, TiO_2, and ethylene-vinyl alcohol copolymer (López-Carballo et al., 2019). In a similar approach Khankaew et al. (2017)_ developed a "consume-within" indicator for O_2 for sensing of fresh and refrigerated foods.

9.8.2 Quality Assessment Due to Unstable Key Ingredients

For the normal functioning of numerous metabolic pathways in our body, micronutrients (vitamins, minerals) and antioxidants that are found in food are vital. These are degraded during processing and their deficiency may cause various disorders, such as carcinogenesis, cardiovascular diseases, and anemia. For the detection of their content in food, the conventional methods used are laborious and time-consuming. Recently, however, nanoparticles have been applied in food products for the detection of these ingredients. Recently, Dadkhah et al. (2021) developed a simple, efficient, and portable nicotinamide-functionalized carbon quantum dot nanosensor for on-site monitoring of vitamin B_{12} detection in triple mode (fluorescence, UV-VIS and smartphone). Vitamin B_2 quantitative detection in orange beverage, milk, and soybean milk was done by an emerging fluorescent nanomaterial of carbon nanodots that have good optical properties (Du et al., 2021). Earlier, folic acid, ascorbic acid, and vitamin detection had been reported by ionc liquid filled single-walled carbon nanotubes; multi-walled carbon nanotubes; and nickel-oxide nanoparticles in wheat flour, fruit juices, and milk (Karimi-Maleh et al., 2014; Wei et al., 2006; Xiao et al., 2008). Similarly, for essential amino acids, tryptophan, and ascorbic acid sensing was done by N-(3,4-dihydroxyphenethyl)-3,5-dinitrobenzamide modified multi-walled carbon nanotubes (Ensafi et al., 2012). In the past, numerous types of nanosensor in conjugation with iron, nickel, silver, nickel-platinum, gold, cuprous oxide, zirconium dioxide, tin dioxide, Prussian blue-gold, and chitosan (Chen et al., 2011; Filippo et al., 2009; Liang and Mu, 2008; Ye et al., 2004) have been employed for monitoring the quality of stored food products for hydrogen peroxide sucrose, fructose, glucose, D-sorbitol, citric acid L-glutamic acid, succinic acid, and L-malic acid (Vermeir et al., 2007; Verstrepen et al., 2004). Recently, molecularly imprinted ratiometric fluorescence nanosensors were employed for visual detection of folic acid in fresh fruits and vegetables (broccoli, spinach, oranges, and tomatoes) (Li et al., 2020). A nano-manganese oxide–based spectrophotometric method was developed by (Can et al., 2020) for indirect measurement of antioxidants, with a limit of detection in the range of 1.23×10^{-9} to 1.71×10^{-7} mol L^{-1}. The method was tested for the detection of hydrophilic and lipophilic antioxidants in green tea and orange juice.

9.9 NANO BARCODES FOR FOOD AUTHENTICITY

Globalization of the food trade with high consumer demand has intensified adulteration of some foods. Adulteration is highly relevant for spices, herbal products, powders, and the seafood industry due to morphological similarities resulting in ease of camouflaging the adulterant. This results in higher public health implications along with growing economic impact. For instance, research claims that 50% of the commercialized fish products are mislabeled in European markets. There was global economic impact of $10 billion losses through food frauds (Valentini et al., 2017). Results of the research suggest DNA barcoding could help identify food products and minimize fraud (Galimberti et al., 2019; Valentini et al., 2017). Barcoding is based on sequencing, by acquiring molecular data at a low-analysis cost with results supported by comprehensive reference sequence libraries, including the barcode of life database (BOLD). Furthermore, new regulatory directives on food labeling, including European Regulation N° 1169/2011, will inevitably force the national institutions to use molecular DNA-based tools to ensure food authenticity. Some technical drawbacks of DNA barcoding, including the need for high-quality DNA extract of food matrices (500 bp) and the need for specialized laboratories with skilled personnel, have escalated the need for nanobarcodes. Use of NanoTracer involves the usage of naked-eye molecular authentication of any food, with low-cost processing and instrumentation involved in the process (Shikha et al., 2017; Tousch and Charlot,

2017). The portability and sequencing-free ability of NanoTracer have simplified the standard barcoding process, easing the analysis (Handy et al., 2011).

Nanosensors have unique attributes to improve food security by detecting pathogenic bacteria and food contaminants, including toxins, adulterants, vitamins, dyes, fertilizers, and pesticides, with variations in taste and smell. Therefore, the invisible nanobarcodes can be used to assess the product authenticity and apply brand protection, along with monitoring time-temperature and oxygen indicators. Other than stable key ingredients, proper storage is also imperative to maintaining the quality of food products. Effective packaging will stop any contaminants, humidity, or anything else from entering, thereby enhancing the shelf life. However, conventional packaging (passive) failed against bioterrorism and food security as compared to intelligent packaging systems that not only protect the shelf life but also contribute to the name of the brand. It uses digital barcodes associated with inventory systems and sensors sensitive to time-temperature and gas indicators along with biosensors composed of nanoparticle-based tags to verify the origin of the product (Kumar et al., 2017).

The encoding system based on nanoparticle and nanodisk codes has been used with Raman scattering–producing disk pairs coupled up with Raman microscope. Previous studies have indicated the usage of different nanobarcodes, including DNA nanobarcodes that use dot blotting, flow cytometry technique, and fluorescence microscopy to determine the presence of Anthrax, Ebola, severe acute respiratory syndrome virus, tularemia, and *E. coli.* (Li et al., 2010). Another nanobarcode used was gold-nickel based with scanning confocal Raman spectroscopy for detection of DNA (Qin et al., 2006; Qin et al., 2007). Silver-gold nanobarcodes also employ the same technique but detect barcodes and DNA as per the research (Banholzer et al., 2010; Liusman et al., 2012). Additionally, with different metal compositions and the functions of different chromophores, the function of nanodisks can be further enhanced. Fluorescent poly(p-phenylene vinylene)–based barcode nanorods are used primarily for labels on single packages. Invisible tags with 7400 and 68,000 unique barcodes are also reported in previous research (Shikha et al., 2017). Therefore, nanoparticle-based robust nanobarcodes are exhibiting improved results as compared to conventional counterparts (Kumar et al., 2017).

9.10 ELECTRONIC NOSE

For automating smell detection, the artificial nose has been effective and is considered quite similar to the human nose in efficiency in accurately matching the food with its smell, as shown in the Figure 9.3. This technology is designed similarly to the human olfactory system, and is equipped with electronic chemical sensors arranged in an array responsible for recognizing simple or complex odor (Song et al., 2013).

Low-cost electronic systems have been used in different ways based on the artificial olfaction system and chemiresistive sensors present inside (Chiu and Tang, 2013). Three different categories are used, mainly including metal oxide semiconductor (MOS) film-based, conducting polymer (CP), and piezoelectric crystal sensors (bulk acoustic wave and surface acoustic wave) (Tan and Xu, 2020). Different nanoparticles have been utilized to produce the electronic nose, including zinc oxide

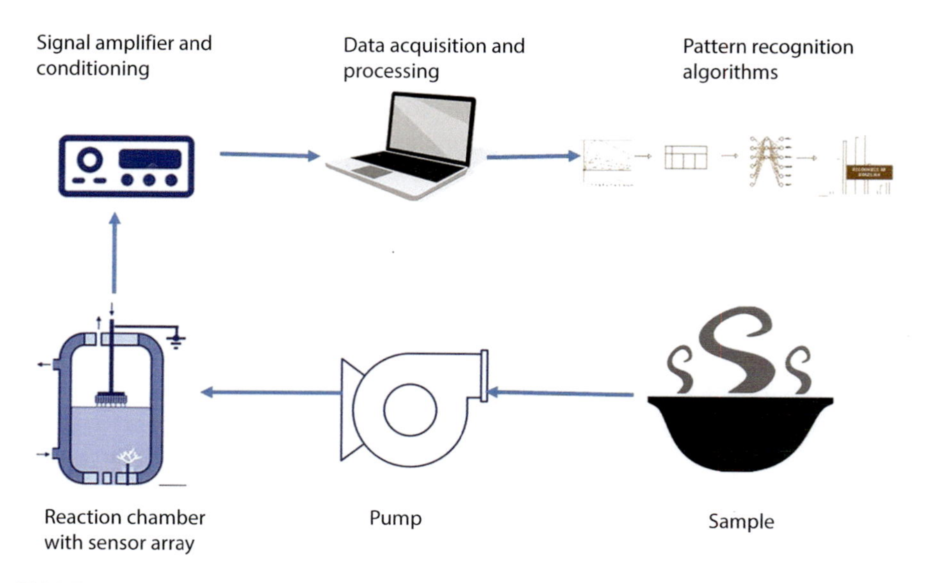

FIGURE 9.3 Schematic representation of e-nose.

effective in determining 17 different commercial kinds of Chinese vinegar attributes by gauging the electrical resistance (Zhang et al., 2006). Similarly, tungsten oxide-tin oxide and silver-tin dioxide also use the electrical resistance measurement technique to determine the ethylene gas analyte (Baik et al., 2010; Pimtong-Ngam et al., 2007). In the case of ethanol gas detection as an analyte, silver-tin oxide, molybdenum trioxide, cobalt tetraoxide, gold-tungsten oxide, tin dioxide, platinum-tin oxide, zinc oxide, and magnesium-zinc oxide nanowires have been used by employing different measurement techniques and methods, including electrical resistance measurement and dynamic headspace extraction analysis, linear discriminant analysis, and electrode less monolithic multichannel quartz crystal microbalance. Ethylene gas detection is effective as a monitor of the harvesting, storage, and processing of fruits and vegetables, but with its excessive release deterioration it can also impact the quality (Calderer, 2012; Ivanov et al., 2004; Kumar et al., 2017; Vickers, 2017; Wang et al., 2014a). For benzaldehyde and olive oil, gold nanoparticles are used with indirect competitive immunoassay and surface plasmon resonance. These gold Nps are used to alter the balance of sensors in an array of quartz crystal used for the formation of e-nose in the detection of virgin, extra virgin, and non-edible lampante olive oil (Del Carlo et al., 2014; Gobi et al., 2008). Additionally, silver carbon nanotubes are used for amyl butyrate, trimethylamine, sucrose, phenylthiocarbaminde, and propylthiouracil detection using field-effect transistors technique (Song et al., 2014; Kim et al., 2011a). Moreover, carboxylated polypyrrole nanotube and carbon nanotubes are also used for helional, hexanal, L-monosodium glutamate, phenylthiocarbamide, and propylthiouracil utilizing electrical source meter and field-effect transistor technique, where they sense gaseous odorants up to parts trillion concentration (Lee et al., 2012; Lee et al., 2015). In the manufacturing industry, e-nose was also used for examining the quality aspects of both the final products and the initial raw materials for their maturity stage, processing, authentic form, freshness level, and the microorganisms present around responsible for the pathogenic examination (Di Rosa et al., 2017). Some applications listed include identification of pork (Kim et al., 2014; Li et al., 2016; Nurjuliana et al., 2011; Tang et al., 2013), cheese (Branciari et al., 2012; Cevoli et al., 2011; Di Rosa et al., 2017; Nateghi et al., 2012), eggs (Di Rosa et al., 2017; Wang et al., 2014b), fish (Du et al., 2015; Zhang et al., 2012), mutton (Tian et al., 2013), butter (Lorenzen et al., 2013), ham (Laureati et al., 2014), yogurt (Li et al., 2014), honey (Huang et al., 2015), poultry (Song et al., 2013; Xiao et al., 2015), and beef (Kodogiannis, 2016; Mohareb et al., 2016).

9.11 ELECTRONIC TONGUE

The electric tongue concept was first proposed by Hayashi et al. (1990) as a "taste sensor" that contained multi-channel electrodes with transducers composed of polymer immobilized lipid membranes that can replicate the gustatory receptors of the human body. The tastes are perceived as sour, salty, sweet, bitter, and umami. Additionally, pungent and astringent are partially associated with pain sense (Magro et al., 2019; Tahara et al., 2013). Sweetness is attributed to the presence of sucrose, glucose, and/or artificial sweeteners. Saltiness is due to the cations mainly represented by sodium ions, while sour taste is due to the dissociated H^+ ions from acetic acid, hydrochloric acid (HCl), and citric acid, etc. Bitter taste is linked to the presence of caffeine, theobromine, quinine, and humulone while umami flavor is associated with disodium inosinate (IMP), monosodium glutamate (MSG), and disodium guanylate (GMP). Similarly, astringent taste is mainly attributed to compounds associated with tannins, and pungent taste having capsaicin, allyl isothiocyanate, and piperine (Toko et al., 2016).

The e-tongue includes non-specified chemical sensors arranged in an array form designated on basis of cross selection. On chemometric processing, it enables quality and quantity differentiation of solutions having multiple species (Ha et al., 2015). A schematic representation of e-tongue is given in the Figure 9.4.

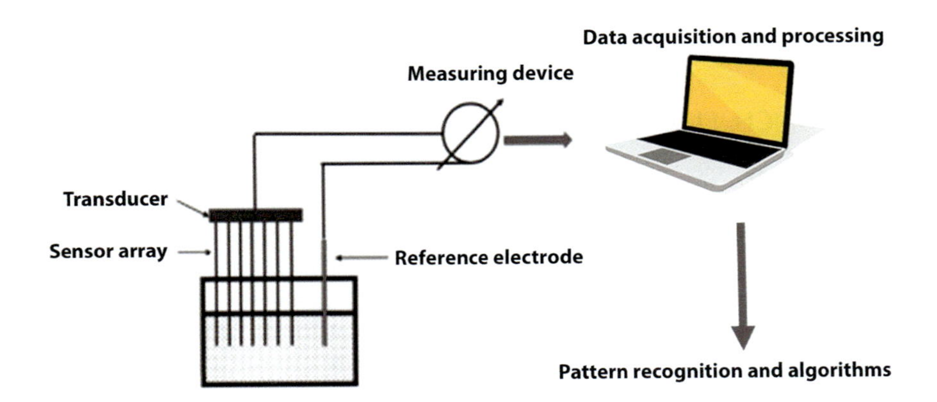

FIGURE 9.4 Schematic representation of e-tongue.

Various sensors are used in the e-tongue, including electrochemical (voltametric, potentiometric, impedimetric, amperometric, conductimetric), optical, or enzymatic/biosensors, etc. Potentiometric sensors are used in the majority of instruments with ion-selective electrodes (ISEs) present in the highest numbers (Ciosek and Wróblewski, 2007). Many applications of the electronic tongue have been recorded, including the identification of residues of antibiotics in milk (Wei and Wang, 2011), in discriminating contaminated skimmed milk powder with urea and melamine (Hilding-Ohlsson et al., 2012), with determination of quality and shelf-life (Wei et al., 2013b), for brand classification (Yu et al., 2015), and for urea quality (Li et al., 2015). Similarly, pork is employed for monitoring chemical, physical, and microbiological changes during the cold storage process (Gil et al., 2011). In poultry, it is used to detect the flavoring constituents of hydrolyzed bones of chicken during Millard reaction using potentiometry. Honey can be easily traced for floral, botanical, and geographical origin (Drivelos et al., 2021; Sobrino-Gregorio et al., 2020). Similarly, the e-tongue has also been used for different salt-based curing in ham (Gil-Sánchez et al., 2015) and the comparisons of umami taste peptides in the water-soluble extractions (Gil-Sánchez et al., 2015). Moreover, yogurt varieties are also evaluated with voltammetry techniques (Wei et al., 2013a). Furthermore, for beef, the electronic tongue used the same technique in flavor assessment, recognition, and chemical composition (Zhang et al., 2015) along with putrescine and ammonia detection (Apetrei and Apetrei, 2016). Its applications are also seen in the estimating stage of fresh quality (Apetrei et al., 2013) and in detection of fish quality (Ruiz-Rico et al., 2013).

9.12 REGULATORY CONCERNS (BIOACCUMULATION)

Numerous explorations have exhibited promising results to ensure food safety and quality; however, the structure reproducibility, biocompatibility, and applications of the nanosensors for biofilm is still a challenge (Pu et al., 2020). Ongoing research is also considering the migration of nanomaterials from food packaging into the food matrix owing to different shortfalls observed in analytical methods. Therefore, selectivity, accuracy, and sensitivity are highly recommended (Caon et al., 2017). These nanosensors are, significantly, detecting the low concentrations of targeted analyte or the pathogens along with multiple mechanisms in case of various targets as per the food matrix complexity (Caon et al., 2017).

9.13 CONCLUSION

Among the wide spectrum of nanoparticle applications, ongoing research studies highlight the precision required for structural reproduction and biocompatibility. A rise in food scandals has strengthened the demand for precise nanosensors that can detect pathogens and minimize the toxins associated with the food industries. As these nanosensors are classified based on their structure, energy sources, and application, further advantages are being associated with their diversity in types. Bioaccumulation or safety concerns associated with nanosensors need further addressing. Furthermore, research gaps in its utilization for biofilms need addressing with more in-depth experimental analysis.

REFERENCES

Actis, P., Jejelowo, O. and Pourmand, N. 2010. Ultrasensitive mycotoxin detection by STING sensors. *Biosensors and Bioelectronics* 26:333–337.

Agrawal, S. and Prajapati, R. 2012. Nanosensors and their pharmaceutical applications: a review. *International Journal of Pharmaceutical Science and Technology* 4:1528–1535.

Albelda, J.A., Uzunoglu, A., Santos, G.N.C. and Stanciu, L.A. 2017. Graphene-titanium dioxide nanocomposite based hypoxanthine sensor for assessment of meat freshness. *Biosensors Bioelectronics* 89:518–524.

Annavaram, V., Chen, M., Kutsanedzie, F.Y., Agyekum, A.A., Zareef, M., Ahmad, W., Hassan, M.M., Huanhuan, L. and Chen, Q. 2019. Synthesis of highly fluorescent RhDCP as an ideal inner filter effect pair for the NaYF4: Yb, Er upconversion fluorescent nanoparticles to detect trace amount of Hg (II) in water and food samples. *Journal of Photochemistry Photobiology A: Chemistry* 382:111950.

Ansari, A.A., Kaushik, A., Solanki, P.R. and Malhotra, B.D. 2010. Nanostructured zinc oxide platform for mycotoxin detection. *Bioelectrochemistry* 77:75–81.

Apetrei, I.M. and Apetrei, C. 2016. Application of voltammetric e-tongue for the detection of ammonia and putrescine in beef products. *Sensors and Actuators B: Chemical* 234:371–379.

Apetrei, I.M., Rodriguez-Mendez, M.L., Apetrei, C. and de Saja, J.A. 2013. Fish freshness monitoring using an E-tongue based on polypyrrole modified screen-printed electrodes. *IEEE Sensors Journal* 13:2548–2554.

Badawy, M.E. and El-Aswad, A.F. 2014. Bioactive paper sensor based on the acetylcholinesterase for the rapid detection of organophosphate and carbamate pesticides. *International Journal of Analytical Chemistry* 2014. doi: 10.1155/2014/536823.

Baik, J.M., Zielke, M., Kim, M.H., Turner, K.L., Wodtke, A.M. and Moskovits, M. 2010. Tin-oxide-nanowire-based electronic nose using heterogeneous catalysis as a functionalization strategy. *ACS Nano* 4:3117–3122.

Banholzer, M.J., Osberg, K.D., Li, S., Mangelson, B.F., Schatz, G.C. and Mirkin, C.A. 2010. Silver-based nanodisk codes. *Acs Nano* 4:446–5452.

Branciari, R., Valiani, A., Trabalza-Marinucci, M., Miraglia, D., Ranucci, D., Acuti, G., Esposto, S. and Mughetti, L. 2012. Consumer acceptability of ovine cheese from ewes fed extruded linseed-enriched diets. *Small Ruminant Research* 106:43–48.

Calderer, J. 2012. Important considerations for effective gas sensors based on metal oxide nanoneedles films. *Sensors and Actuators B: Chemical* 161: 406–413.

Can, K., Üzer, A. and Apak, R. 2020. A manganese oxide (MnO_x)-Based colorimetric nanosensor for indirect measurement of lipophilic and hydrophilic antioxidant capacity. *Analytical Methods* 12:448–455.

Caon, T., Silvia, M.M., and Farayde, M.F. 2017. New trends in the food industry: application of nanosensors in food packaging. In *Nanobiosensors* 773–804. Elsevier.

Cevoli, C., Cerretani, L., Gori, A., Caboni, M.F., Toschi, T.G. and Fabbri, A. 2011. Classification of Pecorino cheeses using electronic nose combined with artificial neural network and comparison with GC–MS analysis of volatile compounds. *Food Chemistry* 129:1315–1319.

Chawla, P., Kaushik, R., Swaraj, V.S. and Kumar, N. 2018. Organophosphorus pesticides residues in food and their colorimetric detection. *Environmental Nanotechnology, Monitoring & Management* 10:292–307.

Chen, L., Remondetto, G.E. and Subirade, M. 2006. Food protein-based materials as nutraceutical delivery systems. *Trends in Food Science & Technology* 17: 272–283.

Chen, S., Ma, L., Yuan, R., Chai, Y., Xiang, Y. and Wang, C. 2011. Electrochemical sensor based on Prussian blue nanorods and gold nanochains for the determination of H_2O_2. *European Food Research and Technology* 232:87–95.

Chen, X.Y., Ha, W. and Shi, Y.P. 2019. Sensitive colorimetric detection of melamine in processed raw milk using asymmetrically PEGylated gold nanoparticles. *Talanta* 194:475–484.

Chen, Z., Lin, Y., Ma, X., Guo, L., Qiu, B., Chen, G. and Lin, Z. 2017. Multicolor biosensor for fish freshness assessment with the naked eye. *Sensors and Actuators B: Chemical* 252:201–208.

Cheng, N., Shi, Q., Zhu, C., Li, S., Lin, Y. and Du, D. 2019. Pt–Ni $(OH)_2$ nanosheets amplified two-way lateral flow immunoassays with smartphone readout for quantification of pesticides. *Biosensors and Bioelectronics* 142:111498.

Chiao, D.J., Shyu, R.H., Hu, C.S., Chiang, H.Y. and Tang, S.S. 2004. Colloidal gold-based immunochromatographic assay for detection of botulinum neurotoxin type B. *Journal of Chromatography B* 809:37–41.

Chiu, S. and Tang, K.T. 2013. Towards a chemiresistive sensor-integrated electronic nose: a review. *Sensors* 13:14214–14247.

Choi, S., Eom, Y., Kim, S.M., Jeong, D.W., Han, J., Koo, J.M., Hwang, S.Y., Park, J. and Oh, D.X.J.A.M. 2020. A self-healing nanofiber-based self-responsive time-temperature indicator for securing a cold-supply chain. *Advanced Materials* 32:1907064.

Chun, H., Kim, B. and Shin, H. 2014. Evaluation of a freshness indicator for quality of fish products during storage. *Food Science Biotechnology* 23:1719–1725.

Ciosek, P. and Wróblewski, W. 2007. Sensor arrays for liquid sensing–electronic tongue systems. *Analyst* 132:963–978.

Dadkhah, S., Mehdinia, A., Jabbari, A. and Manbohi, A. 2021. Nicotinamide-Functionalized Carbon Quantum Dot as New Sensing Platform for Portable Quantification of Vitamin B12 in Fluorescence, Uv-Vis and Smartphone Triple Mode. *Uv-Vis Smartphone Triple Mode* 32: 681–689.

Dasgupta, N., Ranjan, S., Mundekkad, D., Ramalingam, C., Shanker, R. and Kumar, A. 2015. Nanotechnology in agro-food: from field to plate. *Food Research International* 69:381–400.

Del Carlo, M., Fusella, G., Pepe, A., Sergi, M., Di Martino, M., Mascini, M., Martino, G., Cichelli, A., Di Natale, C. and Compagnone, D. 2014. Novel oligopeptides based e-nose for food quality control: application to extra-virgin olive samples. *Quality Assurance and Safety of Crops & Foods* 309–317.

Di Rosa, A.R., Leone, F., Cheli, F. and Chiofalo, V. 2017. Fusion of electronic nose, electronic tongue and computer vision for animal source food authentication and quality assessment–a review. *Journal of Food Engineering* 210:62–75.

Driskell, J. and Tripp, R. 2009. Emerging technologies in nanotechnology-based pathogen detection. *Clinical Microbiology Newsletter* 31:137–144.

Drivelos, S.A., Danezis, G.P., Halagarda, M., Popek, S. and Georgiou, C.A. 2021. Geographical origin and botanical type honey authentication through elemental metabolomics via chemometrics. *Food Chemistry* 338:127936.

Du, F., Cheng, Z., Wang, G., Li, M., Lu, W., Shuang, S. and Dong, C. 2021. Carbon Nanodots as a Multifunctional Fluorescent Sensing Platform for Ratiometric Determination of Vitamin B2 and "Turn-Off" Detection of pH. *Journal of Agricultural Food Chemistry* 69:2836–2844.

Du, L., Chai, C., Guo, M. and Lu, X. 2015. A model for discrimination freshness of shrimp. *Sensing and Bio-Sensing Research* 6:28–32.

Ensafi, A.A., KarimiMaleh, H. and Mallakpour, S. 2012. Simultaneous determination of ascorbic acid, acetaminophen, and tryptophan by square wave voltammetry using N-(3, 4-Dihydroxyphenethyl)-3,5-Dinitrobenzamide-modified carbon nanotubes paste electrode. *Electroanalysis* 24:666–675.

Eom, K.H., Kim, M.C., Lee, S. and Lee, C.W. 2012. The vegetable freshness monitoring system using RFID with oxygen and carbon dioxide sensor. *International Journal of Distributed Sensor Networks* 8:472986.

Erfanian, A., Mirhosseini, H., Rasti, B., Hair-Bejo, M., Mustafa, S.B. and Manap, M.Y.A. 2015. Absorption and bioavailability of nano-size reduced calcium citrate fortified milk powder in ovariectomized and ovariectomized-osteoporosis rats. *Journal of Agricultural and Food Chemistry* 63:5795–5804.

Ezhilarasi, P., Karthik, P., Chhanwal, N. and Anandharamakrishnan, C. 2013. Nanoencapsulation techniques for food bioactive components: a review. *Food and Bioprocess Technology* 6:628–647.

Fan, H., Zhang, M., Bhandari, B. and Yang, C.-h. 2020. Food waste as a carbon source in carbon quantum dots technology and their applications in food safety detection. *Trends in Food Science Technology* 95:86–96.

Fathi, M., Mozafari, M.R. and Mohebbi, M. 2012. Nanoencapsulation of food ingredients using lipid based delivery systems. *Trends in Food Science & Technology* 23:13–27.

Filippo, E., Serra, A. and Manno, D. 2009. Poly (vinyl alcohol) capped silver nanoparticles as localized surface plasmon resonance-based hydrogen peroxide sensor. *Sensors Actuators B: Chemical* 138:625–630.

Forghani, S., Almasi, H. and Moradi, M. 2021. Electrospun nanofibers as food freshness and time-temperature indicators: A new approach in food intelligent packaging. *Innovative Food Science Emerging Technologies* 73: 102804.

Fu, Q., Zhang, C., Xie, J., Li, Z., Qu, L., Cai, X., Ouyang, H., Song, Y., Du, D. and Lin, Y. 2019. Ambient light sensor based colorimetric dipstick reader for rapid monitoring organophosphate pesticides on a smart phone. *Analytica Chimica Acta* 1092:126–131.

Gabr, M.H., Okumura, W., Ueda, H., Kuriyama, W., Uzawa, K. and Kimpara, I. 2015. Mechanical and thermal properties of carbon fiber/polypropylene composite filled with nano-clay. *Composites Part B: Engineering* 69:94–100.

Galimberti, A., Casiraghi, M., Bruni, I., Guzzetti, L., Cortis, P., Berterame, N.M. and Labra, M. 2019. From DNA barcoding to personalized nutrition: the evolution of food traceability. *Current Opinion in Food Science* 28:41–48.

Ghazanfari, Z., Sarhadi, H. and Tajik, S. 2021. Determination of Sudan I and Bisphenol A in Tap Water and Food Samples Using Electrochemical Nanosensor. *Surface Engineering Applied Electrochemistry* 57:397–407.

Gil-Sánchez, L., Garrigues, J., Garcia-Breijo, E., Grau, R., Aliño, M., Baigts, D. and Barat, J.M. 2015. Artificial neural networks (Fuzzy ARTMAP) analysis of the data obtained with an electronic tongue applied to a ham-curing process with different salt formulations. *Applied Soft Computing* 30:421–429.

Gil, L., Barat, J.M., Baigts, D., Martínez-Máñez, R., Soto, J., Garcia-Breijo, E., Aristoy, M.-C., Toldrá, F. and Llobet, E. 2011. Monitoring of physical–chemical and microbiological changes in fresh pork meat under cold storage by means of a potentiometric electronic tongue. *Food Chemistry* 126:1261–1268.

Gilmartin, N. and O'Kennedy, R. 2012. Nanobiotechnologies for the detection and reduction of pathogens. *Enzyme and Microbial Technology* 50:87–95.

Gobi, K.V., Matsumoto, K., Toko, K. and Miura, N. 2008. Highly regenerable and storageable all-chemical based PEG-immunosensor chip for SPR detection of ppt levels of fragrant compounds from beverage samples. *Sensing and Instrumentation for Food Quality and Safety* 2:225–233.

Gori, A., Ferrini, F. and Fini, A. 2019. Reprint of: Growing healthy food under heavy metal pollution load: overview and major challenges of tree based edible landscapes. *Urban Forestry Urban Greening* 45:126292.

Guo, X., Zhang, X., Cai, Q., Shen, T. and Zhu, S. 2013. Developing a novel sensitive visual screening card for rapid detection of pesticide residues in food. *Food Control* 30:15–23.

Ha, D., Sun, Q., Su, K., Wan, H., Li, H., Xu, N., Sun, F., Zhuang, L., Hu, N. and Wang, P. 2015. Recent achievements in electronic tongue and bioelectronic tongue as taste sensors. *Sensors and Actuators B: Chemical* 207:1136–1146.

Hamid, Y. and Fat'hi, M.R. 2018. A colorimetric-dispersive solid-phase extraction method for the sensitive and selective determination of iron using dissolvable bathocuproinedisulfonic acid-intercalated layered double hydroxide nanosheets. *New Journal of Chemistry* 42:5489–5498.

Handy, S.M., Deeds, J.R., Ivanova, N.V., Hebert, P.D., Hanner, R.H., Ormos, A., Weigt, L.A., Moore, M.M. and Yancy, H.F. 2011. A single-laboratory validated method for the generation of DNA barcodes for the identification of fish for regulatory compliance. *Journal of AOAC International* 94:201–210.

He, X., Han, Y., Luo, X., Yang, W., Li, C., Tang, W., Yue, T. and Li, Z. 2020. Terbium (III)-referenced N-doped carbon dots for ratiometric fluorescent sensing of mercury (II) in seafood. *Food Chemistry* 320:126624.

Heo, W., Kim, J.H., Pan, J.H. and Kim, Y.J. 2016. Lecithin-based nano-emulsification improves the bioavailability of conjugated linoleic acid. *Journal of Agricultural and Food Chemistry* 64:1355–1360.

Hilding-Ohlsson, A., Fauerbach, J.A., Sacco, N.J., Bonetto, M.C. and Cortón, E. 2012. Voltamperometric discrimination of urea and melamine adulterated skimmed milk powder. *Sensors* 12:12220–12234.

Hu, Y., Wang, J. and Wu, Y. 2019. A simple and rapid chemosensor for colorimetric detection of dimethoate pesticide based on the peroxidase-mimicking catalytic activity of gold nanoparticles. *Analytical Methods* 11:5337–5347.

Huang, C., Cheng, Y., Gao, Z., Zhang, H., Wei, J.J.S. and Chemical, A.B. 2018. Portable label-free inverse opal photonic hydrogel particles serve as facile pesticides colorimetric monitoring. *Sensors and Actuators B: Chemical* 273:1705–1712.

Huang, L., Liu, H., Zhang, B. and Wu, D. 2015. Application of electronic nose with multivariate analysis and sensor selection for botanical origin identification and quality determination of honey. *Food and Bioprocess Technology* 8:359–370.

Ivanov, P., Llobet, E., Vilanova, X., Brezmes, J., Hubalek, J. and Correig, X. 2004. Development of high sensitivity ethanol gas sensors based on Pt-doped SnO2 surfaces. *Sensors and Actuators B: Chemical* 99:201–206.

Jiang, X., Valdeperez, D., Nazarenus, M., Wang, Z., Stellacci, F., Parak, W.J. and Del Pino, P. 2015. Future perspectives towards the use of nanomaterials for smart food packaging and quality control. *Particle Particle Systems Characterization* 32:408–416.

Jose, T., George, S.C., Maria, H.J., Wilson, R. and Thomas, S. 2014. Effect of bentonite clay on the mechanical, thermal, and pervaporation performance of the poly (vinyl alcohol) nanocomposite membranes. *Industrial & Engineering Chemistry Research* 53:16820–16831.

Joye, I.J. and McClements, D.J. 2014. Biopolymer-based nanoparticles and microparticles: Fabrication, characterization, and application. *Current Opinion in Colloid & Interface Science* 19:417–427.

Karimi-Maleh, H., Moazampour, M., Yoosefian, M., Sanati, A.L., Tahernejad-Javazmi, F. and Mahani, M. 2014. An electrochemical nanosensor for simultaneous voltammetric determination of ascorbic acid and Sudan I in food samples. *Food Analytical Methods* 7:2169–2176.

Kaur, J. and Singh, P.K. 2020. Enzyme-based optical biosensors for organophosphate class of pesticide detection. *Physical Chemistry Chemical Physics* 22:15105–15119.

Kaushik, A., Solanki, P.R., Ansari, A.A., Ahmad, S. and Malhotra, B.D. 2009. A nanostructured cerium oxide film-based immunosensor for mycotoxin detection. *Nanotechnology* 20:055105.

Khankaew, S., Mills, A., Yusufu, D., Wells, N., Hodgen, S., Boonsupthip, W. and Suppakul, P. 2017. Multifunctional anthraquinone-based sensors: UV, O2 and time. *Sensors Actuators B: Chemical* 238:76–82.

Khanna, V.K. 2008. New-generation nano-engineered biosensors, enabling nanotechnologies and nanomaterials. *Sensor Review* 28: 39–45.

Kim, H.-W., Choi, J.-H., Choi, Y.-S., Kim, H.-Y., Lee, M.-A., Hwang, K.-E., Song, D.-H., Lee, J.-W. and Kim, C.-J. 2014. Effects of kimchi and smoking on quality characteristics and shelf life of cooked sausages prepared with irradiated pork. *Meat science* 96:548–553.

Kim, T.H., Song, H.S., Jin, H.J., Lee, S.H., Namgung, S., Kim, U.-k., Park, T.H. and Hong, S. 2011a. "Bioelectronic super-taster" device based on taste receptor-carbon nanotube hybrid structures. *Lab on a Chip* 11:2262–2267.

Kim, Y.A., Lee, E.-H., Kim, K.-O., Lee, Y.T., Hammock, B.D. and Lee, H.-S. 2011b. Competitive immunochromatographic assay for the detection of the organophosphorus pesticide chlorpyrifos. *Analytica Chimica Acta* 693:106–113.

Kodogiannis, V.S. A rapid detection of meat spoilage using an electronic nose and fuzzy-wavelet systems. Proceedings of SAI Intelligent Systems Conference, 2016. Springer, 521–539.

Kumar, P., Kumar, P., Manhas, S. and Navani, N.K. 2016. A simple method for detection of anionic detergents in milk using unmodified gold nanoparticles. *Sensors Actuators B: Chemical* 233:157–161.

Kumar, V., Guleria, P. and Mehta, S.K. 2017. Nanosensors for food quality and safety assessment. *Environmental Chemistry Letters* 15:165–177.

Kuswandi, B., Futra, D. and Heng, L. 2017. Nanosensors for the detection of food contaminants. *Nanotechnology Applications in Food*. Elsevier.

Kuswandi, B., Oktaviana, R., Abdullah, A. and Heng, L.Y. 2014. A novel on-package sticker sensor based on methyl red for real-time monitoring of broiler chicken cut freshness. *Packaging Technology and Science* 27:69–81.

Laureati, M., Buratti, S., Giovanelli, G., Corazzin, M., Fiego, D.P.L. and Pagliarini, E. 2014. Characterization and differentiation of Italian Parma, San Daniele and Toscano dry-cured hams: a multi-disciplinary approach. *Meat science* 96:288–294.

Lee, M., Jung, J.W., Kim, D., Ahn, Y.-J., Hong, S. and Kwon, H.W. 2015. Discrimination of umami tastants using floating electrode-based bioelectronic tongue mimicking insect taste systems. *ACS Nano* 9:11728–11736.

Lee, S.H., Kwon, O.S., Song, H.S., Park, S.J., Sung, J.H., Jang, J. and Park, T.H. 2012. Mimicking the human smell sensing mechanism with an artificial nose platform. *Biomaterials* 33:1722–1729.

Lee, S.J. and Rahman, A.M. 2014. Intelligent packaging for food products. *Innovations in Food Packaging*. Elsevier.

Li, C., Yang, Q., Wang, X., Arabi, M., Peng, H., Li, J., Xiong, H. and Chen, L. 2020. Facile approach to the synthesis of molecularly imprinted ratiometric fluorescence nanosensor for the visual detection of folic acid. *Food Chemistry* 319:126575.

Li, L.-a., Yu, Y., Yang, J., Yang, R., Dong, G. and Jin, T. 2015. Voltammetric electronic tongue for the qualitative analysis of milk adulterated with urea combined with multi-way data analysis. *International Journal of Electrochemical Sciences* 10:5970–5980.

Li, M., Wang, H., Sun, L., Zhao, G. and Huang, X. 2016. Application of electronic nose for measuring total volatile basic nitrogen and total viable counts in packaged pork during refrigerated storage. *Journal of Food Science* 81:M906–M912.

Li, S., Ma, C., Liu, Z., Gong, G., Xu, Z., Xu, A. and Hua, B. 2014. Flavour analysis of stirred yoghurt with Cheddar cheese adding into milk. *Food Science and Technology Research* 20:939–946.

Li, X., Wang, T., Zhang, J., Zhu, D., Zhang, X., Ning, Y., Zhang, H. and Yang, B. 2010. Controlled fabrication of fluorescent barcode nanorods. *ACS Nano* 4:4350–4360.

Li, Y., Liu, L., Song, S. and Kuang, H. 2018. Development of a gold nanoparticle immunochromatographic assay for the on-site analysis of 6-benzylaminopurine residues in bean sprouts. *Food Agricultural Immunology* 29:14–26.

Liang, K.-Z. and Mu, W.-J. 2008. ZrO_2/DNA-derivated polyion hybrid complex membrane for the determination of hydrogen peroxide in milk. *Ionics* 14:533–539.

Lim, T.-C. and Ramakrishna, S. 2006. A conceptual review of nanosensors. *Zeitschrift Für Naturforschung A* 61:402–412.

Lisa, M., Chouhan, R., Vinayaka, A., Manonmani, H. and Thakur, M. 2009. Gold nanoparticles based dipstick immunoassay for the rapid detection of dichlorodiphenyltrichloroethane: an organochlorine pesticide. *Biosensors Bioelectronics* 25:224–227.

Liu, M., Khan, A., Wang, Z., Liu, Y., Yang, G., Deng, Y. and He, N. 2019. Aptasensors for Pesticide Detection. *Biosensors Bioelectronics* 130:174–184.

Liusman, C., Li, H., Lu, G., Wu, J., Boey, F., Li, S. and Zhang, H. 2012. Surface-enhanced Raman scattering of Ag–Au nanodisk heterodimers. *The Journal of Physical Chemistry C* 116:10390–10395.

López-Carballo, G., Muriel-Galet, V., Hernández-Muñoz, P. and Gavara, R. 2019. Chromatic sensor to determine oxygen presence for applications in intelligent packaging. *Sensors* 19:4684.

Lorenzen, P.C., Walte, H.-G. and Bosse, B. 2013. Development of a method for butter type differentiation by electronic nose technology. *Sensors and Actuators B: Chemical* 181:690–693.

Lyu, Y., Tao, Z., Lin, X., Qian, P., Li, Y., Wang, S. and Liu, Y. 2019. A MnO_2 nanosheet-based ratiometric fluorescent nanosensor with single excitation for rapid and specific detection of ascorbic acid. *Analytical Bioanalytical Chemistry* 411:4093–4101.

Magro, C., Mateus, E.P., Raposo, M. and Ribeiro, A.B. 2019. Overview of electronic tongue sensing in environmental aqueous matrices: potential for monitoring emerging organic contaminants. *Environmental Reviews* 27:202–214.

Mak, A.C., Osterfeld, S.J., Yu, H., Wang, S.X., Davis, R.W., Jejelowo, O.A. and Pourmand, N. 2010. Sensitive giant magnetoresistive-based immunoassay for multiplex mycotoxin detection. *Biosensors and Bioelectronics* 25:1635–1639.

Martínez-Olmos, A., Fernández-Salmerón, J., Lopez-Ruiz, N., Rivadeneyra Torres, A., Capitan-Vallvey, L. and Palma, A. 2013. Screen printed flexible radiofrequency identification tag for oxygen monitoring. *Analytical Chemistry* 85:11098–11105.

McClements, D.J. and Xiao, H. 2012. Potential biological fate of ingested nanoemulsions: influence of particle characteristics. *Food & Function* 3:202–220.

Mohammadalinejhad, S., Almasi, H. and Moradi, M. 2020. Immobilization of Echium amoenum anthocyanins into bacterial cellulose film: A novel colorimetric pH indicator for freshness/spoilage monitoring of shrimp. *Food Control* 113:107169.

Mohammadpour, Z. and Naghib, S.M. 2021. Smart nanosensors for intelligent packaging. *Nanosensors for Smart Manufacturing*. 1st ed.: Elsevier.

Mohareb, F., Papadopoulou, O., Panagou, E., Nychas, G.-J. and Bessant, C. 2016. Ensemble-based support vector machine classifiers as an efficient tool for quality assessment of beef fillets from electronic nose data. *Analytical Methods* 8:3711–3721.

Morsin, M., Umar, A.A., Salleh, M.M. and Majlis, B.Y. 2012. High sensitivity localized surface plasmon resonance sensor of gold nanoparticles: Surface density effect for detection of boric acid. 2012 10th IEEE International Conference on Semiconductor Electronics (ICSE), IEEE, 352–356.

Murphy, C., Carroll, C. and Jordan, K. 2006. Environmental survival mechanisms of the foodborne pathogen Campylobacter jejuni. *Journal of Applied Microbiology* 100: 623–632.

Narenderan, S., Meyyanathan, S. and Babu, B. 2020. Review of pesticide residue analysis in fruits and vegetables. Pre-treatment, extraction and detection techniques. *Food Research International* 133:109141.

Nateghi, L., Roohinejad, S., Totosaus, A., Rahmani, A., Tajabadi, N., Meimandipour, A., Rasti, B. and Manap, M. 2012. Physicochemical and textural properties of reduced fat Cheddar cheese formulated with xanthan gum and/or sodium caseinate as fat replacers. *Journal of Food, Agriculture and Environment* 10:59–63.

Noor Aini, B., Siddiquee, S. and Ampon, K. 2016. Development of formaldehyde biosensor for determination of formalin in fish samples; malabar red snapper (*Lutjanus malabaricus*) and longtail tuna (*Thunnus tonggol*). *Biosensors* 6:32.

Norouzi, P., Larijani, B. and Ganjali, M. 2012. Ochratoxin A sensor based on nanocomposite hybrid film of ionic liquid-graphene nano-sheets using coulometric FFT cyclic voltammetry. *International Journal of Electrochemical Science* 7:7313–7324.

Nurjuliana, M., Man, Y.C., Hashim, D.M. and Mohamed, A. 2011. Rapid identification of pork for halal authentication using the electronic nose and gas chromatography mass spectrometer with headspace analyzer. *Meat Science* 88:638–644.

Patel, A., Patra, F., Shah, N. and Khedkar, C. 2018. Application of nanotechnology in the food industry: Present status and future prospects. *Impact of Nanoscience in the Food Industry*. Elsevier.

Patel, H., Rawtani, D. and Agrawal, Y.K. 2019. A newly emerging trend of chitosan-based sensing platform for the organophosphate pesticide detection using Acetylcholinesterase-a review. *Trends in Food Science Technology* 85:78–91.

Pimtong-Ngam, Y., Jiemsirilers, S. and Supothina, S. 2007. Preparation of tungsten oxide–tin oxide nanocomposites and their ethylene sensing characteristics. *Sensors and Actuators A: Physical* 139:7–11.

Potyrailo, R.A., Nagraj, N., Tang, Z., Mondello, F.J., Surman, C. and Morris, W. 2012. Battery-free radio frequency identification (RFID) sensors for food quality and safety. *Journal of Agricultural Food Chemistry* 60:8535–8543.

Prapainop, K., Mekseriwattana, W., Siangproh, W., Chailapakul, O. and Songsrirote, K. 2019. Successive detection of benzoic acid and total parabens in foodstuffs using mercaptosuccinic acid capped cadmium telluride quantum dots. *Food Control* 96:508–516.

Pu, H., Xu, Y., Sun, D.-W., Wei, Q. and Li, X. 2020. Optical nanosensors for biofilm detection in the food industry: principles, applications and challenges. *Critical Reviews in Food Science and Nutrition* 61:1–18.

Qin, L., Banholzer, M.J., Millstone, J.E. and Mirkin, C.A. 2007. Nanodisk codes. *Nano letters* 7:3849–3853.

Qin, L., Zou, S., Xue, C., Atkinson, A., Schatz, G.C. and Mirkin, C.A. 2006. Designing, fabricating, and imaging Raman hot spots. *Proceedings of the National Academy of Sciences* 103:13300–13303.

Qureshi, M., Karthikeyan, S., Karthikeyan, P., Khan, P., Uprit, S. and Mishra, U. 2012. Application of nanotechnology in food and dairy processing: an overview. *Pakistan Journal of Food Sciences* 22:23–31.

Radoi, A., Targa, M., Prieto-Simon, B. and Marty, J.L. 2008. Enzyme-Linked Immunosorbent Assay (ELISA) based on superparamagnetic nanoparticles for aflatoxin M1 detection. *Talanta* 77:138–143.

Ranjan, S., Dasgupta, N., Chakraborty, A.R., Samuel, S.M., Ramalingam, C., Shanker, R. and Kumar, A. 2014. Nanoscience and nanotechnologies in food industries: opportunities and research trends. *Journal of Nanoparticle Research* 16:1–23.

Rao, H., Liu, W., Lu, Z., Wang, Y., Ge, H., Zou, P., Wang, X., He, H., Zeng, X. and Wang, Y. 2016. Silica-coated carbon dots conjugated to CdTe quantum dots: a ratiometric fluorescent probe for copper (II). *Microchimica Acta* 183:581–588.

Ravichandran, R. 2010. Nanotechnology applications in food and food processing: innovative green approaches, opportunities and uncertainties for global market. *International Journal of Green Nanotechnology: Physics and Chemistry* 1:P72–P96.

Realini, C.E. and Marcos, B. 2014. Active and intelligent packaging systems for a modern society. *Meat Science* 98:404–419.

Roy, S. and Rhim, J. 2021. Anthocyanin food colorant and its application in pH-responsive color change indicator films. *Critical Reviews in Food Science and Nutrition* 61: 2297–2325.

Ruiz-Rico, M., Fuentes, A., Masot, R., Alcañiz, M., Fernández-Segovia, I. and Barat, J.M. 2013. Use of the voltammetric tongue in fresh cod (Gadus morhua) quality assessment. *Innovative Food Science & Emerging Technologies* 18:256–263.

Saini, R., Bagri, L. and Bajpai, A. 2017. *New Pesticides and Soil Sensors*. Elsevier.

Salavati-Niasari, M., Davar, F. and Mir, N. 2008. Synthesis and characterization of metallic copper nanoparticles via thermal decomposition. *Polyhedron* 27:3514–3518.

Sen, L., Hyun, K.H., Kim, J.W., Shin, J.W. and Eom, K.H. 2013. The design of smart RFID system with gas sensor for meat freshness monitoring. *Advanced Science Technology Letters* 41:17–20.

Sessa, M., Balestrieri, M.L., Ferrari, G., Servillo, L., Castaldo, D., D'Onofrio, N., Donsì, F. and Tsao, R. 2014. Bioavailability of encapsulated resveratrol into nanoemulsion-based delivery systems. *Food Chemistry* 147:42–50.

Shah, A. 2020. A novel electrochemical nanosensor for the simultaneous sensing of two toxic food dyes. *ACS Omega* 5:6187–6193.

Shaikh, J., Ankola, D., Beniwal, V., Singh, D. and Kumar, M.R. 2009. Nanoparticle encapsulation improves oral bioavailability of curcumin by at least 9-fold when compared to curcumin administered with piperine as absorption enhancer. *European Journal of Pharmaceutical Sciences* 37:223–230.

Sharma, A., Matharu, Z., Sumana, G., Solanki, P.R., Kim, C.G. and Malhotra, B.D. 2010. Antibody immobilized cysteamine functionalized-gold nanoparticles for aflatoxin detection. *Thin Solid Films* 519:1213–1218.

Shikha, S., Salafi, T., Cheng, J. and Zhang, Y. 2017. Versatile design and synthesis of nano-barcodes. *Chemical Society Reviews* 46:7054–7093.

Shukla, K. 2012. Nanotechnology and emerging trends in dairy foods: the inside story to food additives and ingredients. *International Journal of Nanoscience and Nanotechnology* 1:41–58.

Singh, H. 2016. Nanotechnology applications in functional foods; opportunities and challenges. *Preventive Nutrition and Food Science* 21:1.

Singh, R., Kumar, N., Mehra, R., Kumar, H. and Singh, V.P. 2020. Progress and challenges in the detection of residual pesticides using nanotechnology based colorimetric techniques. *Trends in Environmental Analytical Chemistry* 26:e00086.

Smits, E., Schram, J., Nagelkerke, M., Kusters, R., Heck,van, van Acht, G., Koetse, V., van den Brand, M., Gerlinck J., G. Development of printed RFID sensor tags for smart food packaging. Proceedings of the 14th international meeting on chemical sensors, Nuremberg, Germany 2012. 20–23.

Sobrino-Gregorio, L., Tanleque-Alberto, F., Bataller, R., Soto, J. and Escriche, I. 2020. Using an automatic pulse voltammetric electronic tongue to verify the origin of honey from Spain, Honduras, and Mozambique. *Journal of the Science of Food and Agriculture* 100:212–217.

Sohail, M., Sun, D.-W. and Zhu, Z. 2018. Recent developments in intelligent packaging for enhancing food quality and safety. *Critical Reviews in Food Science and Nutrition* 58:2650–2662.

Song, H.S., Jin, H.J., Ahn, S.R., Kim, D., Lee, S.H., Kim, U.-K., Simons, C.T., Hong, S. and Park, T.H. 2014. Bioelectronic tongue using heterodimeric human taste receptor for the discrimination of sweeteners with human-like performance. *ACS Nano* 8:9781–9789.

Song, J., Li, J., Guo, Z., Liu, W., Ma, Q., Feng, F. and Dong, C. 2017. A novel fluorescent sensor based on sulfur and nitrogen co-doped carbon dots with excellent stability for selective detection of doxycycline in raw milk. *RSC Advances* 7:12827–12834.

Song, S., Yuan, L., Zhang, X., Hayat, K., Chen, H., Liu, F., Xiao, Z. and Niu, Y. 2013. Rapid measuring and modelling flavour quality changes of oxidised chicken fat by electronic nose profiles through the partial least squares regression analysis. *Food chemistry* 141:4278–4288.

Sulaiman, I.C., Chieng, B., Osman, M., Ong, K., Rashid, J., Yunus, W.W., Noor, S., Kasim, N., Halim, N. and Mohamad, A. 2020. A review on colorimetric methods for determination of organophosphate pesticides using gold and silver nanoparticles. *Microchimica Acta* 187:1–22.

Sun, Z., Tian, L., Guo, M., Xu, X., Li, Q. and Weng, H. 2017. A double-film screening card for rapid detection of organophosphate and carbamate pesticide residues by one step in vegetables and fruits. *Food Control* 81:23–29.

Tahara, Y., Nakashi, K., Ji, K., Ikeda, A. and Toko, K. 2013. Development of a portable taste sensor with a lipid/polymer membrane. *Sensors* 13:1076–1084.

Taherkhani, E., Moradi, M., Tajik, H., Molaei, R. and Ezati, P. 2020. Preparation of on-package halochromic freshness/spoilage nanocellulose label for the visual shelf life estimation of meat. *International Journal of Biological Macromolecules* 164:2632–2640.

Tan, J. and Xu, J. 2020. Applications of electronic nose (e-nose) and electronic tongue (e-tongue) in food quality-related properties determination: a review. *Artificial Intelligence in Agriculture* 4:104–115.

Tang, X., Sun, X., Wu, V.C., Xie, J., Pan, Y., Zhao, Y. and Malakar, P.K. 2013. Predicting shelf-life of chilled pork sold in China. *Food Control* 32:334–340.

Thandavan, K., Gandhi, S., Nesakumar, N., Sethuraman, S., Rayappan, J.B.B. and Krishnan, U.M. 2015. Hydrogen peroxide biosensor utilizing a hybrid nano-interface of iron oxide nanoparticles and carbon nanotubes to assess the quality of milk. *Sensors Actuators B: Chemical* 215:166–173.

Tian, X., Wang, J. and Cui, S. 2013. Analysis of pork adulteration in minced mutton using electronic nose of metal oxide sensors. *Journal of Food Engineering* 119:744–749.

Toko, K., Tahara, Y., Habara, M., Kobayashi, Y., Ikezaki, H. and Nakamoto, T. 2016. Taste sensor: electronic tongue with global selectivity. *Essentials of Machine Olfaction and Taste* 87–174.

Tousch, D. and Charlot, B. 2017. Biocaptors and Barcodes. *Food Traceability and Authenticity: Analytical Techniques* 332–351.

Tripathy, S., Reddy, M.S., Vanjari, S.R.K., Jana, S. and Singh, S.G. 2019. A step towards miniaturized milk adulteration detection system: smartphone-based accurate pH sensing using electrospun halochromic nanofibers. *Food Analytical Methods* 12: 612–624.

Tsai, M., Chen, R., Bai, S. and Chen, W. 2011. The storage stability of chitosan/tripolyphosphate nanoparticles in a phosphate buffer. *Carbohydrate Polymers* 84:756–761.

Tümay, S.O., Şenocak, A., Sarı, E., Şanko, V., Durmuş, M. and Demirbas, E. 2021. A new perspective for electrochemical determination of parathion and chlorantraniliprole pesticides via carbon nanotube-based thiophene-ferrocene appended hybrid nanosensor. *Sensors Actuators B: Chemical* 345:130344.

Umapathi, R., Sonwal, S., Lee, M.J., Rani, G.M., Lee, E., Jeon, T., Kang, S., Oh, M. and Huh, Y.S. 2021. Colorimetric based on-site sensing strategies for the rapid detection of pesticides in agricultural foods: new horizons, perspectives, and challenges. *Coordination Chemistry Reviews* 446:214061.

Valentini, P., Galimberti, A., Mezzasalma, V., De Mattia, F., Casiraghi, M., Labra, M. and Pompa, P.P. 2017. DNA barcoding meets nanotechnology: development of a universal colorimetric test for food authentication. *Angewandte Chemie International Edition* 56:8094–8098.

Vermeir, S., Nicolai, B., Jans, K., Maes, G. and Lammertyn, J. 2007. High-throughput microplate enzymatic assays for fast sugar and acid quantification in apple and tomato. *Journal of Agricultural Food Chemistry* 55:3240–3248.

Verstrepen, K.J., Iserentant, D., Malcorps, P., Derdelinckx, G., Van Dijck, P., Winderickx, J., Pretorius, I.S., Thevelein, J.M. and Delvaux, F.R. 2004. Glucose and sucrose: hazardous fast-food for industrial yeast? *Trends in biotechnology* 22:531–537.

Vickers, N.J. 2017. Animal communication: when I'm calling you, will you answer too? *Current Biology* 27:R713–R715.

Wang, J.J., Liu, B.H., Hsu, Y.T. and Yu, F.Y. 2011. Sensitive competitive direct enzyme-linked immunosorbent assay and gold nanoparticle immunochromatographic strip for detecting aflatoxin M1 in milk. *Food Control* 22:964–969.

Wang, L., Gao, P., Bao, D., Wang, Y., Chen, Y., Chang, C., Li, G. and Yang, P. 2014a. Synthesis of crystalline/amorphous core/shell MoO_3 composites through a controlled dehydration route and their enhanced ethanol sensing properties. *Crystal Growth & Design* 14:569–575.

Wang, Q., Jin, G., Jin, Y., Ma, M., Wang, N., Liu, C. and He, L. 2014b. Discriminating eggs from different poultry species by fatty acids and volatiles profiling: comparison of SPME-GC/MS, electronic nose, and principal component analysis method. *European Journal of Lipid Science and Technology* 116:1044–1053.

Wang, Y.-C., Lu, L. and Gunasekaran, S. 2015. Gold nanoparticle-based thermal history indicator for monitoring low-temperature storage. *Microchimica Acta* 182:1305–1311.

Wang, Y.-C., Mohan, C., Guan, J., Ravishankar, C. and Gunasekaran, S. 2018. Chitosan and gold nanoparticles-based thermal history indicators and frozen indicators for perishable and temperature-sensitive products. *Food Control* 85:186–193.

Wei, S., Zhao, F., Xu, Z. and Zeng, B. 2006. Voltammetric determination of folic acid with a multi-walled carbon nanotube-modified gold electrode. *Microchimica Acta* 152:285–290.

Wei, Z. and Wang, J. 2011. Detection of antibiotic residues in bovine milk by a voltammetric electronic tongue system. *Analytica Chimica Acta* 694:46–56.

Wei, Z., Wang, J. and Jin, W. 2013a. Evaluation of varieties of set yogurts and their physical properties using a voltammetric electronic tongue based on various potential waveforms. *Sensors and Actuators B: Chemical* 177:684–694.

Wei, Z., Wang, J. and Zhang, X. 2013b. Monitoring of quality and storage time of unsealed pasteurized milk by voltammetric electronic tongue. *Electrochimica Acta* 88:231–239.

Wu, Q., Long, Q., Li, H., Zhang, Y. and Yao, S. 2015. An upconversion fluorescence resonance energy transfer nanosensor for one step detection of melamine in raw milk. *Talanta* 136:47–53.

Xiao, F., Ruan, C., Liu, L., Yan, R., Zhao, F. and Zeng, B. 2008. Single-walled carbon nanotube-ionic liquid paste electrode for the sensitive voltammetric determination of folic acid. *Sensors Actuators B: Chemical* 134:895–901.

Xiao, Z., Wu, M., Niu, Y., Chen, F., Zhang, X., Zhu, J., Song, S. and Zhu, G. 2015. Contribution of chicken base addition to aroma characteristics of Maillard reaction products based on gas chromatography-mass spectrometry, electronic nose, and statistical analysis. *Food Science and Biotechnology* 24:411–419.

Xiulan, S., Xiaolian, Z., Jian, T., Zhou, J. and Chu, F.S. 2005. Preparation of gold-labeled antibody probe and its use in immunochromatography assay for detection of aflatoxin B1. *International Journal of Food Microbiology* 99:185–194.

Yadav, S., Nair, S.S., Sai, V. and Satija, J. 2019. Nanomaterials based optical and electrochemical sensing of histamine: Progress and perspectives. *Food Research International* 119:99–109.

Yang, X., Luo, Y., Zhu, S., Feng, Y., Zhuo, Y. and Dou, Y. 2014. One-pot synthesis of high fluorescent carbon nanoparticles and their applications as probes for detection of tetracyclines. *Biosensors Bioelectronics* 56:6–11.

Yao, Y., Wu, H. and Ping, J. 2019. Simultaneous determination of Cd (II) and Pb (II) ions in honey and milk samples using a single-walled carbon nanohorns modified screen-printed electrochemical sensor. *Food Chemistry* 274:8–15.

Ye, J.-S., Wen, Y., De Zhang, W., Gan, L.M., Xu, G.Q. and Sheu, F.-S. 2004. Nonenzymatic glucose detection using multi-walled carbon nanotube electrodes. *Electrochemistry Communications* 6:66–70.

Yu, Y., Zhao, H., Yang, R., Dong, G., Li, L., Yang, J., Jin, T., Zhang, W. and Liu, Y. 2015. Pure milk brands classification by means of a voltammetric electronic tongue and multivariate analysis. *International Journal of Electrochemical Science* 10:4381–4392.

Zhang, M., Wang, X., Liu, Y., Xu, X. and Zhou, G. 2012. Species discrimination among three kinds of puffer fish using an electronic nose combined with olfactory sensory evaluation. *Sensors* 12:12562–12571.

Zhang, C., Yin, A.X., Jiang, R., Rong, J., Dong, L., Zhao, T., Sun, L.D., Wang, J., Chen, X. and Yan, C.H. 2013. Time–temperature indicator for perishable products based on kinetically programmable Ag overgrowth on Au nanorods. *ACS Nano* 7:4561–4568.

Zhang, Q., Zhang, S., Xie, C., Zeng, D., Fan, C., Li, D. and Bai, Z. 2006. Characterization of Chinese vinegars by electronic nose. *Sensors and Actuators B: Chemical* 119:538–546.

Zhang, W.-H. and Zhang, W.-D. 2008. Fabrication of SnO2–ZnO nanocomposite sensor for selective sensing of trimethylamine and the freshness of fishes. *Sensors Actuators B: Chemical* 134:403–408.

Zhang, X., Zhang, Y., Meng, Q., Li, N. and Ren, L. 2015. Evaluation of beef by electronic tongue system TS-5000Z: Flavor assessment, recognition and chemical compositions according to its correlation with flavor. *PLoS One* 10:e0137807.

Zhao, L., Li, H., Xu, Y., Liu, H., Zhou, T., Huang, N., Li, Y. and Ding, L. 2018. Selective detection of copper ion in complex real samples based on nitrogen-doped carbon quantum dots. *Analytical Bioanalytical Chemistry* 410:4301–4309.

Zhao, S., Huang, J., Lei, J., Huo, D., Huang, Q., Tan, J., Li, Y., Hou, C. and Tian, F. 2021. A portable and automatic dual-readout detector integrated with 3D-printed microfluidic nanosensors for rapid carbamate pesticides detection. *Sensors Actuators B: Chemical* 346:130454.

Zhou, G. 2013. Nanotechnology in the food system: consumer acceptance and willingness to pay. *Theses and Dissertations--Agricultural Economics*. 10. https://uknowledge.uky.edu/a_etds/10.

Zhou, Y., Pan, F.G., Li, Y.S., Zhang, Y.Y., Zhang, J.H., Lu, S.Y., Ren, H.L. and Liu, Z.S. 2009. Colloidal gold probe-based immunochromatographic assay for the rapid detection of brevetoxins in fishery product samples. *Biosensors and Bioelectronics* 24:2744–2747.

Zhu, N., Zou, Y., Huang, M., Dong, S., Wu, X., Liang, G., Han, Z. and Zhang, Z. 2018. A sensitive, colorimetric immunosensor based on Cu-MOFs and HRP for detection of dibutyl phthalate in environmental and food samples. *Talanta* 186:104–109.

10 Silver Nanoparticles as Food Packaging Additives for Shelf-Life Extension

Kashif Ameer[1], *Guihun Jiang*[2], *Chang-Cheng Zhao*[3], *Muhammad Nadeem*[1], *Mian Anjum Murtaza*[1], *Ghulam Mueen-ud-Din*[1], *and Shahid Mahmood*[1]
[1]Institute of Food Science and Nutrition, University of Sargodha, Sargodha, Pakistan
[2]School of Public Health, Jilin Medical University, Jilin, China
[3]School of Life Science, Zhengzhou University, Henan, China

CONTENTS

10.1 INTRODUCTION

Conventionally, the food packaging materials are employed for various purposes, such as protection, communication, containment, and convenience purposes. However, in the usual scenario of food processing, the recent modalities convert the traditional packaging into active packaging materials. In recent times, conventional food packaging encompasses the active food packaging of foodstuffs as its extension (Carbone et al., 2016). Usually, the packaging materials work to incorporate the active food substances for shelf-life extension and characteristic improvement of packaged foods. Packaging materials are improved through deliberate incorporation of active components (Ahari and Lahijani, 2021). These components allow to-and-fro absorption or discharge of substances from the outer environment or packaged foods. In active packaging, incorporation of various types of active substances is

DOI: 10.1201/9781003207641-12

carried out, including of antimicrobials, antioxidants, oxygen scavengers, ethylene or free radicals absorbers, as well as carbon dioxide emitters (Anvar et al., 2021). Antimicrobial food packaging technology is regarded as one of the effective methods for delaying or averting decay and microbial decontamination and for augmenting of food products' shelf life (Ahmad et al., 2021). Silver nanoparticle fusion with food packaging materials is widely known to impart bactericidal activity and it plays critical role in achieving food packaging materials that allow microbial decontamination of viruses, bacteria, fungi, yeasts, and mold (Ahari and Lahijani, 2021).

Nanomaterials can be categorized as particles having a size range of 1 to 100 nm. These exhibit a larger surface-to-volume ratio and are extensively employed in manipulation, measurement, and manufacturing of nanoscale objects (Lyons et al., 2018; Usman et al., 2020). When materials are handled at nanoscale levels, the result is creation of particular properties with newly emerged quantum mechanical effects which play vital role in diffusion at elevated temperatures (Scott et al., 2018). The particles responsible for enhanced materials characteristics are forced through nanoscale manipulation to produce a higher degree of association with between particles at neon level, resulting in enhancement of materials characteristics (Prasad et al., 2017). Nanoparticles are also manufactured from semiconductors, and nanomaterials made up of nanoparticles derived from metal oxides, including ($CoFe_2O_4$, $ZnCoFe_2O_4$, $CuFe_2O_4$, Fe_3O_4, FeOH etc.), metals and semiconductors exhibiting high potential of electromagnetic and physicochemical properties (Misra et al., 2013; Mittal et al., 2020). Nanomaterials are also produced from the carbon allotropes, such as grapheme, which is further processed to form spheres and tubes possessing a high degree of electromechanical strength (Dasgupta et al., 2015).

Nanomaterial classification might also be carried out in terms of the constituents making up the nanomaterials, such as inorganic and hybrid nanomaterials (Mitter and Hussey, 2019), and organics. In published literature, the efficiency of the hybrid system has been widely reported and could be utilized in case of advance oxidation processes (AOPs), which play vital role in degradation of organic materials in a catalytic manner (Kumari and Yadav, 2014). Researchers are interested in and emphasize enhancing nanoparticle manipulation for technological purposes (Scott and Chen, 2013). This includes the application of nanomaterials and utilization of nanoparticles in systemizing and designing particle control mechanisms, innovative structures, and waste and water treatments. As an example, hybrid nanoparticles have been utilized for water treatment by causing the removal of cyfluthrin from water (Chhipa, 2019; He et al., 2019). It was also concluded by researchers that degradation and mineralization efficiency of nanomaterials are 88.5 and 52.4%, respectively (Kim et al., 2018). In a similar study, magnetic nanoparticles derived from the hybrid system were employed in conjunction with ultraviolet and ultrasound irradiations as a mechanistic approach to degrade catalytically and produce scavengers to treat wastewater of petrochemical byproducts (Ndlovu et al., 2020; Tripathi et al., 2018). Efficiency of about 87% was reported in terms of pollutant removal over short time interval. Moreover, application of nanomaterials has shown rapid rise in recent years in manufacturing, on a commercial scale according to McIntyre (2012). It has also been endorsed by Maynard et al. (2006) that a four-fold increase was seen from 2010 to 2015 in the manufacturing of a large number of products with improved properties because of integrated nanomaterials. The majority of the products consisted of nanomaterials and the most commonly reported and employed nanomaterials in commercial product manufacturing are metal oxides, carbon-based nanorods, and nanotubes as well as silver-based nanoparticles (Ndukwu et al., 2020). Nanoscience application has influenced almost every facet of human advancement in various technological sectors, such as natural sciences, agriculture, medicine, and engineering, etc., with significant market potential worth more than $1.6 trillion USD in 2013 already (Mohammad et al., 2012).

Despite the innovative benefits of nanotechnology and nanomaterials, they also exhibit potential drawbacks which limit their practical applications (Sastry et al., 2011). As far as the magnetic nanoparticles are concerned, these exhibit low surface catalytic activity accompanied by a highly significant tendency of agglomeration and lower degradation rates because of low surface-to-volume ratio (Eze et al., 2016). These may be released into consumables, which may lead to detrimental toxic effects. These metal oxides may lead to increased generation of reactive oxygen species (ROS) at the cellular levels and may exert damaging effect on DNA (Mfon et al., 2017).

Hence, various countries have implemented framework policies and regulatory guidelines for research, innovation, and manipulation of nanoscience, as nanoscience could impact military as well as humanitarian applications significantly (Huang et al., 2015). Moreover, the incidence and outbreak of coronavirus pandemic has signaled the importance of global food supply chains and of the agricultural and food sectors for societies (Kim et al., 2018). To improve the growth rate of agricultural and food sectors, exploitation of nanotechnology and nano-biomaterials has been carried out in various research domains, such as renewable energy systems development for agriculture, bichar improvement to enrich soils, soil bioremediation, crop protection, coating of the crop processing system to protect from corrosion to achieve crop detoxification and improved efficiency, disease control, and biosensors applications (Reichert et al., 2020; Sekhon, 2014; Shuttleworth et al., 2014). However, the real contributory role of nanotechnology is toward agricultural production. Various types of nanomaterials employed for agricultural purposes have been developed; however, the main issue is the scaling up of these research developments at farmer and agro-processor levels to get direct benefits (García et al., 2010; Mahato et al., 2021). The majority of the nanomaterials applications in agriculture revolve around the treatment of agricultural produce, irrigation water purification, improvement of the yield and vegetative growth, disease control in cattle and crop plants, gene expression, and improvement of nutritional content (Reichert et al., 2020; Sekhon, 2014). Much of the published literature regarding agricultural application of nanomaterials has focused on migration properties of the metallic nanoparticles in soil and plants and their interaction, precision farming, seed science, biosensors, nutritional improvement, evaluation of the effect of polymeric nano-carriers for potential agricultural applications, fertilizer use, and disease control (Ajayi et al., 2018; Reichert et al., 2020; Sekhon, 2014). However, several other areas of nanotechnology application need to be explored and this chapter examines the use of nanoparticles in food packaging (Huang et al., 2015).

10.2 FOOD PACKAGING THROUGH USE OF AgNPs-BASED NANOMATERIALS

AgNPs-based antimicrobial packaging has been reported in published literature and has been in use as a promising type of active food packaging that plays a decisive role in reducing the risk of food-borne pathogens in conjunction with the shelf-life extension of foods (Sharma et al., 2017). With respect to the type of polymer matrix employed for hosting AgNPs, there are two groups.: 1) an edible coating of biodegradable nature synthesized through the polymer or usage of a stabilizer and 2) a non-degradable polymeric packaging material (Qiu et al., 2020). It is worth mentioning that the addition of AgNPs into both types of polymeric matrices affects film permeability and, hence, product quality characteristics.

10.3 AgNPs-BASED NON-DEGRADABLE POLYMERIC MATRIX

There are various non-degradable polymers which are utilized for hosting silver nanoparticles to manufacture food packaging materials. These non-degradable polymers include ethylene vinyl alcohol (EVOH), polyvinyl chloride (PVC), and polyethylene (PE) (Venkateshwarlu and Nagalakshmi, 2013). PE is the most commonly utilized plastic material, and it is synthesized by adding ethylene during the polymerization process. Two basic forms of PE are employed: low-density PE (LDPE) and high-density PE (HDPE). The former is found to be more flexible and exhibits material properties of transparency and resistance to moisture (Hamad et al., 2018; Martínez-Abad et al., 2012). For this reason, this LDPE is suitable for polymeric covering films intended for fresh foods for long-term storage. PVC as the thermoplastic material; it exhibits material properties of stiffness, transparency, and ductility and it is obtained through a process of polymerization of vinyl chloride polymer. PVS has excellent barrier properties to restrict acids, oils, and bases and it is largely employed to manufacture packaging films and food containers (Rossi et al., 2017; Valipoor Motlagh et al., 2013). EVOH is obtained as the copolymer made by reaction of vinyl alcohol and ethylene. EVOH is categorized as a thin film and exhibits the ability to resist oxygen, fats, and oils. Nevertheless, EVOH has a sensitivity to moisture content and, hence, cannot be employed in the case of liquid foods which will come into direct contact with the EVOH (Carbone et al., 2016). The usage of these polymers by incorporation with AgNPs for various food packaging applications is given in detail here.

LDPE polymer matrix is comprised of ZnO and Ag nanoparticles, and was used for extending orange juice shelf life (Ahari and Lahijani, 2021). The effectiveness of active-nanocomposite is obvious in terms of antimicrobial efficacy in conjunction with thermal treatment at a pasteurization temperature range of 55 to 65°C (Mahdi et al., 2012). LDPE polymeric material with AgNPs showed high effectiveness in inhibiting the growth of fungi, including yeasts and molds as compared to polymeric matrices of ZnO-based nanoparticles (Anvar et al., 2021; Carbone et al., 2016). The antimicrobial effect of AgNPs decreases by 10°C change in pasteurization temperature of orange juice. AgNPs-based packages were also evaluated for their effects regarding preservation of sensorial and quality characteristics of stored berries (Ahmad et al., 2021). It was evident from the above research that LDPE in conjunction with AgNPs was successful for the maintenance and preservation of physicochemical and physiological properties of blue berries and strawberries as compared to normal packaging with that of polythene bags (Carbone et al., 2016; Scott et al., 2018).

Refrigeration is necessary in the case of ready-to-use foods, for assurance of quality properties and for retention of nutritional wholesomeness. The raw material type and quality are factors determining the length of shelf life of meat and its derivatives (Lamba and Garg, 2018). In a report by Metak (2015), PVC in conjunction with AgNPs was evaluated for its antimicrobial efficacy as nano-packaging material for minced beef which was subjected to storage at refrigeration temperature (4°C). After completion of seven days of storage, the inhibitory effect of AgNPs packaging on microbial growth was obvious (Mlalila et al., 2016). Particularly, the AgNPs exhibited better antimicrobial effects against growth of *Escherichia coli* as compared to the effect on *Staphylococcus aureus* growth. The PVC-AgNPs nanocomposite packaging showed significant antimicrobial effect by retarding the bacterial growth and, consequently, helping to extend the shelf life of minced meat which otherwise has short shelf life of two days when stored in common food packaging materials (Babu, 2021). Metak (2015) also investigated the effects of silver and TiO_2 nanoparticles used to synthesize the PE polymeric packaging materials on liquid and solid food samples as well as on highly acidic and fat-containing foodstuffs, also comparing the efficacy of Ag-NPs based materials with that of conventionally employed containers (Carbone et al., 2016). It was indicated by the studies, using energy-dispersive X-ray spectroscopy (EDX), that nanoparticles made from Ag and TiO_2 were aligned in layers in a distinct manner and were found embedded in the bulk polymer instead of as a coating on the outer surface of polymer (Carbone et al., 2016). The containers made by the AgNPs exhibit significantly higher antimicrobial activity by causing inhibition of microbial growth in containers subjected to storage for several days. The deterioration rate was quite a bit lower in samples observed after ten days of storage, due to antimicrobial effects of TiO_2 and AgNPs in comparison with that of conventional packaging (Carbone et al., 2016; Toker et al., 2013).

The nanocomposites were also assessed for their antimicrobial activity, and in order to achieve effective antimicrobial activity, the food samples should be differentiated with respect to their protein content, as greater presence of proteins might lead to inactivation of silver (Carbone et al., 2016; Youssef and Abdel-Aziz, 2013). In a similar work, incorporation of EVOH polymer matrix was carried out with silver ions, and their antibacterial efficacy was assessed when in contact with several kinds of food materials (Toker et al., 2013). Specifically, two experiments were carried out. The first experiment concerned foods with low protein content, including fresh cut vegetables, fruits, and juices. The second group consisted of high protein foods, such as cheese and meat (Carbone et al., 2016). It was evident from the results that low protein foods samples exhibited a lower degree of inactivation of silver ions, and the highest bactericidal effects were achieved in the case of 10% Ag-EVOH films. However, it was also observed that even with the food samples with low protein content, 0.1% content of silver ions embedded in EVOH films was not enough to surpass the European Food Safety Authority (EFSA) restriction limits regarding microbial contamination (Youssef and Abdel-Aziz, 2013). Other polymer types were also employed for preparing new nanocomposites for food packaging applications. Toker et al. (2013) reported that AgNPs could be incorporated in polyurethane hybrid coating which exhibited potent antimicrobial activity against *S. aureus* and *E. coli*. Youssef and Abdel-Aziz (2013) also carried out soaking of polystyrene matrix into AgNPs to form nanocomposite termed as Ag/PS composite. Characterization of the nanocomposite was performed through instrumental analyses by means of FTIR, SEM and TEM and it was implied that growth of pathogenic gram-positive and gram-negative bacteria, including *Bacillus subtilis* and *Enterococcus faecalis* as well as *E. coli*, yeast *(Candida albicans)* and *Salmonella typhimurium* was significantly inhibited (Carbone et al., 2016).

10.4 NANOCOMPOSITE PACKAGING BASED ON AgNPs/BIODEGRADABLE EDIBLE MATRIX

Biodegradable films made from polymeric materials have been utilized as an alternative for conventionally used materials for food packaging purposes. These polymeric materials are obtained from the renewable sources at an economical cost, and these do not contribute to environmental pollution owing to their biodegradable nature (Mazzuca et al., 2014). Because of their eco-friendly nature, the biodegradable polymers have been employed extensively in various fields and garnered the attention of researchers as a viable option for utilization in food packaging materials (Carbone et al., 2016). Among biodegradable materials, the most commonly employed are the polysaccharides, including chitosan, starch, agarose, pullulan, and cellulose (Valencia-Chamorro et al., 2011). The electropositive transition metal ions are usually attracted to cellulose by means of electrostatic interactions. As a result, the silver ions are absorbed when they are immersed in the silver nitrate and the porosity of cellulose usually helps in stabilization and synthesis of silver nanoparticles (Carbone et al., 2016; Elsabee and Abdou, 2013). Absorbent pads find their wide applications in synthesis of modern food packaging materials and fabric of polymeric material intended for fresh produce. The cost of production of cellulose is usually quite low and it is found abundantly in nature; therefore, it is considered biocompatible and has been utilized for food material packaging as an edible form of biodegradable material (Carbone et al., 2016; Fernández et al., 2010).

Pullalan is also categorized as an edible form of polysaccharide material comprising maltotriose units in its structural configuration (Carbone et al., 2016). The polymeric films of pullulan are characterized as tasteless and colorless, and it has excellent barrier properties against oil with a low degree of permeability against oxygen (Khalaf et al., 2013). However, these are highly susceptible to moisture, a disadvantage in the case of food storage, with respect to microbial growth, but an advantage during the process of cooking (Morsy et al., 2014).

Red algae is also a wonderful source of polysaccharides, and agarose is usually obtained from red algae. The agarose is chemically defined as the linear polymer, made up of agarobiose as its basic unit. These units are found in repeated fashion in the agarose configuration (Carbone et al., 2016; Incoronato et al., 2011). Agarobiose is a form of disaccharides made by the polymerization of two basic sugars, such as 3,6-anhydro- L-galactopyranose and D-galactose. Agarose is the raw material of agar-agar gel, a jelly-like substance. Agar gel is reported to have properties of water-insolubility, non-toxicity, non-immunogenicity, and biodegradability (Carbone et al., 2016; Fernández et al., 2010). Due to its peculiar material properties, agar-agar gel is used to synthesize coating materials for packaging ready-to-eat foods and it has also been proposed for the gel packaging of liquid foodstuffs (De Moura et al., 2012).

The absorbent pads made up of the cellulose are employed as the matrix for embedding silver nanoparticles synthesized in situ through physicochemical reduction methods. Beef meat was analyzed and it was evident that there was a significant reduction in microbial growth during an extended period of storage (Carbone et al., 2016). Microbial analysis included enumeration of total aerobic count as well as yeast and mold counts by the plate count agar method at a temperature range of 25–30°C. The food-borne pathogens included *E. coli (E. coli) O157:H7, Salmonella, S. aureus*, and *Listeria monocytogenes*. Significant reduction of bacterial load was observed during the storage interval, whereas lactic acid bacteria showed a lower degree of sensitivity. As compared to control samples, the samples wrapped in AgNPs/cellulose polymeric films exhibited lower degree of *Pseudomonas* spp. and *Enterobacteriaceae* counts (<1 log CFU/g) (Carbone et al., 2016; Fernández et al., 2010). Fresh-cut melon samples demonstrated similar results, implying that AgNPs-loaded absorbent pads resulted in reduction of microbial loads in the fruit. In a study conducted by De Moura et al. (2012), a distinct type of cellulose was employed in which AgNPs was incorporated in the matrix of hydroxypropyl methyl-cellulose (HPMC) for possible applications in food packaging. The HPMC matrix embedded with the AgNPs exhibited good antimicrobial activity against food-borne pathogens, like *E. coli* and *S. aureus* (De Moura et al., 2012). Spices have been known as to have several antimicrobial compounds in their chemical compositions, and rosemary and oregano were processed to extract essential oils which exhibit antibacterial agents. With the use of these in conjunction with the AgNPs the quality and safety of the foodstuffs could be improved (Costa et al., 2012). Edible pullulan films were prepared by incorporation with the nanoparticles, including ZnO NPs and AgNPs, and studied for their antimicrobial efficacy. The essential oils, like oregano or rosemary oils, played a significant role in conserving the physicochemical and sensory properties of turkey deli meat (Mastromatteo et al., 2015). It was evident from the results that AgNPs in conjunction with the rosemary essential oils proved to be more effective for microbial safety of foodstuffs in comparison with the combination of rosemary oil with ZnO-NPs. The antimicrobial effectiveness of edible polymeric film was studied against gram-positive bacteria, such as *L. monocytogenes* and *S. aureus*, at different storage temperatures of 4°C, 25°C, 37°C, and 55°C for extended storage intervals up to seven weeks (Abreu et al., 2015). The foodstuffs wrapped in the pullulan edible films/AgNPs/EOs were stored at the optimum temperatures of 4°C and 25°C. The food samples exhibited marked decline in the growth of both gram-positive and gram-negative bacteria (Carbone et al., 2016).

Similarly, Morsy et al. (2014) also reported the effectiveness of pullulan films in conjunction with the essential oils, and authors have endorsed the efficacy of pullulan films made with AgNPs incorporated to control food-borne pathogens in poultry and meat products. Agar hydrogels were also employed as the hosting matrix to facilitate embedding of AgNPs. Silver montmorillonite (Ag-MMT) were also reported in the published literature, and these nanoparticles were obtained by replacing Na+ ions in the natural silver montmorillonite with silver ions (Maizura et al., 2007). These Ag-MMT were loaded into the agar hydrogels for prolonging the shelf life of *Fior di Latte* cheese. The antimicrobial effect was evident from the inhibition of pathogens, and *Pseudomonas* spp. growth specifically slowed owing to antimicrobial efficacy of Ag-MMT (Carbone et al., 2016).

The researchers also prepared edible film materials by incorporating AgNPs within a biopolymer matrix, processing this using stabilizing agents like salts of sodium alginate (anionic polysaccharide alginic acid). The gel-based films made with the sodium alginate exhibited good material properties, such as flexibility, tensile strength, and resistance to wear and tear, as well as barrier properties against oil (Khalaf et al., 2013). The alginate gel synthesis for coating purposes was carried out at low temperature. The synthesis of coating at low temperature resulted in minimization of inactivation of antimicrobial agent and helped to achieve the preservation of food quality characteristics during extended storage (Gammariello et al., 2011). The silver nanoparticles were incorporated in the sodium alginate films, and these polymeric alginate films exhibited antimicrobial efficacy and retarded the decay rate of fruits and vegetables subjected to several days of storage period. The gram-positive and gram-negative bacteria grew at reduced rate during storage (Carbone et al., 2016). In a similar research work, calcium ions were employed as active coating for sodium alginate-based films loaded with AgNPs and Ag-MMT. These films, when used as packaging material, caused improvement in the shelf life of fresh-cut carrots. Sodium alginate films in the form of bio-composite coating were also utilized for quality preservation of *Fior di latte* cheese (Gammariello et al., 2011).

Starch is also a commonly employed edible film for biodegradable packaging materials (Carbone et al., 2016). Abreu et al. (2015) prepared nanostructured starch–based film which contained AgNPs and clay. In native form, starches have their structural configurations organized into granules. Usually, starches exhibit poor mechanical properties with high affinity for water. AgNPs were modified with numerous hydroxyl groups to form a biopolymer (Khalaf et al., 2013). The physical modification of the biopolymer resulted in excellent gaseous and mechanical barrier properties and caused significant microbial reduction (Carbone et al., 2016). In other research, starch and sodium alginate films were mixed to form an edible film, and the mixing led to improvements in mechanical properties. Biopolymer films have been made with the alginate, and alginate has peculiar colloidal properties and has been shown to demonstrate poor water resistance owing to their hydrophilic nature (Carbone et al., 2016), but mixing the biopolymers resulted in improved mechanical properties of packaging film. In another study, film was prepared by adding lemongrass oil which proved effective in inhibiting *E. coli* growth (Khalaf et al., 2013). The antimicrobial effect of films were enhanced using glycerol when compared to films which did not include glycerol in its chemical composition (Carbone et al., 2016).

10.5 RELEASE CHARACTERISTICS OF SILVER NANOPARTICLES IN FOODSTUFFS

There has been a lot of research carried out in the last years regarding assessment of risks and potential health hazards to consumers associated with the migration of packaging constituents into foodstuffs (Carbone et al., 2016). Echegoyen and Nerín (2013) carried out research into the migration properties of silver nanoparticles, in a very extensive manner, using various nanocomposites, such as polypropylene and LDPE, in food materials. Specifically, the authors reported on the migration of silver across the packaging material in the form of particles or ions. The authors not only reported the migration of silver into food simulants (Simbine et al., 2019), but the rate of migration was found to be dependent on the type of food and thermal constraints applied on foodstuffs (Cushen et al., 2014). Acidic foods as well as the foods which were oven dried exhibited the highest levels of migrations of silver (Simbine et al., 2019). The authors have implied two separate mechanisms; 1) first was related to the release of nanoparticles from composites after detachment, 2) second was about silver ions dissolution after exposure to oxidation (Carbone et al., 2016; Simbine et al., 2019). In another study, it was found that silver exhibited migration below the maximum migration limits permissible by the legislation according to the European Union (Simbine et al., 2019). Among the published review papers in recent years, a study was found being conducted by Metak (2015) who carried out migration assessments of titanium dioxide and silver nanoparticles across the food packages in wide variety of foodstuffs. The instrumental analysis involved inductive-coupled plasma coupled with mass spectrometry (ICP-MS) was utilized to carry out assessment studies. The results showed the significant migration of silver and TiO2 NPs from the packaging materials in which foodstuffs were packed during extended periods of storage after 7 to 10 days. The authors concluded that the orange juice samples exhibited the highest levels of migration levels followed by apple juice samples. Among all samples, bread samples showed the least level of NPs migration (Simbine et al., 2019). Irrespective of the food type, the amounts of migrations of NPs of both silver and titanium oxide were below the prescribed concentration limit of 10 mg/mL (2002/72/EC) (Carbone et al., 2016).

10.6 CONCLUSIONS

The NPs synthesized from the silver and copper has fostered the attention of the researchers owing to their appreciable physicochemical properties and good antimicrobial effects. Bionanopolymers have faced several issues of mechanical and barrier properties for many years, and exploitation of AgNPs has led to improve fragility, gaseous barrier properties and decreased toxicity of bionanopolymers. Among the various particles evaluated for their properties, silver played the crucial role in manufacturing of NPs because of requirement of simple equipment, good antimicrobial effects, and cost-effectiveness. Hence, it is of prime significance to measure and monitor the NPs release potential under the conditions in which the modifications in the surface area of food packaging occur. However, food packaging materials prepared from the AgNPs not only caused significant improvement in shelf life of foodstuffs by retarding the growth of microorganisms but also may also exert hazardous effects on human health. Therefore, scientists have increased their efforts to study the migration of AgNPs through use of various migration tests.

Although, NPs have antimicrobial effects, however, toxic nature of AgNPs and their inherent toxicity may could have repercussions in terms of onset of disease and potential toxicity for humans. Therefore, it is dire need to carry out more studies regarding migration of AgNPs into foodstuffs. It is also justifiable to assume that to and fro migration of nanoparticles into the foodstuffs may cause accumulation of NPs and affect health of consumers in negative manner. In terms of scientific principles, organic components migration from plastic materials to foodstuffs or simulants follows the diffusion as per the Fick's law. Such migration phenomenon are governed by two phases of migration: 1) AgNP release from the surface and 2) oxidative dissolution of silver, titanium dioxide, and copper nanoparticles. The amount and rate of migration of AgNPs are usually dependent on various factors, such as film thickness, pH, type of coating/film, food simulants, temperature, and time. Regarding NPs migration, more research studies are recommended to explore the mechanical details related to the migration of NPs from polymeric packaging films into foodstuff.

REFERENCES

Abreu, A. S., Oliveira, M., de Sá, A., Rodrigues, R. M., Cerqueira, M. A., Vicente, A. A., and Machado, A. V. 2015. Antimicrobial nanostructured starch based films for packaging. *Carbohydrate Polymers* 129: 127–134.

Ahari, H., and Lahijani, L. K. 2021. Migration of silver and copper nanoparticles from food coating. *Coatings* 11(4): 380.

Ahmad, S. S., Yousuf, O., Islam, R. U., and Younis, K. 2021. Silver nanoparticles as an active packaging ingredient and its toxicity. *Packaging Technology and Science* 34(11–12): 653–663.

Ajayi, O. O., Omowa, O. F., Omotosho, O. A., Abioye, O. P., Akinlabi, E. T., Akinlabi, S. A., Abioye, A. A., Owoeye, F. T., and Afolalu, S. A. 2018. Experimental investigation of the effect of ZnO-Citrus sinensis nano-additive on the electrokinetic deposition of zinc on mild steel in acid chloride. In *TMS Annual Meeting & Exhibition* (pp. 35–40). Springer, Cham.

Anvar, A. A., Ahari, H., and Ataee, M. 2021. Antimicrobial properties of food nanopackaging: A new focus on foodborne pathogens. *Frontiers in Microbiology* 12. DOI: 10.3389/fmicb.2021.690706

Babu, P. J. 2021. Nanotechnology mediated intelligent and improved food packaging. *International Nano Letters* 12: 1–14.

Carbone, M., Donia, D. T., Sabbatella, G., and Antiochia, R. 2016. Silver nanoparticles in polymeric matrices for fresh food packaging. *Journal of King Saud University-Science* 28(4): 273–279.

Chhipa, H. 2019. Mycosynthesis of nanoparticles for smart agricultural practice: A green and eco-friendly approach. In *Green synthesis, characterization and applications of nanoparticles* (pp. 87–109). Elsevier Radarweg, Amsterdam, Netherlands.

Costa, C., Conte, A., Buonocore, G. G., Lavorgna, M., and Del Nobile, M. A. 2012. Calcium-alginate coating loaded with silver-montmorillonite nanoparticles to prolong the shelf-life of fresh-cut carrots. *Food Research International* 48(1): 164–169.

Cushen, M., Kerry, J., Morris, M., Cruz-Romero, M., and Cummins, E. 2014. Evaluation and simulation of silver and copper nanoparticle migration from polyethylene nanocomposites to food and an associated exposure assessment. *Journal of Agricultural and Food Chemistry* 62(6): 1403–1411.

Dasgupta, N., Ranjan, S., Mundekkad, D., Ramalingam, C., Shanker, R., and Kumar, A. 2015. Nanotechnology in agro-food: From field to plate. *Food Research International* 69: 381–400.

De Moura, M. R., Mattoso, L. H., and Zucolotto, V. 2012. Development of cellulose-based bactericidal nanocomposites containing silver nanoparticles and their use as active food packaging. *Journal of Food Engineering* 109(3): 520–524.

Echegoyen, Y., and Nerín, C. 2013. Nanoparticle release from nano-silver antimicrobial food containers. *Food and Chemical Toxicology* 62: 16–22.

Elsabee, M. Z., and Abdou, E. S. 2013. Chitosan based edible films and coatings: A review. *Materials Science and Engineering: C* 33(4): 1819–1841.

Eze, S. C., Umeh, S. I., Onyeke, C. C., Ameh, G. I., and Ugwuoke, K. I. 2016. Preliminary investigations on the control of yam *Dioscorea rotundata Poir* tuber rot through nanoscience. *Nanotechnology Reviews* 56: 499–505.

Fernández, A., Picouet, P., and Lloret, E. 2010. Cellulose-silver nanoparticle hybrid materials to control spoilage-related microflora in absorbent pads located in trays of fresh-cut melon. *International Journal of Food Microbiology* 142(1–2): 222–228.

Gammariello, D., Conte, A., Buonocore, G. G., and Del Nobile, M. A. 2011. Bio-based nanocomposite coating to preserve quality of Fior di latte cheese. *Journal of Dairy Science* 94: 5298–5304.

García, M., Forbe, T., and Gonzalez, E. 2010. Potential applications of nanotechnology in the agro-food sector. *Food Science and Technology* 30(3): 573–581.

Hamad, A. F., Han, J. H., Kim, B. C., and Rather, I. A. 2018. The intertwine of nanotechnology with the food industry. *Saudi Journal of Biological Sciences* 25: 27–30.

He, X., Deng, H., and Hwang, H. M. 2019. The current application of nanotechnology in food and agriculture. *Journal of Food and Drug Analysis* 27: 1–21.

Huang, S., Wang, L., Liu, L., Hou, Y., and Li, L. 2015. Nanotechnology in agriculture, livestock, and aquaculture in China. A review. *Agronomy for Sustainable Development* 35: 369–400.

Incoronato, A. L., Conte, A., Buonocore, G. G., and Del Nobile, M. A. 2011. Agar hydrogel with silver nanoparticles to prolong the shelf life of fior di latte cheese. *Journal of Dairy Science* 94(4): 1697–1704.

Khalaf, H. H., Sharoba, A. M., El-Tanahi, H. H., and Morsy, M. K. 2013. Stability of antimicrobial activity of pullulan edible films incorporated with nanoparticles and essential oils and their impact on turkey deli meat quality. *Journal of Food and Dairy Sciences* 4(11): 557–573.

Kim, D. Y., Kadam, A., Shinde, S., Saratale, R. G., Patra, J., and Ghodake, G. 2018. Recent developments in nanotechnology transforming the agricultural sector: A transition replete with opportunities. *Journal of the Science of Food and Agriculture* 98(3): 849–864.

Kumari, A., and Yadav, S. K. 2014. Nanotechnology in agri-food sector. *Critical Reviews in Food Science and Nutrition* 54(8): 975–984.

Lamba, A., and Garg, V. 2018. Nanotechnology approach in food science: A review. *Nanotechnology* 3(2): 183–186.

Lyons, K., Scrinis, G., and Whelan, J. 2018. Nanotechnology, Agriculture, and Food. In *Nanotechnology and Global Sustainability* (pp. 146–169). CRC Press, Boca Raton, Florida, United States.

Mahato, D. K., Mishra, A., and Kumar, P. 2021. Nanoencapsulation for agri-food application and associated health and environmental concerns. *Frontiers in Nutrition* 8: 146.

Mahdi, S. S., Vadood, R., and Nourdahr, R. 2012. Study on the antimicrobial effect of nanosilver tray packaging of minced beef at refrigerator temperature. *Global Veterinaria* 9(3): 284–249.

Maizura, M., Fazilah, A., Norziah, M. H., and Karim, A. A. 2007. Antibacterial activity and mechanical properties of partially hydrolyzed sago starch–alginate edible film containing lemongrass oil. *Journal of Food Science* 72(6): C324–C330.

Martínez-Abad, A., Lagaron, J. M., and Ocio, M. J. 2012. Development and characterization of silver-based antimicrobial ethylene–vinyl alcohol copolymer EVOH films for food-packaging applications. *Journal of Agricultural and Food Chemistry* 60(21): 5350–5359.

Mastromatteo, M., Conte, A., Lucera, A., Saccotelli, M. A., Buonocore, G. G., Zambrini, A. V., and Del Nobile, M. A. 2015. Packaging solutions to prolong the shelf life of fiordilatte cheese: Bio-based nanocomposite coating and modified atmosphere packaging. *LWT-Food Science and Technology* 60(1): 230–237.

Maynard, A. D., Aitken, R. J., Butz, T., Colvin, V., Donaldson, K., Oberdörster, G., Philbert, M. A., Ryan, J., Seaton, A., Stone, V., and Tinkle, S. S. 2006. Safe handling of nanotechnology. *Nature* 444(7117): 267–269.

Mazzuca, C., Micheli, L., Carbone, M., Basoli, F., Cervelli, E., Iannuccelli, S., Sotgiu, S., and Palleschi, A. 2014. Gellan hydrogel as a powerful tool in paper cleaning process: A detailed study. *Journal of Colloid and Interface Science* 416: 205–211.

McIntyre, R. A. 2012. Common nano-materials and their use in real world applications. *Science Progress* 95(1): 1–22.

Metak, A. M. 2015. Effects of nanocomposite based nano-silver and nano-titanium dioxide on food packaging materials. *International Journal of Applied Science and Technology* 5(2): 26–40.

Mfon, R. E., Odiaka, N. I., and Sarua, A. 2017. Interactive effect of colloidal solution of zinc oxide nanoparticles biosynthesized using *Ocimum gratissimum* and *Vernonia amygdalina* leaf extracts on the growth of *Amaranthus cruentus* seeds. *African Journal of Biotechnology* 16(26): 1481–1489.

Misra, A. N., Misra, M., and Singh, R. 2013. Nanotechnology in agriculture and food industry. *International Journal of Pure and Applied Sciences and Technology* 16(2): 1–9.

Mittal, D., Kaur, G., Singh, P., Yadav, K., and Ali, S. A. 2020. Nanoparticle-based sustainable agriculture and food science: Recent advances and future outlook. *Frontiers in Nanotechnology* 2: 10. DOI: 10.3389/fnano.2020.579954

Mitter, N., and Hussey, K. 2019. Moving policy and regulation forward for nanotechnology applications in agriculture. *Nature Nanotechnology* 14(6): 508–510.

Mlalila, N., Kadam, D. M., Swai, H., and Hilonga, A. 2016. Transformation of food packaging from passive to innovative via nanotechnology: Concepts and critiques. *Journal of Food Science and Technology* 53(9): 3395–3407.

Mohammad, A. W., Lau, C. H., Zaharim, A., and Omar, M. Z. 2012. Elements of nanotechnology education in engineering curriculum worldwide. *Procedia-Social and Behavioral Sciences* 60: 405–412.

Morsy, M. K., Khalaf, H. H., Sharoba, A. M., El, Tanahi, H. H., and Cutter, C. N. 2014. Incorporation of essential oils and nanoparticles in pullulan films to control foodborne pathogens on meat and poultry products. *Journal of Food Science* 79(4): M675–M684.

Ndlovu, N., Mayaya, T., Muitire, C., and Munyengwa, N. 2020. Nanotechnology applications in crop production and food systems. *International Journal of Plant Breeding and Crop Science* 7(1): 624–634.

Ndukwu, M. C., Ikechukwu-Edeh, C. E., Nwakuba, N. R., Okosa, I., Horsefall, I. T., and Orji, F. N. 2020. Nanomaterials application in greenhouse structures, crop processing machinery, packaging materials and agro-biomass conversion. *Materials Science for Energy Technologies* 3: 690–699.

Prasad, R., Bhattacharyya, A., and Nguyen, Q. D. 2017. Nanotechnology in sustainable agriculture: Recent developments, challenges, and perspectives. *Frontiers in Microbiology* 8: 1014.

Qiu, L., Zhang, M., Bhandari, B., and Yang, C. 2020. Shelf life extension of aquatic products by applying nanotechnology: A review. *Critical Reviews in Food Science and Nutrition* 62(6): 1–15.

Reichert, C. L., Bugnicourt, E., Coltelli, M. B., Cinelli, P., Lazzeri, A., Canesi, I., Braca, F., Martínez, B. M., Alonso, R., Agostinis, L., and Verstichel, S. 2020. Bio-based packaging: Materials, modifications, industrial applications and sustainability. *Polymers* 12(7): 1558.

Rossi, M., Passeri, D., Sinibaldi, A., Angjellari, M., Tamburri, E., Sorbo, A., and Dini, L. 2017. Nanotechnology for food packaging and food quality assessment. *Advances in Food and Nutrition Research* 82: 149–204.

Sastry, R. K., Rashmi, H. B., and Rao, N. H. 2011. Nanotechnology for enhancing food security in India. *Food Policy* 36: 391–400.

Scott, N., and Chen, H. 2013. Nanoscale science and engineering for agriculture and food systems. *Industrial Biotechnology* 9(1): 17–18.

Scott, N. R., Chen, H., and Cui, H. 2018. Nanotechnology applications and implications of agrochemicals toward sustainable agriculture and food systems. *Journal of Agricultural and Food Chemistry* 66(26): 6451–6456.

Sekhon, B. S. 2014. Nanotechnology in agri-food production: An overview. *Nanotechnology, Science and Applications* 7: 31–53.

Sharma, C., Dhiman, R., Rokana, N., and Panwar, H. 2017. Nanotechnology: An untapped Resource for food packaging. *Frontiers in Microbiology* 8: 1735.

Shuttleworth, P. S., Parker, H. L., Hunt, A. J., Budarin, V. L., Matharu, A. S., and Clark, J. H. 2014. Applications of nanoparticles in biomass conversion to chemicals and fuels. *Green Chemistry* 16(2): 573–584.

Simbine, E. O., Rodrigues, L. D. C., Lapa-Guimaraes, J., Kamimura, E. S., Corassin, C. H., and Oliveira, C. A. F. D. 2019. Application of silver nanoparticles in food packages: A review. *Food Science and Technology* 39(4): 793–802.

Toker, R. D., Kayaman-Apohan, N. L. H. A. N., and Kahraman, M. V. 2013. UV-curable nano-silver containing polyurethane based organic–inorganic hybrid coatings. *Progress in Organic Coatings* 76(9): 1243–1250.

Tripathi, M., Kumar, S., Kumar, A., Tripathi, P., and Kumar, S. 2018. Agro-nanotechnology: A future technology for sustainable agriculture. *International Journal of Current Microbiology and Applied Sciences* 7: 196–200.

Usman, M., Pan, L., Farid, A., Khan, A. S., Yongpeng, Z., Khan, M. A., and Hashim, M. 2020. Carbon nanocoils-nickel foam decorated with silver nanoparticles/sheets using a novel stirring assisted electrodeposition technique for non-enzymatic glucose sensor. *Carbon* 157: 761–766.

Valencia-Chamorro, S. A., Palou, L., Del Río, M. A., and Perez-Gago, M. B. 2011. Antimicrobial edible films and coatings for fresh and minimally processed fruits and vegetables: A review. *Critical Reviews in Food Science and Nutrition* 51(9): 872–900.

Valipoor Motlagh, N., Hamed Mosavian, M. T., and Mortazavi, S. A. 2013. Effect of polyethylene packaging modified with silver particles on the microbial, sensory and appearance of dried barberry. *Packaging Technology and Science* 26(1): 39–49.

Venkateshwarlu, G., and Nagalakshmi, K. 2013. Developments in bionanocomposite films: Prospects for eco-Friendly and smart food packaging. *Asian Biotechnology and Development Review* 15(3): 51–66.

Youssef, A. M., and Abdel-Aziz, M. S. 2013. Preparation of polystyrene nanocomposites based on silver nanoparticles using marine bacterium for packaging. *Polymer-Plastics Technology and Engineering* 52(6): 607–613.

11 Nanoencapsulation of Essential Oil-Based Packaging for Shelf-Life Extension of Foods

Suleyman Polat[1] *and Ahmet Aygun*[2]
[1]Food Engineering Department, Cukurova University, Adana, Turkey
[2]Finike Vocational School, Akdeniz University, Antalya, Turkey

CONTENTS

DOI: 10.1201/9781003207641-13

11.1 INTRODUCTION

Essential oils are volatile compounds obtained from a variety of parts of aromatic plants. The composition and amounts of essential oils vary depending on the type of spice, production method, climate, and geographical structure of the region where it is grown. It is known that the antimicrobial and antioxidant properties of essential oils highly depend on the type and composition of the essential oils. Distillation and extraction methods are used to obtain essential oils, and methods such as water distillation, supercritical liquid extraction, microwave extraction, and solid-phase microextraction are commonly used (Reyes-Jurado et al., 2015).

Recently, consumer demand for natural ingredients have been increased due to the potential dangers of synthetic additives. Thus, the use of essential oils in the food industry is increasing due to their antimicrobial and antioxidant properties as well as the flavour and colour they bring to food. In parallel with food production technologies, food packaging technology is developing day by day. Packaging produced with nanotechnology, such as biodegradable packaging, active packaging, and smart packaging, are new packaging techniques that have been frequently mentioned in the recent decade. Essential oils or their components are very suitable compounds to be used in active packaging, thanks to their properties, such as inhibiting pathogenic microorganisms and protecting food products. The essential oils or components can enable packaging materials to gain antioxidant activity with the help of their capability to act as oxygen scavengers. Also, it can delay or prevent the microbial spoilage of foods by providing antibacterial properties to the packaging material. Recent applications of essential oils for active packaging consist of their usage as films or coatings applied to packaging materials for different food groups, such as fruits and vegetables, and to fish, meat, dairy, and bakery products.

Essential oil or its components can affect the microstructure as well as optical (colour, gloss, and transparency), mechanical, and barrier properties of the final packaging material. For these effects, the type and concentration of essential oils or its components play a crucial role. In general, positive effects of using essential oils in packaging studies have been reported, but some factors, such as high volatility, low solubility, strong aroma, and the possibility of adversely affecting the sensory properties of foods, may limit their usage. Also, some of them are sensitive to heat and light. These disadvantages limit the widespread usage of essential oils in active packaging material production. Nanoemulsification and encapsulation technology (microencapsulation or nanoencapsulation) increase the use of essential oils in applications. These technologies have advantages like reducing volatility, ensuring controlled release, and increasing the stability of essential oils or components (Sharma et al., 2021).

In this chapter, the most used essential oils in food packaging studies are introduced, extraction and encapsulation techniques are explained, and the use of nanoencapsulated essential oils in active packaging technology for extending the shelf life of foods are discussed from the light of recently available study consequences.

11.2 ESSENTIAL OILS

Essential oils are liquid, volatile, clear or rarely coloured, and are soluble in oil and often in organic solvents which have a lower density than water. They are widely used in food, agriculture, pharmaceutics, sanitary, and cosmetic industries with their antimicrobial, antiparasitic, insecticide properties. They are mostly distilled from aromatic plants and contain various volatile molecules, such as terpenes and terpenoids, aromatic compounds derived from phenol, and aliphatic compounds. The production method is one of the factors affecting the amount and concentration of these volatile compounds. Distillation and extraction methods are applied to obtain essential oils. Methods such as water distillation, supercritical liquid extraction, microwave extraction, and solid-phase microextraction are processes that have been applied in recent years. Due to their bactericidal and fungicidal properties, pharmaceutical and food uses are becoming more common as an alternative to synthetic chemical products to maintain ecological balance (Angioni et al., 2006; Masotti et al., 2003).

Essential oils or their components are used in the cosmetic industry, such as for perfume and make-up products, as well as in the fields of medicine and alternative medicine, agricultural pesticides, and food additives, due to their various properties. For example, geranyl acetate, d-carvone or d-limonene are used in products that are already on the market, such as perfumes, creams, soaps, and household cleaning products, and as flavouring additives in foods (Bakkali et al., 2008).

The application purposes of essential oils or their components in food industry can be summarized: to give aroma, to increase the quality of food, and to minimize the deterioration that may occur as a result of oxidation and microbial activity. The composition of essential oils largely depends on the type of plant, the ecology and soil structure in which it grows, the part of the plant used in the extraction (leaf, seed, flower, fruit, stem, etc.), and the extraction method. Thus, the biological functions of the essential oils can vary depending on these factors (Reyes-Jurado et al., 2015). The most used and/or investigated essential oils and their main components are given in Table 11.1.

TABLE 11.1
Some Essential Oils and Their Major Compounds

Essential Oil Source	Main Component	Picture
Oregano (*Origanum vulgare* L.)	Thymol, carvacrol, γ-terpinene	
Thyme (*Thymus vulgaris* L.)	Thymol	
Winter savory (*Santureja Montana* L.)	Thymol, p-cymene, γ-terpinene, carvacrol	
Cuneate Turkish savory (*Santureja* cuneifolia Ten.)	Linalool, germacrene D, α-pinene, thymol	
Clove (*Syzygium aromaticum* L.)	Eugenol	
Sage (*Salvia officinalis* L.)	α-thujone, camphor, viridiflorol	
Rosemary (*Rosmarinus officinalis* L.)	α-pinene, limonene, camphor	

(Continued)

TABLE 11.1 (*Continued*)
Some Essential Oils and Their Major Compounds

Essential Oil Source	Main Component	Picture
Sweet basil (*Ocimum basilicum* L.)	Linalool, estragole, methyl cynnamate, eugenol	
Cumin (*Cumin cyminum* L.)	Cuminaldehyde	
Black cumin (*Nigella sativa* L.)	Thymoquinone, carvacrol	
Caraway (*Carum carvi* L.)	Carvone	
Mint (*Mentha piperita* L.)	Isomenthone, neomenthol, 1,8-cineole	
Green anise (*Pimpinella anisum* L.)	Trans-anethol, estragole	
Celery (*Apium graveolens* L.)	β-selinene, phellandral, limonene	

(*Continued*)

TABLE 11.1 ***(Continued)***
Some Essential Oils and Their Major Compounds

Essential Oil Source	Main Component	Picture
Cinnamon (*Cinnamomum zeylanicum*)	*E*-Cinnamaldehyde	
Eucalyptus globulus	Eucalyptol, α-pinene	
Orange peel	Limonene, citral	
Lemon peel (*Citrus limonum*)	Limonene, citral	

11.3 PRODUCTION OF ESSENTIAL OILS

Different production techniques are widely used to produce essential oils, such as distillation, extraction, and mechanic methods (Figure 11.1). Essential oils can be produced using different techniques, and choosing the appropriate technique significantly affects the yield and composition of essential oils. In the selection of the method, the source of the essential oil, the process time and cost, and the possible impacts on environment should be considered.

11.3.1 Hydrodistillation

Hydrodistillation is a traditional method widely used in obtaining volatile compounds. Clevenger-type apparatus is used for small-scale production, while a large distilling apparatus (still) is used in industrial applications. The method is based on the principle of boiling the water and plant material for a length of time, condensing the oil molecules that become volatile with water vapour during boiling and separating them from the water.

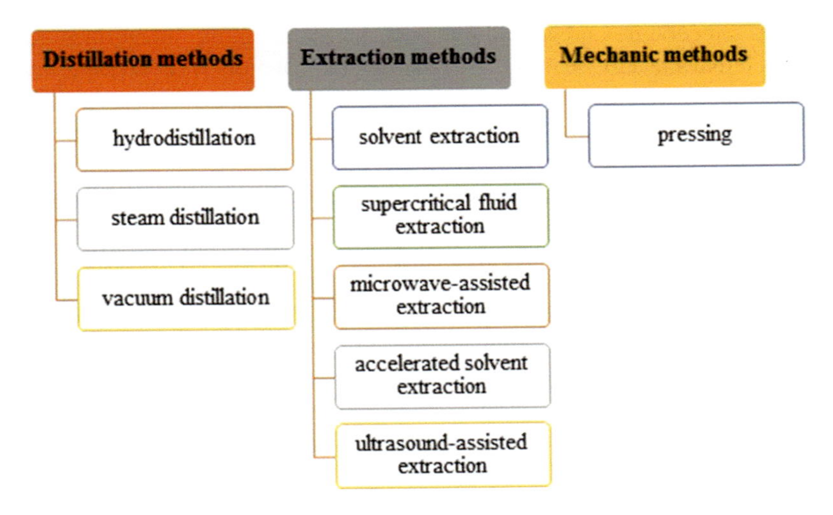

FIGURE 11.1 Production methods of essential oils.

11.3.2 Steam Distillation

In this method, the steam applied to the fresh plant material is placed in the container with the help of pressure, the oil droplets are dragged along with it to the collection container, where the oil is condensed and separated from the water.

11.3.3 Vacuum Distillation

Some compounds have very high boiling points. To obtain these compounds, the sources of these compounds must be boiled at low temperature to prevent thermal degradation. When the pressure value is lowered below the vapour pressure of the compound, the distillation process takes place at a lower temperature.

11.3.4 Solvent Extraction

This is the traditional extraction method, where the plant material is immersed directly in the solvent and treated at appropriate temperatures. Hexane and ethanol are mostly used as organic solvents in industrial applications. The advantage of this method over steam distillation is the use of low temperatures during extraction. The low temperature ensures that the essential oil obtained has more natural content than when extracted through the steam distillation process by avoiding hydrolysis and isomerization reactions. Nevertheless, the drawbacks of the method are connected to the usage of a high volume of solvents and loss of low molecular weight volatile compounds during the condensation process.

11.3.5 Supercritical Fluid Extraction (SFE)

In this method, substances with supercritical fluid properties are used as solvents instead of organic solvents. A substance exhibits supercritical fluid properties above the critical temperature and critical pressure. At this point, the supercritical fluid is between a liquid and a gas in terms of its thermophysical properties. While many substances can be dissolved owing to the dissolving power of liquid solvents, the fact that the diffusion coefficient is closer to gases in supercritical fluids makes the dissolution process more effective. Carbon dioxide (CO_2) is the most used solvent in this method due to its low cost, high purity rate, easy usage, and minimal environmental impact.

11.3.6 Accelerated Solvent Extraction

In this method, higher temperatures, and pressures (usually 50–200°C and 10–15 MPa) are used to decrease the extraction time compared to solvent extraction. Thus, this process can be called a type of pressurized solvent extraction that is quite similar to supercritical fluid extraction. Extraction is performed under pressure to keep the solvent in a liquid state at high temperatures. Due to extreme operation conditions, it has limited industrial applications to produce essential oils.

11.3.7 Microwave-Assisted Extraction

This method provides a quick energy supply for the extraction medium (solvent and solid plant matrix). Efficient and homogeneous heating of the extraction medium occurs. Mostly, two different types of microwave-assisted extraction techniques are used. One of them is closed-system extraction, which is done in a closed vessel where temperature and pressure can be controlled. The other is carried out in an open container under atmospheric pressure. When compared to supercritical fluid extraction, this method offers process simplicity and low cost. Compared to the distillation methods, it provides lower extraction time and solvent amount. However, additional operations such as filtration or centrifugation should be applied to remove the solid plant residues in this method.

11.3.8 Ultrasound-Assisted Extraction

Ultrasound extraction treatments mainly can be applied with an ultrasound water bath or with a probe sonicator. The ultrasonic baths have more than one transducer at the base part of the device and the probe sonicators have one transducer. Generally, probe systems can give higher energy directly to the extraction medium than the ultrasonic bath is able to. During processing, electrical energy is transformed into mechanical vibration by a transducer. While some of the electrical energy is converted to heat, the generated ultrasound energy causes a phenomenon in the medium called "cavitation." The cavitation contributes to penetration of the solvent into cellular materials, disrupting biological cell walls and improving mass transfer.

11.3.9 Pressing

Some plants' essential oils decompose when higher temperatures are applied in the extraction. Pressing method is generally used for the peels of some citrus fruits. Essential oils can be obtained by putting the peels of citrus fruits in a cloth bag and squeezing them in a cold hydraulic press.

Among the methods outlined, the distillation method has been widely used in the production of most essential oils for a long time. With developing technologies, research has shown that newer techniques, such as microwave-assisted, ultrasound-assisted, or supercritical fluid extraction, can be used in the production of essential oils. Studies have shown that these methods provide some advantages compared to the distillation methods which are accepted as traditional methods (Reyes-Jurado et al., 2015). Possible advantages provided by the new methods can be listed:

- shortened extraction time
- increased extraction yield
- enhanced essential oil quality
- reduced operation cost
- less environmental effect compared to solvent extraction

The microwave method can be combined with solvent extraction (solvent-less microwave) (Lucchesi et al., 2004) or hydrodistillation method (microwave-assisted hydrodistillation) (Stashenko et al., 2004; Wang et al., 2006) to take advantage of microwave heating. The combination methods significantly reduce extraction time and have less effect on essential oil quality and quantity. Although laboratory studies report positive results for the new extraction techniques to produce essential oils, these have limited use in industrial applications. Nevertheless, these combined techniques have the potential to be widely used in commercial productions soon.

11.4 POTENTIAL USAGE OF ESSENTIAL OILS IN THE FOOD INDUSTRY

The potential applications of essential oils in the food industry are given in Figure 11.2. They are generally used in small amount to give flavour to some foods, such as hard candies, bakery products, and beverages. As they are produced in concentrated form,

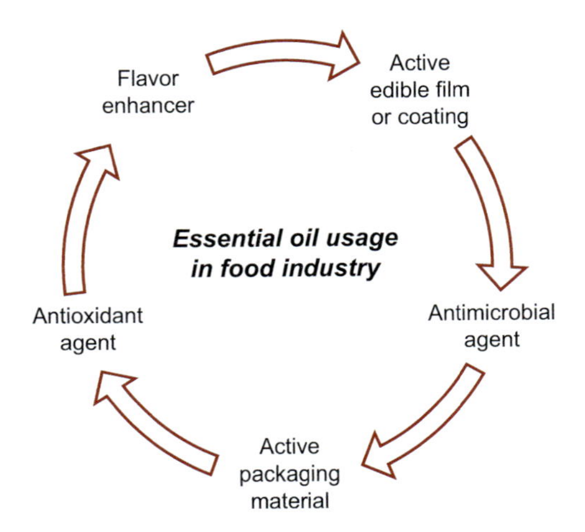

FIGURE 11.2 Potential applications of essential oils in the food industry.

they have a higher intensity of flavour. Thus, they cannot be used in large amounts in formulations of foods because they can affect the sensory properties of the product. Different kinds of essential oil, such as spearmint oil, grapefruit oil, peppermint oil, lemon oil, cinnamon oil, and lemongrass oil are commercially available in small bottles.

The Food and Drug Administration (FDA) and the US Department of Agriculture (USDA) authorize food additive regulations in the United States. The FDA classifies the essential oils as generally recognized as safe (GRAS) when used for their intended purpose (FDA, 2020). The European Commission (EC) announced that the essential oils may be used as flavouring agents in and on food products. According to Commission regulations, essential oils which will be used as a flavouring agent should meet various requirements that show the safety of food flavourings and that they are registered by the Commission. There are different regulations in the evaluation of the safety of food packaging materials but only that labelled with (EC) No. 1935/2004 was applied for the evaluation of the safety of the substances used in active packaging. This evaluation considers only the amount of the active substances found in the food materials and that they were within the maximum limit of the regulation, whatever the sources of these compounds.

11.4.1 Antimicrobial Usage of the Essential Oils

The increasing trend of consumers toward products without synthetic preservatives has guided food researchers to develop alternative foods, food additives, and techniques. Different types of essential oils or their components are presented as naturally antimicrobial substances for restricting microbial spoilage and also as antioxidant substances. Generally, the phenolic components of the essential oils (carvacrol, eugenol, cinnamaldehyde, thymol, or citral) show significant activity. Also terpenes (citronellol, limonene, linalool, α-pinene, terpineol) and ketones (α-thujone, geranyl acetate, or β-myrcene) have shown antimicrobial activity (Dhifi et al., 2016). Most studies showed that bacteria, especially gram-positive strains, are more sensitive than gram-negative bacteria to the essential oils or their bioactive compounds. Moreover, researchers concluded that the different possible essential oils have different component action mechanisms on microorganisms. Those action mechanisms were summarized by Khorshidian et al. (2018) as membrane protein damage, cell wall destruction, cytoplasm coagulation, adenosine triphosphate (ATP) hydrolysis, decrease in the synthesis of ATP, proton motive force reduction, cytoplasmic membrane damage, increasing cell permeability and leakage of the cell content, and membrane protein damage. Antifungal activity of most essential oils or their main compounds have also been proven (Basak and Guha, 2017; Belasli et al., 2020; Boubaker et al., 2016; Chaudhari et al., 2020). Generally, the studies on antimicrobial activity of essential oils showed that higher concentrations of essential oils are required in food systems than for in vitro experiments. The explanation for this is that foods with rich composition serve as nutrient sources for microorganisms. The higher concentration of essential oils created concerns about changing organoleptic properties of the foodstuff (Fisher and Phillips, 2008).

11.4.2 Antioxidant Usage of the Essential Oils

Undesirable changes in foods occur because of oxidation reactions. Foods containing oils and fats are subjected to oxidation, resulting in rancidity, nutrient loss, discolouration, and changes in texture due to peroxidation. Antioxidant substances are generally used to prevent oxidation in food products. The antioxidant potential of the essential oils and health concerns about the most common synthetic antioxidants, butylated hydroxyanisole (BHA) and butylated hydroxytoluene (BHT), applied in the food industry have accelerated the investigations of essential oil antioxidant properties. The antioxidant properties are mostly derived from their chemical composition, particularly of phenols, some terpenoids, and other volatile constituents. Those molecules can react with radicals and stop or delay the aerobic oxidation of organic matter. Thymol and carvacrol (monoterpenes) are found in different types of essential oils and their antioxidant properties are comparable to those of synthetic antioxidants. These compounds are detected in oils extracted from medicinal and pharmacological plants, such as oregano, nutmeg, parsley, thyme, cinnamon, basil, and clove. Moreover, the other components, such as some alcohols, ethers, ketones, aldehydes, and terpenes (1,8-Cineole, geranial/neral, citronellal, linalool, isomenthone, menthone), contribute to the antioxidant properties (Amorati et al., 2013). Studies showed that botanical sources, environmental factors, and extraction methods affect the antioxidant activity properties.

11.4.3 Usage in Food Packaging Applications

It has been stated that essential oils with antioxidant and antimicrobial properties can be used for active food packaging materials development. Thus, the essential oils have been used in different ways in food packaging applications by researchers (Figure 11.3). They can be applied as a component of washing solution before packaging for post-harvest disease control of fresh fruit and vegetables, used as an active compound in edible film and coatings, or incorporated with existing food packaging materials.

The inhibitory and synergistic effects of essential oils (e.g., marjoram, thyme, rosemary, and cinnamon oils) were shown in pear fruits against *Botrytis cinerea* and *Penicillum expansum* (Nikkhah et al., 2017). Nikkhah and Hashemi (2020), applied selected essential oils (e.g., thyme, cinnamon, thyme/cinnamon, and cinnamon/rosemary/thyme) to jujube fruit and the efficiency of those essential oils to control and regulate the post-harvest changes were detected. Also, the application of dual or triple combinations of oils was recommended to control fungal spoilage.

FIGURE 11.3 Potential food packaging applications of essential oils.

Active packaging methods can be defined as the addition of certain active substances into the package to preserve or extend the product quality and shelf life. As the essential oils have high amounts of antioxidant and antimicrobial properties, various studies have been conducted on the use of essential oils as films and coatings in active packaging (Sharma et al., 2021).

An edible film or coating is a natural material used for encasing (i.e., coating, packaging, or wrapping) food, to extend the shelf life and to improve product quality (Pavlath and Orts, 2009). While edible films are identified as thin-layer structures, coatings are a particular form of edible films, directly applied to the surface of materials (Han and Aristippos, 2005). Edible films and coatings are generally produced from polysaccharides (starch and cellulosic derivatives), proteins, and lipids. Also, plasticizers and other additives (e.g., colourants, flavours, fortifying nutrients) can be added to the film and coating solutions. Edible films and coatings can constitute a good protective environment to essential oils or their components due to their carrier and encapsulation properties. They may improve the ability of edible films to retard moisture, oxygen, aromas, and solute transport (Sharma et al., 2021). The properties of the edible film are highly dependent on the type and concentration of the essential oil added.

Zheng et al. (2019) produced edible chitosan-based film that contains eugenol and acorn starch. It was determined that the antioxidant activity increased with increased amounts of eugenol in the film. Cinnamon oil, ginger essential oil, rosemary extracts, and lavender essential oils were used in the different polymer matrices, and the resulting films showed antioxidant properties (Sharma et al., 2021). In addition to studies of films, several studies are available on edible coatings that contain essential oils. Chitosan-hydroxypropyl methylcellulose–based coatings with bergamot oil were applied to grapes as a post-harvest preservation method (Sánchez-González et. al., 2011). The coating showed effective antimicrobial activity and minimal impact on grape sensory properties.

Apples coated with a solution containing oregano, lemongrass essential oil, and vanillin (incorporated into puree-alginate) significantly reduced the growth of microorganisms (Rojas-Graü et al., 2007). Chitosan-clove essential oil coating was applied to cooked pork sausages and inhibition of microbial development, delayed lipid oxidation, and shelf-life extension were reported (Lekjing, 2016). Gelatin-orange essential oil coating was applied on pink shrimp and the coating extended the shelf life by nearly ten days (Alparslan et al., 2016). Cinnamon or lemongrass essential oil were incorporated in an arabic gum-sodium caseinate–based coating solution and applied on guava fruit. The coating resulted in browning and related enzyme decreases and higher antioxidant activity, and a high content of phenolic compounds were detected in coated samples (Murmu and Mishra, 2018). Gelatin-oregano essential oil–based coating applied to refrigerated rainbow trout fillets resulted in lower microbial growth, thiobarbituric acid, peroxide value, and total volatile basic nitrogen (Hosseini et al., 2016).

The other technique for producing essential oil–based active food packaging is by incorporating the essential oils or their components on existing packaging materials, such as plastic films (polyethylene, polypropylene, etc.) or papers. Hu et al. (2015) used cinnamon essential oil incorporated with chitosan-based emulsion sprayed on the surface of low-density polyethylene (LDPE) films. Depending on the chitosan nanoparticle size, active LDPE film increased the shelf life of the chilled pork. Rodriguez et al., (2008) produced an active paper packaging that contained cinnamon essential oil and solid wax paraffin. The active paper used for packaging slices of bread showed good antifungal activity against *R. stolonifer.* The same researchers tried active paper (containing cinnamon essential oil) for packaging strawberries and the protective effect of the active paper lasted for seven days of storage.

Chemically unstable essential oils degrade very rapidly under air, light, humidity, and high temperature (Beristain et al., 2001). Also, incorporation of some essential oils or their components into the packaging materials for active food packaging is highly difficult due to their strong odour and high volatility. With the encapsulation technology, developed as a solution for those problems, essential oils are coated with layers, thus protecting their components and delaying their volatility. In addition, it is possible to reduce potential problems that may be encountered in the processing, storage, and transportation processes of essential oils and to extend their shelf life by converting them from liquid state to dry powder form (Jackson and Lee, 1991).

11.5 NANOENCAPSULATION METHODS AND WALL MATERIALS FOR ENCAPSULATION OF ESSENTIAL OILS OR COMPOUNDS

Nanoencapsulation is a technique developed to retain bioactive substances in nanoscale shells made of biodegradable polymers (polysaccharides and proteins) and lipids. Various methods and/or equipment is used for nanoencapsulation of essential oils. Below are explanations and some case studies about the equipment and nanoencapsulation techniques, as well as about how to encapsulate with wall materials and the advantages and disadvantages of these wall materials.

11.5.1 Types of Equipment

Some equipment, especially homogenizers, grinders, mixing equipment, and etc., are used to encapsulate bioactive compounds and essential oils. Compounds such as nanofibres can only be produced with specially developed equipment such as electro-spinning and electro-spray (Bhushani and Anandharamakrishnan, 2014). Electro-spinning and electro-spray are electrohydrodynamic processes used to produce fibres or particles by spraying polymer solutions (Liao et al., 2021).

11.5.2 Emulsification

Emulsification can be defined as the process of mixing two immiscible liquids by a method such as ultrasonication, high pressure homogenization, microfluidization, and high shear mixing. The resulting product is called an emulsion. In addition, emulsions can be obtained by methods such as spontaneous nano-emulsification and phase inversion (Liao et al., 2021).

11.5.3 Spray Drying

Spray drying can be defined as spraying the liquid (to be encapsulated) and the coating material together into a drying chamber and encapsulating the liquid that has become small droplets during drying. During drying, hot air is fed to the chamber. Nano or microcapsules are obtained by passing through the sieve integrated to the exit of the drying cell.

Encapsulation with the nano-spray drying technique is done using a drying gas and a coating agent during the drying of the liquid, like conventional spray drying. The formation of droplets during drying and the surrounding of this droplet with the help of the agent used is based on the principle of nanoemulsion formation by turning the hydrophilic parts of the used agent outwards and the lipophilic parts inwards while the water in the liquid moves away from the droplet during drying. For the sprayed liquid to be nano-sized, an atomizer different from conventional drying is used, and these nano-droplets must pass through a vibrating sieve consisting of 4.0, 5.5, and 7.0 μm sized holes. Additionally, it was reported that the particle sizes of the samples obtained by spray drying vary depending on the liquid concentration, spraying rate, drying gas temperature, and droplet size (Jafari et al., 2008).

11.5.4 Ionotropic Gelation

Ionotropic gelation is another technique used to produce micro- or nano-sized particles by electrostatic interaction between two ionic species, at least one of which is polymer, under certain conditions. When a bioactive compound is added to the reaction, this compound joins between the chains and takes place in the polymeric structure in nanoparticle size. Such a formulation provides for controlled release of the bioactive compound and opens other potential uses, such as co-encapsulation of compounds, site-specific functionalization of particles, and increased times of compound bioactivity.

The most common species used for the ionic gelation method is 83.18% of chitosan and/or alginate. Also, carboxymethyl cellulose, hyaluronic acid, gelatine, gellan gum, fibrin, collagen, dextran, and pectin have been used as polymeric species for the ionic gelation method (Pedroso-Santana and Fleitas-Salazar, 2020).

11.5.5 Complex Coacervation

Coacervation is a method for the encapsulation of essential oils and other hydrophobic bioactive compounds. The encapsulated essential oil is created by the electrical attraction of molecules, one of which is positively charged and the other negatively charged (Zhu et al., 2009).

11.5.6 Electro-Spinning and Electro-Spraying

Electro-spinning and electro-spray are two different types of electrohydrodynamic processes in which a charged jet is used to spin or spray a prepared polymer solution to produce fibres or particles.

The electro-spinning is a method used to produce dry micro and nanoparticles by passing electrically charged fluids, high-voltage electricity, through a polymeric fluid. The device used in electro-spinning consists of three main parts: a syringe pump, an electric needle, and a collecting plate in a chamber. The liquid for producing nanoparticles is fed to the needle under high voltage electric current at a constant speed, and the resulting nanofibres are collected on the electric collector plate.

Electro-spray is an alternative to common encapsulation techniques and is commonly known as electrohydrodynamic atomization (EHDA). Like electro-spinning, it operates on a high-voltage current at room temperature. The only difference between these technologies is the cohesion between the liquid polymer molecules, which is very low in the case of electro-spray. Due to this cohesive force, the resultant liquid disperses as fine droplets and forms a spherical shape with shear surface tension in the air (Rostami et al., 2019).

These techniques can be used alone or in combination to increase the effectiveness of nanoencapsulated essential oils or to achieve a smaller size.

11.5.7 Biopolymer-Based Nanoencapsulation

Biofilm-forming agents are widely used as wall materials to encapsulate antibacterial agents. There are various sources of biopolymer, which can be used in its original form or modified to obtain some of the desired properties. In general, essential oils are encapsulated as with food-grade biofilm-forming agents, such as polysaccharides (starch, cyclodextrin, pectin, chitosan, etc.), and proteins (zein, casein, whey protein, etc.) and their synthetic derivatives (Zhang et al., 2017). Nanocapsules are based on biofermentation, which can form nanocapsules or nanoparticles. In nanocapsules, the oil core is surrounded by polymer walls, and the nanoparticles are a matrix systems in which the core is dispersed in a polymer matrix (Bilia et al., 2014). Several different technologies are commonly used to produce nanocapsules or nanospheres, such as complex coagulation (Penalva et al., 2015), freeze drying (Hill et al., 2013), spray drying (Pérez-Masiá et al., 2015), and ionic gelation (Pan et al., 2014; Penalva et al., 2015).

11.5.7.1 Polysaccharide-Based Nanoencapsulation

Polysaccharide-based biopolymers are divided into two groups, depending on the type and the number of monosaccharide units that make up the chain. If they consist of the same type of monosaccharides they are named homopolysaccharides, and if they consist of different types of monosaccharides they are named heteropolysaccharides. The different chemical structure of each monosaccharide unit allows polysaccharides to have different degrees of polymerization (number of monosaccharide units), hydrophobicity/hydrophilicity, molecular weight, electrical charge and electrical properties, and viscosities. Polysaccharides used for encapsulation are generally expected to have properties such as low toxicity, low cost, stability over a wide pH range, and good biodegradability (Hill et al., 2013).

Starch is a renewable polysaccharide that is also the main energy source of the human body. In addition, modifying it by some physical and chemical method can change some properties of starch and give this polymer some unique properties. Thus, starch or its derivatives are widely used for encapsulation of bioactive compounds, especially essential oils and carotenoids (Fathi et al., 2014). Chin et al. (2011) dissolved sago starch in NaOH/urea, added the obtained starch solution drop by drop to analytic grade ethyl alcohol on a stirrer, and centrifuged the resulting suspension to obtain starch nanoparticles (300–400 nm).

Cyclodextrins (CD) are ring-shaped oligosaccharides which are a derivative of starch and are the most widely used wall materials to prepare nanocapsules (Karathanos et al., 2007). In the nanoencapsulation process, the inner wall of the β-cyclodextrin—that is, the inner wall of the ring structure—is the combination of a hydrophobic epoxy group and the C-H bond, and the outer surface of the ring is the hydroxyl group of the primary alcohol at C6. This structure ensures that the inner surface of the β-cyclodextrin ring is hydrophobic or lipophilic, and that the outer surface is hydrophilic. This allows the complex formed with cyclodextrin to be added to water-insoluble compounds (for example, essential fatty acids) into aqueous solutions. Many different host molecules can be added to the gap portion and remain stable for a longer period due to the control of light and oxygen or the inability to interact directly with the compound. In addition, β-cyclodextrin can mask the distinctive potent odour of some oils more than other coating materials. In a study, it was stated that eugenol encapsulated with β-cyclodextrin protected against cold and frost damage (Lee et al., 2009).

Chitosan is a cationic polysaccharide formed by reduction of chitin by a chemical process in nature. Chitosan is very suitable for use as a nanocarrier due to its biodegradability, sensitivity to pH, and easy chemical modification (Wang et al., 2014). Ionotropic gelation is one of the most widely used techniques using chitosan. The technique is quite simple and at the same time the activity of the encapsulated compounds can be preserved to a maximum. In a study, Zataria multiflora essential oil–loaded chitosan nanoparticles (125–175 nm) were prepared by this method, and it was reported that the essential oils stability and antifungal activity against *Botrytis cinerea* were increased (Mohammadi et al., 2015).

11.5.7.2 Protein-Based Nanoencapsulation

Proteins are very suitable biopolymers for the encapsulation of essential oils due to their functional properties. They change the texture, taste, and colour of the foods they are incorporated into. Factors such as amino acid sequence, pH, ionic bonds, and temperature of the molecular structures of proteins can significantly affect their emulsifying capacity, solubility, and stability (Zhang et al., 2017).

Casein, which makes up about 80% of milk proteins, is a protein in phospholipid structure, while the carbon end (N-terminal), constituting two-thirds, is hydrophobic; the carbon end (C-terminal), constituting the remaining third, is strongly hydrophilic. This structural feature is very important for the stability and properties of casein micelles. In this way, the use of casein molecules in nanoencapsulation of essential oils allows the formation of strong capsule structures. In the literature, gold and iron oxide nanoparticles have been used to produce casein-based nanoparticles. The average sizes of produced nanoparticles were measured as 200 and 38 nm, respectively (Liu and Guo, 2009; Sangeetha and Philip, 2012).

Zein, which is a corn prolamin, has attracted attention among the materials used in nano and microcapsule production in recent years due to its amphiphilic structure, its ability to form microspheres spontaneously in ethanol-water binary mixtures, its biodegradability, and its being on the GRAS (generally recognized as safe) list. Due to its high hydrophobicity in the food industry, zein can be successfully used in the protection, transportation, controlled release, and microencapsulation of lipophilic core materials (Muthuselvi and Dhathathreyan, 2006, Wu et al., 2012).

The combination of different biofilm-forming agents provides several promising functions and properties compared to a single biofilm-forming agent (Hosseini et al., 2015). Biofilm-forming nanoparticles can be synthesized from biofilm-forming food preparations by self-association of single biofilms or by phase separation of biofilm-forming mixtures (Gupta and Gupta, 2005). The ability to control and the interactions involved lead to the design of new molecular structures and development of foods with more desirable structural properties.

11.5.8 Lipid-Based Nanoencapsulation

Although carbohydrate and protein-based nanocapsules have many advantages, they are difficult to fully control due to their different complexity. In addition, negative results may occur in cases where it is necessary to apply heat treatments. However, lipid-based nanocarriers have the advantage of more widespread industrial production, greater encapsulation efficiency, and lower toxicity compared to carbohydrate and protein-based nanocapsules. Lipid-based nanocapsules, such as nanoemulsions, nanoliposomes, and solid lipid nanoparticles, are widely used in the pharmaceutical and food industries due to their biocompatibility (Fathi et al., 2012).

11.5.8.1 Nanoemulsions

Nanoemulsions are multiphase colloidal droplets formed by two immiscible liquids, one of them physically breaking down into nano size and dispersing into the other liquid (Solans et al. 2003). They are classified as oil-in-water (O/W) and/or water-in-oil (W/O), depending on the liquid that is the greater amount. Depending on the hydrophobicity or hydrophilicity of the bioactive material, O/W or W/O emulsion systems can be used to stabilize water- or oil-soluble formulations. The nanoemulsions are used as carrier systems for the industries such as nutraceutical or pharmaceutical, and as for foods like sweeteners, antioxidants, and antimicrobial agents (Jiahui et al., 2004; McClements et al., 2007; Kesisoglou et al., 2007). The nanoemulsions can be obtained using low and high energy methods. High-pressure homogenization and ultrasonication are widely used high energy methods to produce nanoemulsions. Among the low-energy methods, self-assembly and reversed-phase temperature methods are frequently used (Acosta, 2009; Leong et al., 2009; Tadros et al., 2004; Piorkowski and McClements, 2013). Nonionic surfactants are generally used in high concentrations (>6%) to create emulsions in these methods. The concentration of surfactant used severely affects the particle size of emulsions. Jun et al. (2015) produced grapefruit essential oils, oil-in-water–type nanoemulsion using Span80, Tween80, and glycerine as surfactants. The approximate particle size was 21 nm and the stability of the emulsion was good (Jun et al., 2015). Xue et al. (2015) obtained a nanoemulsion using thyme oil with casein and soy lecithin. It was stated that the nanoemulsion obtained in the study showed a superior antibacterial effect than free thyme oil.

11.5.8.2 Nanoliposomes

Nanoliposomes (also called lipid vesicles) are one of the most widely used carriers and nanoemulsion systems. It can be used in any area (food, cosmetic, agricultural, and pharmaceutical industries) where controlled release is required, thanks to its similarity to the biomembrane. Liposomes are colloidal, vesicular structures consisting of one or more phospholipid layers surrounding an equal number of aqueous compartments, which can contain different molecules, such as proteins. Phospholipids, which are abundant in structures such as egg yolk, soybean, and cholesterol, have biocompatibility because they are the main components of the cell membrane. Depending on the phospholipid layer number they contain, they can be single or multilayered and contain hydrophilic, lipophilic, and amphiphilic compounds. In a study, liposome-integrated essential oils were prepared using the membrane dispersion ultrasonic method. According to the results of the study, it has been reported that liposomes can be used in the preservation of foods, especially meat (Nieto et al., 2011). In another study, Bai et al. (2011) produced spray-dried, liposome-encapsulated coix seed oil with ethanol injection. According to the results of the study, they found that the coix seed oil is more stable and better soluble in water.

11.5.8.3 Solid Lipid Nanoparticles

Solid lipid nanoparticles are generally produced by exposing solid lipids in a hot lipid phase to high energy (shearing, ultrasound, or high-pressure homogenization) in a hot aqueous solution containing surfactant, and then cooling it for recrystallization (Souto et al., 2007). As the name suggests, solid lipid nanoparticles are prepared from solid lipids. For example, fatty acids (e.g., palmitic acid), triglycerides (e.g., trilaurin), steroids (e.g., cholesterol), and partial glycerides (e.g. glyceryl monostearate) are used (Liao et al., 2021). In a study, loading the cholesterol with curcumin solid lipid nanoparticles, which were prepared using the high-pressure homogenization technique and had the dimensions between 112 and 163 nm, besides releasing curcumin for a longer time, increased the antibacterial effect by providing a better adhesion of cholesterol to the bacterial cell and exposed the bacterial cell to curcumin for a longer time (Jourghanian et al., 2016).

11.6 USE OF NANOENCAPSULATED ESSENTIAL OILS IN FOOD PACKAGING

Active packaging changes or modifies the content or the environment inside of the package to extend the shelf life of foods by stopping or slowing down the spoilage reactions. It can be defined as the addition of active ingredients to the packaging in order to help the desired interactions occur between the food and the packaging and to extend the shelf life. Oxygen scavengers (usually iron and iron oxide), carbon dioxide scavengers/releasers (calcium hydroxide, sodium hydroxide, potassium hydroxide, or silica gel), ethylene scavengers (e.g., potassium permanganate embedded in silica), moisture regulators (silica gel, natural clay (montmorillonite), calcium oxide, calcium chloride, and modified starch), and various antimicrobial components are used to extend the shelf life and quality parameters of foods in active packaging (Aday, 2021). Studies have reported that essential oils have significant antimicrobial effects (Škrinjar and Nemet, 2009). Using essential oils in food packaging by covering them with a wall material by nanoencapsulation provides a longer and more effective use of the antimicrobial effect with the controlled release of the essential oil. In the literature, there are studies on extending the shelf life of various foods by adding nanoencapsulated essential oils to the chitosan-based coating material; those are summarized in Table 11.2.

TABLE 11.2
The Effects of Nanonencapsulated Essential Oils-Based Food Packaging on Some Foods

Essential Oil Type	Nanoencapsulation Technique	Wall Material	Application Type	Highlights	Reference
Carvacrol/cinnamon/ oregano (70/20/10 w/w/w)	Kneading	β-cyclodextrin	Coating of the cardboard tray	The shelf life of cherry tomatoes was extended by using a coated cardboard tray from 20 to 24 days at 8 °C-90% RH. Also, physicochemical quality was not affected with the effect of the active tray.	Buendía–Moreno et al., 2019
Carvacrol/cinnamon/ oregano (70/20/10 w/w/w)	Kneading	β-cyclodextrin	Coating of the cardboard box	Reduced the decay of peppers and the firmness was better protected after 18 days of storage.	Buendía–Moreno et al., 2020
Thyme essential oil	Homogenization	Chitosan	Edible coating	The coating extended the shelf life of Karish cheese from 2 to 4 weeks	Al-Moghazy et al., 2021
Tarragon essential oil	Ionic gelation	Chitosan-Gelatin	Edible coating	The edible packaging composed of chitosan-gelatin incorporated with tarragon essential oil nanoparticles was thought to preserve fresh meat products by inhibiting quality deterioration.	Zhang et al., 2020
Clove essential oil	Ultrasonic homogenization	Chitosan Nanoparticles	Edible coating	Using clove essential oil–loaded chitosan nanoparticle coating extends the shelf life of fresh-cut fruits and vegetables.	Hasheminejad and Khodaiyan, 2020
Thyme essential oil	Electro-spinning	β-cyclodextrin	Aluminium foil loaded with thyme essential oil/β-cyclodextrin ε-polylysine nanoparticles	The chicken meat was thought to have longer shelf life by the antimicrobial activity against *campylobacter jejuni* of thyme essential oil–loaded β-cyclodextrin and ε-polylysine nanoparticles.	Lin et al., 2018

(Continued)

TABLE 11.2 *(Continued)*
The Effects of Nanonencapsulated Essential Oils-Based Food Packaging on Some Foods

Essential Oil Type	Nanoencapsulation Technique	Wall Material	Application Type	Highlights	Reference
Lemon essential oil	High-pressure homogenization	Modified Chitosan	Edible coating	The use of nanoencapsulated lemon essential oil and modified chitosan in combination with dip coating on rucola leaves was found to extend shelf life from a minimum of 3 days to a maximum of 7 days, compared to uncoated samples.	Sessa et al., 2015
d-limonene	Liposome	Dimyristoyl phosphatidylcholine, Polydiacetylene and N-hydroxysuccinimide	Edible coating	Liposomes were found to be effective in keeping the strawberry quality, and limonene liposomes were effective in controlling fungal rot during longer storage of strawberries.	Dhital et al., 2017
Carvacrol	Homogenization	Chitosan-Tripolyphosphate	Edible coating	Carvacrol-loaded chitosan-tripolyphosphate nanoparticles were used as a washing solution for carrot slices. The solution was found highly effective in maintaining the quality of carrot slices during cold storage (5 °C).	Martínez-Hernández et al., 2017
d-limonene	Liposome	Soy-based lecithin and diacetylene	Edible Coating	Strawberries coated with d-limonene liposomes had significantly lower respiratory rates and pH values, and higher anthocyanin content compared to control and alginate coatings.	Dhital et al., 2018
d-limonene	Liposome	Dimyristoyl phosphatidylcholine, polydiacetylene and N-hydroxysuccinimide	Edible Coating	Blueberry liposome coatings were better protection against degradation after storing for 9 weeks at 4 °C. Liposome coating reduced blueberry storage loss by 33% after 9 weeks.	Umagiliyage et al., 2017

(Continued)

TABLE 11.2 *(Continued)*
The Effects of Nanonencapsulated Essential Oils-Based Food Packaging on Some Foods

Essential Oil Type	Nanoencapsulation Technique	Wall Material	Application Type	Highlights	Reference
Morin	Freeze drying	β-cyclodextrin	Edible coating	The thermal properties of the film were improved, and better dispersion was achieved.	Yuan et al., 2019
Tea tree oil	Liposome/ Electro-spinning	soy lecithin, cholesterol, trichloromethane	Nanofibre Membrane	Microbial growth of Salmonella was been reduced and the quality of chicken was maintained.	Cui et al., 2018
Thymol	Electro-spinning	γ-cyclodextrin	Edible coating	This reduced the microbial count on meat stored for 5 days at 4 °C.	Aytac et al., 2017
Palmarosa Essential oil- Star anise essential oil	Homogenization	β-cyclodextrin	Packaging material	Reduced respiration rate and ethylene production, better textural properties, and restricted growth of *P. expansum* in apples have been reported.	da Rocha Neto et al., 2019
Cinnamon essential oil	Electro-spinning	zein/ethyl cellulose	Edible coating	The films reduced the weight loss and kept the tightness of the mushrooms. The results showed that the film significantly extended the shelf life of mushrooms stored at 20 ± 3°C.	Niu et al., 2020
Satureja plant essential oil	Liposome	lecithin: cholesterol	Edible Coating	It has been reported that coating lamb meat using nanoencapsulated Satureja plant essential oil reduces lipid oxidation and microbial growth, and improves sensory properties.	Pabast et al., 2018
Cinnamaldehyde	Electro-spinning	Polyvinyl alcohol/ permutit	Packaging film	The film was found to delay the rapid deterioration of strawberries during storage.	Shao et al., 2018

A mixture of carvacrol/cinnamon/oregano (70/20/10 (w/w/w)) essential oils were encapsulated in β-cyclodextrin by kneading. Then, it was sprayed onto a cardboard tray and cherry tomatoes were packed with that cardboard tray. The tray extended the shelf life of tomatoes by four days at 8°C and 90% relative humidity RH, without the physicochemical quality changes (Buendía-Moreno et al., 2019). Coating karish cheese with thyme essential oil containing chitosan extended its shelf life from two weeks to four weeks (Al-Moghazy et al., 2021). It has been stated that coatings based on a chitosan-gelatin mixture combined with tarragon essential oil–loaded nanoparticles can be developed as edible packaging to protect fresh meat products by preventing quality losses (Zhang et al., 2020). Clove essential oil–loaded chitosan nanoparticles are considered a promising protective coating to extend the shelf life of freshly cut fruits and vegetables (Hasheminejad and Khodaiyan, 2020). In another study, lemon essential oil–loaded chitosan-based edible coatings were successfully used to extend the shelf life of rucola leaves from three to seven days, compared to untreated samples (Sessa et al., 2015). In another study noted that washing the fresh-cut carrot slices with a solution containing carvacrol-loaded chitosan-tripolyphosphate nanoparticles significantly maintained the quality (Martínez-Hernández et al., 2017). Dhital et al. (2017) have reported that liposomes are effective in preserving strawberry quality, and strawberries coated with limonene liposomes are more resistant to fungal spoilage and can be stored longer. Also, it was reported that blueberries coated with limonene-loaded liposomes were better protected against degradation after nine weeks of storage at 4°C, with reduced storage loss of one-third (Umagiliyage et al., 2017). In a different study, chicken was packaged in aluminium foil loaded with thyme essential oil/β-cyclodextrin ε-polylysine nanoparticles. It was noted that the packaged chicken showed significantly better antimicrobial activity against *C. jejuni* compared to the control (Lin et al., 2018). Yuan et al. (2019) produced an edible film containing morin-loaded hydroxypropyl-β-cyclodextrin and compared its properties with gelatin films. The film exhibited better thermal stability and UV barrier property, and the thickness of the films was significantly improved compared to the gelatin film. Chicken meat samples were coated with tea tree oil liposomes/chitosan nanofibre membrane, and it was stated that, after four days of storage the quality of chicken meat has maintained better (Cui et al., 2018). Aytac et al. (2017) produced thymol encapsulated γ-cyclodextrin, zein nanofibres film and coated the meat. It was found that the microbial count reduced after five days of storage at 4°C. In a study on the development of packaging material for packaging apples, palmarosa or star anise essential oils were encapsulated with ß-cyclodextrin. The resulting packaging material reduced ethylene production, respiration rate, and hardness loss of apples (da Rocha Neto et al., 2019). Also, a cardboard box has been developed by spraying nanoencapsulated essential oil mix (carvacrol/cinnamon/oregano 70/10/20 (w/w/w)) in β-cyclodextrin for bulk packaging of fresh bell pepper. The package has reduced the decay of peppers, and the firmness of the peppers was better protected after 18 days of storage. Pabast et al., (2018) nanoencapsulated the Satureja plant essential oil in lecithin-cholesterol and coated lamb meat. They reported that the quality characteristics and the oxidative stability of the meat were better protected during chilled storage. In a study by Niu et al. (2020), zein-ethylcellulose nanofibres were loaded with cinnamon essential oil via an electro-spinning process. It was stated that the shelf life of mushroom samples coated with the nanofibres was significantly prolonged compared to the uncoated ones. In a different study, poly(vinyl alcohol)/permutite fibrous film were loaded with cinnamaldehyde and strawberries were covered with that film. Deterioration of the strawberries was delayed by the film (Shao et al., 2018).

11.7 CONCLUSION

Studies for better preservation of foods are increasing day by day. The higher food demand due to increasing world population pressures the food industry and researchers to improve the food packaging materials and techniques to preserve agricultural products. Some of the packaging materials developed use essential oils for this purpose. Those oils are used in food preservation because of their antioxidant and antimicrobial effects. However, due to their volatility, they are separated from packaging material or food surface shortly after incorporation. Also, their potent odour can affects food sensory properties and consumer preferences. For those reasons, the most important method to prevent the volatility of essential oils and mask their potent odour is encapsulation. In addition, nanoencapsulation stands out for its longer effect and controlled release. Essential oils nanoencapsulated by various methods are used in studies as a promising method to extend the storage period of foods such as fruits and vegetables or meat and cheese, and to preserve their quality properties as much as possible during this period. Advances in active packaging technology are also expected to expand the use of nanoencapsulated essential oils for better preservation of foods.

REFERENCES

Acosta, E. 2009. Bioavailability of nanoparticles in nutrient and nutraceutical delivery. *Current Opinion in Colloid & Interface Science* 14(1): 3–15.

Aday, M. S. 2021. The use of active packaging technology in fruits and vegetables. *European Journal of Science and Technology* 21: 122–130.

Al-Moghazy, M., El-sayed, H. S., Salama, H. H., and Nada, A. A. 2021. Edible packaging coating of encapsulated thyme essential oil in liposomal chitosan emulsions to improve the shelf life of Karish cheese. *Food Bioscience* 43: 101230.

Alparslan, Y., Yapıcı, H. H., Metin, C., Baygar, T., Günlü, A., and Baygar, T. 2016. Quality assessment of shrimps preserved with orange leaf essential oil incorporated gelatin. *LWT-Food Science and Technology* 72: 457–466.

Amorati, R., Foti, M. C., and Valgimigli, L. 2013. Antioxidant activity of essential oils. *Journal of Agricultural and Food Chemistry* 61(46): 10835–10847.

Angioni, A., Barra, A., Coroneo, V., Dessi, S., and Cabras, P. 2006. Chemical composition, seasonal variability, and antifungal activity of Lavandula Stoechas l. Ssp. Stoechas essential oils from stem/leaves and flowers. *Journal of Agricultural and Food Chemistry* 54: 4364–4370.

Aytac, Z., Ipek, S., Durgun, E., Tekinay, T., and Uyar, T. 2017. Antibacterial electrospun zein nanofibrous web encapsulating thymol/cyclodextrin-inclusion complex for food packaging. *Food Chemistry* 233: 117–124.

Bai, C., Peng, H., Xiong, H., Liu, Y., Zhao, L., and Xiao, X. 2011. Carboxymethylchitosan-coated proliposomes containing coix seed oil: Characterisation, stability and in vitro release evaluation. *Food Chem* 129: 1695–1702.

Bakkali, F., Averbeck, S., Averbeck, D., and Idaomar, M. 2008. Biological effects of essential oils-a review. *Food and Chemical Toxicology* 46(2): 46–475.

Basak, S., and Guha, P. 2017. Use of predictive model to describe sporicidal and cell viability efficacy of betel leaf (*Piper betle* L.) essential oil on *Aspergillus flavus* and *Penicillium expansum* and its antifungal activity in raw apple juice. *LWT Food Science & Technology* 80: 510–516.

Belasli, A., Ben Miri, Y., Aboudaou, M., Aït Ouahioune, L., Montañes, L., and Ariño, A., et al. 2020. Antifungal, antitoxigenic, and antioxidant activities of the essential oil from laurel (*Laurus nobilis* L.): Potential use as wheat preservative. *Food Science Nutrition* 8: 4717–4729.

Beristain, C. I., Garcıa, H. S., and Vernon-Carter, E. J. 2001. Spray-dried encapsulation of cardamom (*Elettaria cardamomum*) essential oil with mesquite (*Prosopis juliflora*) gum. *LWT-Food Science and Technology* 34(6): 398–401.

Bhushani, J. A., and Anandharamakrishnan, C. 2014. Electrospinning and electrospraying techniques: Potential food based applications. *Trends in Food Science & Technology* 38: 21–33.

Bilia, A. R., Guccione, C., Isacchi, B., Righeschi, C., Firenzuoli, F., and Bergonzi, M. C. 2014. Essential oils loaded in nanosystems: A developing strategy for a successful therapeutic approach. *Evidence-Based Complementary and Alternative Medicine* https://doi.org/10.1155/2014/651593

Boubaker, H., Karim, H., El Hamdaoui, A., Msanda, F., Leach, D., and Bombarda, I., et al. 2016. Chemical characterization and antifungal activities of four *Thymus* species essential oils against postharvest fungal pathogens of citrus. *Industrial Crops and Products* 86: 95–101.

Buendía-Moreno, L., Ros-Chumillas, M., Navarro-Segura, L., Sánchez-Martínez, M. J., Soto-Jover, S., Antolinos, V., Martínez-Hernández, G. B., and López-Gómez, A.2019. Effects of an active cardboard box using encapsulated essential oils on the tomato shelf life. *Food and Bioprocess Technology* 12(9): 1548–1558.

Buendía Moreno, L., Soto Jover, S., Ros Chumillas, M., Antolinos López, V., Navarro Segura, L., Sánchez Martínez, M. J., Martínez-Hernández, G. B.,and López Gómez, A. 2020. An innovative active cardboard for box bulk packaging of fresh bell pepper. *Postharvest Biology and Technology,* 164, 111171.

Chaudhari, A. K., Singh, V. K., Das, S., Prasad, J., Dwivedy, A. K., and Dubey, N. K. 2020. Improvement of in vitro and in situ antifungal, AFB1 inhibitory and antioxidant activity of Origanum majorana L. essential oil through nanoemulsion and recommending as novel food preservative. *Food and Chemical Toxicology* 143: 111536.

Chin, S. F., Pang, S. C., and Tay, S. H. 2011. Size controlled synthesis of starch nanoparticles by a simple nanoprecipitation method. *Carbohydrate Polymers* 86: 1817–1819.

Cui, H., Bai, M., Li, C., Liu, R., and Lin, L. 2018. Fabrication of chitosan nanofibers containing tea tree oil liposomes against Salmonella spp. in chicken. *LWT Food Science and Technology* 96: 671–678.

da Rocha Neto, A. C., Beaudry, R., Maraschin, M., Di Piero, R. M., and Almenar, E. 2019. Double-bottom antimicrobial packaging for apple shelf-life extension. *Food Chemistry* 279: 379–388.

Dhifi, W., Bellili, S., Jazi, S., Bahloul, N., and Mnif, W. 2016. Essential oils' chemical characterization and investigation of some biological activities: A critical review. *Medicines* 3(4): 25.

Dhital, R., Joshi, P., Becerra-Mora, N., Umagiliyage, A., Chai, T., Kohli, P., and Choudhary, R. 2017. Integrity of edible nano-coatings and its effects on quality of strawberries subjected to simulated in-transit vibrations. *LWT Food Science and Technology* 80: 257–264.

Dhital, R., Mora, N. B., Watson, D. G., Kohli, P., and Choudhary, R. 2018. Efficacy of limonene nano coatings on post-harvest shelf life of strawberries. *LWT Food Science and Technology* 97: 124–134.

Fathi, M., Martin, A., and McClements, D. J. 2014. Nanoencapsulation of food ingredients using carbohydrate based delivery systems. *Trends in Food Science and Technology* 39: 18–39.

Fathi, M., Mozafari, M.-R., and Mohebbi, M. 2012. Nanoencapsulation of food ingredients using lipid based delivery systems. *Trends in Food Science and Technology* 23: 13–27.

FDA. 2020. Accessed via https://www.accessdata.fda.gov/SCRIPTs/cdrh/cfdocs/cfcfr/CFRSearch.cfm?fr=182.20 (accessed June, 2021)

Fisher, K., and Phillips, C. 2008. Potential antimicrobial uses of essential oils in food: Is citrus the answer? *Trends in Food Science & Technology* 19(3): 156–164.

Gupta, A. K., and Gupta, M. 2005. Synthesis and surface engineering of iron oxide nanoparticles for biomedical applications. *Biomaterials* 26: 3995–4021.

Han, J. H., and Aristippos, G. 2005. Edible films and coatings: A review. In Han J. H. (Ed.), *Innovations in Food Packaging*, Elsevier Academic Press, London, UK, 239–262

Hasheminejad, N., and Khodaiyan, F. 2020. The effect of clove essential oil loaded chitosan nanoparticles on the shelf life and quality of pomegranate arils. *Food Chemistry* 309: 125520.

Hill, L. E., Gomes, C., and Taylor, T. M. 2013. Characterization of beta-cyclodextrin inclusion complexes containing essential oils (transcinnamaldehyde, eugenol, cinnamon bark, and clove bud extracts) for antimicrobial delivery applications. *LWT Food Science and Technology* 51: 86–93.

Hosseini, S. F., Rezaei, M., Zandi, M., and Ghavi, F. F. 2016. Effect of fish gelatin coating enriched with oregano essential oil on the quality of refrigerated rainbow trout fillet. *Journal of Aquatic Food Product Technology* 25(6): 835–842.

Hosseini, S. M. H., Emam-Djomeh, Z., Sabatino, P., and Van der Meeren, P. 2015. Nanocomplexes arising from protein-polysaccharide electrostatic interaction as a promising carrier for nutraceutical compounds. *Food Hydrocolloids* 50: 16–26.

Hu, J., Wang, X., Xiao, Z., and Bi, W. 2015. Effect of chitosan nanoparticles loaded with cinnamon essential oil on the quality of chilled pork. *LWT-Food Science and Technology* 63(1): 519–526.

Jackson, L. S., and Lee, K. 1991. Microencapsulation and the food industry. *Lebensmittel-Wissenschaft & Technologie* 24(4): 289–297.

Jafari, S. M., Assadpoor, E., Bhandari, B., and He, Y. 2008. Nano-particle encapsulation of fish oil by spray drying. *Food Research International* 41(2): 172–183.

Jiahui, H., Johnston, K. P., and Williams III, R. O. 2004. Nanoparticle engineering processes for enhancing the dissolution rates of poorly water soluble drugs. *Drug Development and Industrial Pharmacy* 30(3): 233–245.

Jourghanian, P., Ghaffari, S., Ardjmand, M., Haghighat, S., and Mohammadnejad, M. 2016. Sustained release curcumin loaded solid lipid nanoparticles. *Advanced Pharmaceutical Bulletin* 6(1): 17.

Jun, H. E., Yao, X. L., Feng, G. R., Wang, R., and Yang, B. L. 2015. Preparations of the nano-emulsion grapefruit essential oil and its quality evaluation. *Food Research Development* 36: 3–6.

Karathanos, V. T., Mourtzinos, I., Yannakopoulou, K., and Andrikopoulos, N. K. 2007. Study of the solubility, antioxidant activity and structure of inclusion complex of vanillin with β-cyclodextrin. *Food Chemistry* 101: 652–658.

Kesisoglou, F., Panmai, S., and Wu, Y. H. 2007. Application of nanoparticles in oral delivery of immediate release formulations. *Current Nanoscience* 3: 183–190.

Khorshidian, N., Yousefi, M., Khanniri, E., and Mortazavian, A. M. 2018. Potential application of essential oils as antimicrobial preservatives in cheese. *Innovative Food Science & Emerging Technologies* 45: 62–72.

Lee, M. Y., Min, S. G., Bourgeois, S., and Choi, M. J. 2009. Development of a novel nanocapsule formulation by emulsion-diffusion combined with high hydrostatic pressure. *Journal of Microencapsulation* 26: 122–129.

Lekjing, S. 2016. A chitosan-based coating with or without clove oil extends the shelf life of cooked pork sausages in refrigerated storage. *Meat Science* 111: 192–197.

Leong, T. S. H., Wooster, T. J., Kentish, S. E., and Ashokkumar, M. 2009. Minimising oil droplet size using ultrasonic emulsification. *Ultrasonics Sonochemistry* 16(6): 721–727.

Liao, W., Badri, W., Dumas, E., Ghnimi, S., Elaissari, A., Saurel, R., and Gharsallaoui, A. 2021. Nanoencapsulation of essential oils as natural food antimicrobial agents: An overview. *Applied Science*. https://doi.org/10.3390/app11135778

Lin, L., Zhu, Y., and Cui, H. 2018. Electrospun thyme essential oil/gelatin nanofibers for active packaging against Campylobacter jejuni in chicken. *LWT Food Science & Technology* 97: 711–718.

Liu, Y., and Guo, R. 2009. The interaction between casein micelles and gold nanoparticles. *Journal of Colloid and Interface Science* 332: 265–269.

Lucchesi, M. E., Chemat, F., and Smadja, J. 2004. Solvent-free microwave extraction of essential oil from aromatic herbs: Comparison with conventional hydro-distillation. *Journal of Chromatography A* 1043(2): 323–327.

Martínez-Hernández, G. B., Amodio, M. L., and Colelli, G. 2017. Carvacrol-loaded chitosan nanoparticles maintain quality of fresh-cut carrots. *Innovative Food Science and Emerging Technologies* 41: 56–63.

Masotti, V., Juteau, F., Bessie`re, J. M., and Viano, J. 2003. Seasonal and phenological variations of the essential oil from the narrow endemic species Artemisia molinieri and its biological activities. *Journal of Agricultural and Food Chemistry* 51: 7115–7121.

McClements, D. J., Decker, E., and Weiss, J. 2007. Emulsion-based delivery systems for lipophilic bioactive components. *Journal of Food Science* 72(8): 109–124.

Mohammadi, A., Hashemi, M., and Hosseini, S. M. 2015. Nanoencapsulation of Zataria multiflora essential oil preparation and characterization with enhanced antifungal activity for controlling Botrytis cinerea, the causal agent of gray mould disease. *Innovative Food Science and Emerging Technologies* 28: 73–80.

Murmu, S. B., and Mishra, H. N. 2018. The effect of edible coating based on Arabic gum, sodium caseinate and essential oil of cinnamon and lemon grass on guava. *Food Chemistry* 245: 820–828.

Muthuselvi, L., and Dhathathreyan, A. 2006. Simple coacervates of zein to encapsulate Gitoxin. *Colloids and Surfaces B Biointerfaces* 51: 39–43.

Nieto, G., Huvaere, K., and Skibsted, L. H. 2011. Antioxidant activity of rosemary and thyme by-products and synergism with added antioxidant in a liposome system. *European Food Research & Technology* 233: 11–18.

Nikkhah, M., and Hashemi, M. 2020. Boosting antifungal effect of essential oils using combination approach as an efficient strategy to control postharvest spoilage and preserving the jujube fruit quality. *Postharvest Biology and Technology* 164: 111159.

Nikkhah, M., Hashemi, M., Najafi, M. B. H., and Farhoosh, R. 2017. Synergistic effects of some essential oils against fungal spoilage on pear fruit. *International Journal of Food Microbiology* 257: 285–294.

Niu, B., Zhan, L., Shao, P., Xiang, N., Sun, P., Chen, H., and Gao, H. 2020. Electrospinning of zein-ethyl cellulose hybrid nanofibers with improved water resistance for food preservation. *International Journal of Biological Macromolecules* 142: 592–599.

Pabast, M., Shariatifar, N., Beikzadeh, S., and Jahed, G. 2018. Effects of chitosan coatings incorporating with free or nano-encapsulated Satureja plant essential oil on quality characteristics of lamb meat. *Food Control* 91: 185–192.

Pan, K., Chen, H., Davidson, P. M., and Zhong, Q. J. 2014. Thymol nanoencapsulated by sodium caseinate: Physical and antilisteria properties. *Journal of Agricultural and Food Chemistry* 62: 1649–1657.

Pavlath, A. E., and Orts, W. 2009. Edible films and coatings: Why, what, and how. In Embuscado M. E., and Huber K. C. (Eds.), *Edible Films and Coatings for Food Applications*, Springer, New York, 1–23.

Pedroso-Santana, S., and Fleitas-Salazar, N. 2020. Ionotropic gelation method in the synthesis of nanoparticles/microparticles for biomedical purposes. *Polymer International* 69(5): 443–447.

Penalva, R., Esparza, I., Agüeros, M., Gonzalez-Navarro, C. J., Gonzalez-Ferrero, C., and Irache, J. M. 2015. Casein nanoparticles as carriers for the oral delivery of folic acid. *Food Hydrocolloids* 44: 399–406.

Pérez-Masiá, R., López-Nicolás, R., Periago, M. J., Ros, G., Lagaron, J. M., and López-Rubio, A. 2015. Encapsulation of folic acid in food hydrocolloids through nanospray drying and electrospraying for nutraceutical applications. *Food Chemistry* 168: 124–133.

Piorkowski, D. T., and McClements, D. J. 2013. Beverage emulsions: Recent developments in formulation, production, and applications. *Food Hydrocolloids* 42: 5–41.

Reyes-Jurado, F., Franco-Vega, A., Ramírez-Corona, N., Palou, E., and López-Malo, A. 2015. Essential oils: Antimicrobial activities, extraction methods, and their modeling. *Food Engineering Reviews* 7(3): 275–297.

Rodriguez, A., Nerín, C., and Batlle, R. 2008. New cinnamon-based active paper packaging against *Rhizopus stolonifer* food spoilage. *Journal of Agricultural and Food Chemistry* 56(15): 6364–6369.

Rojas-Graü, M. A., Raybaudi-Massilia, R. M., Soliva-Fortuny, R. C., Avena-Bustillos, R. J., McHugh, T. H., and Martín-Belloso, O. 2007. Apple puree-alginate edible coating as carrier of antimicrobial agents to prolong shelf-life of fresh-cut apples. *Postharvest Biology and Technology* 45(2): 254–264.

Rostami, M., Yousefi, M., Khezerlou, A., Mohammadi, M. A., and Jafari, S. M. 2019. Application of different biopolymers for nanoencapsulation of antioxidants via electrohydrodynamic processes. *Food Hydrocolloids* 97: 105170.

Sánchez-González, L., Pastor, C., Vargas, M., Chiralt, A., González-Martínez, C., and Cháfer, M. 2011. Effect of hydroxypropylmethylcellulose and chitosan coatings with and without bergamot essential oil on quality and safety of cold-stored grapes. *Postharvest Biology and Technology* 60(1): 57–63.

Sangeetha, J., and Philip, J. 2012. The interaction, stability and response to an external stimulus of iron oxide nanoparticle-casein nanocomplexes. *Colloids and Surfaces A: Physicochemical and Engineering Aspects* 406: 52–60.

Sessa, M., Ferrari, G., and Donsì, F. 2015. Novel edible coating containing essential oil nanoemulsions to prolong the shelf life of vegetable products. *Chemical Engineering Transactions* 43: 55–60.

Shao, P., Yan, Z., Chen, H., and Xiao, J. 2018. Electrospun poly (vinyl alcohol)/permutite fibrous film loaded with cinnamaldehyde for active food packaging. *Journal of Applied Polymer Science* 135(16): 46117.

Sharma, S., Barkauskaite, S., Jaiswal, A. K., and Jaiswal, S. 2021. Essential oils as additives in active food packaging. *Food Chemistry* 343: 128403.

Škrinjar, M. M., and Nemet, N. T. 2009. Antimicrobial effects of spices and herbs essential oils. *Acta Periodica Technologica* (40): 195–209.

Solans, C., Esquena, J., Forgiarini, A. M., Uson, N., Morales, D., Izquierdo, P., Azemar, N., and Garcia-Celma, M. J. 2003. Nano-emulsions: formation, properties, and applications. *Surfactant science series*, 525–554.

Souto, E. B., Almeida, A. J., and Müller, R. H. 2007. Lipid nanoparticles (SLN®, NLC®) for cutaneous drug delivery: Structure, protection and skin effects. *Journal of Biomedical Nanotechnology* 3: 317–331.

Stashenko, E. E., Jaramillo, B. E., and Martınez, J. R. 2004. Analysis of volatile secondary metabolites from Colombian Xylopia aromatica (Lamarck) by different extraction and headspace methods and gas chromatography. *Journal of Chromatography A* 1025(1): 105–113.

Tadros, T., Izquierdo, P., Esquena, J., and Solans, C. 2004. Formation and stability of nano-emulsions. *Advances in Colloid and Interface Science* 108: 303–318.

Umagiliyage, A. L., Becerra-Mora, N., Kohli, P., Fisher, D. J., and Choudhary, R. 2017. Antimicrobial efficacy of liposomes containing d-limonene and its effect on the storage life of blueberries. *Postharvest Biology and Technology* 128: 130–137.

Wang, X., Chen, Y., Dahmani, F. Z., Yin, L., Zhou, J., and Yao, J. 2014. Amphiphilic carboxymethyl chitosan-quercetin conjugate with P-gp inhibitory properties for oral delivery of paclitaxel. *Biomaterials* 35: 7654–7665.

Wang, Z., Ding, L., Li, T., Zhou, X., Wang, L., Zhang, H., and He, H. 2006. Improved solvent-free microwave extraction of essential oil from dried Cuminum cyminum L. and Zanthoxylum bungeanum Maxim. *Journal of Chromatography A* 1102(1–2): 11–17.

Weıss, J., Decker, E., Mcclements, D., Krıstbergsson, K., Helgason, T., and Awad, T. 2008. Solid lipid nanoparticles as delivery systems for bioactive food components. *Food Biophysics* 3(2): 146–154.

Wu, Y., Luo, Y., and Wang, Q. 2012. Antioxidant and antimicrobial properties of essential oils encapsulated in zein nanoparticles prepared by liquid-liquid method. *LWT-Food Science and Technology* 48: 283–290.

Xue, J., Davidson, P. M., and Zhong, Q. 2015. Antimicrobial activity of thyme oil co-nanoemulsified with sodium caseinate and lecithin. *International Journal of Food Microbiology* 210: 1–8.

Yuan, L., Li, S., Zhou, W., Chen, Y., Zhang, B., and Guo, Y. 2019. Effect of morin-HP-β-CD inclusion complex on anti-ultraviolet and antioxidant properties of gelatin film. *Reactive and Functional Polymers* 137: 140–146.

Zhang, H., Liang, Y., Li, X., and Kang, H. 2020. Effect of chitosan-gelatin coating containing nano-encapsulated tarragon essential oil on the preservation of pork slices. *Meat Science* 166: 108137.

Zhang, Y., Chen, H., Pan, K. 2017. Chapter 5—Nanoencapsulation of food antimicrobial agents and essential oils. In Jafari, S.M. (Ed.), *Nanoencapsulation of Food Bioactive Ingredients*, Academic Press, Cambridge, MA, 183–221.

Zheng, K., Xiao, S., Li, W., Wang, W., Chen, H., Yang, F., and Qin, C. 2019. Chitosan-acorn starch-eugenol edible film: Physico-chemical, barrier, antimicrobial, antioxidant and structural properties. *International Journal of Biological Macromolecules* 135: 344–352.

Zhu, Y., An, X., Li, S., and Yu, S. 2009. Nanoencapsulation of β-Cypermethrin by Complex Coacervation in a Microemulsion. *Journal of surfactants and detergents* 12(4), 305–311.

12 Nanoparticles Designed for Shelf-Life Extension and Food Safety

Franklin Ore Areche[1], *Denis Dante Corilla Flores*[1], *Alfonso Ruiz Rodriguez*[1], *Jovencio Ticsihua Huaman*[1], *Miguel Angel Quispe Solano*[2], *Yash D. Jagdale*[3], *and Mohammad Javed Ansari*[4]
[1]Professional School of Agroindustrial Engineering, National University of Huancavelica, Huancavelica, Peru
[2]Faculty of Food Industry Engineering, National University of the Center of Peru, Huancayo, Peru
[3]MIT School of Food Technology, MIT ADT University, Pune, Maharashtra, India
[4]Department of Botany, Hindu College Moradabad (Mahatma Jyotiba Phule Rohilkhand University, Bareilly), UP, India

CONTENTS

DOI: 10.1201/9781003207641-14

12.1 INTRODUCTION

Diseases caused by contaminated food have increased enormously due to local, national, and international trade and commerce. These illnesses are caused by fungus, viruses, and bacteria, and they have raised significant worries about food safety not just in one region but globally (Bahrami, 2020). According to the World Health Organization (WHO), contaminated food has harmed 600 million people's health in 2015 alone and resulted in 420,000 fatalities (Hoelzer, 2018). As a result, new food packing solutions must be proposed to lower these food borne illness (Bahrami, 2020). Food packaging enables us to increase the shelf life of food and maintain the food chain by facilitating foods' handling, flexibility, and ergonomics, and minimizing physical or physicochemical changes in food (Jafari, 2018; Khaneghah, 2018).

Nanotechnology is the science that manipulates and employs nanometric structures (1 to 100 nm in length), hence affecting the physicochemical properties of minor to significant components (Garofalo, 2018). Nanomaterials may be generated naturally (from biomolecules), purposefully, or accidentally (El Sheikha et al., 2018). The top-down technique involves shrinking the materials (grinding, laser abrasion, and chemistry); conversely, the bottom-up method involves building nanomaterials from individual molecules or atoms (metal salt) (Mohajerani, 2019). Nano packaging is characterized as something new in the food sector that can replace conventional packaging because of its many advantages (antioxidants and antimicrobials) (Pandey, 2020).

Nanomaterials as an additive in the manufacture of packaging increase the storage period of standard packaging (Biswas et al., 2019). Along the same lines, nanomaterials can prevent the growth of various microorganisms, such as *Lactobacillus plantarum*, *Escherichia Coli*, and *Staphylococcus aureus*, also of *Salmonella*, including aerobic psychrotrophs, molds, and yeasts (Karimi, 2020).

A wide range of foods may benefit from the usage of antimicrobial nanocomposites such as Ag and ZnO, which can help to increase foods' shelf life (Sharma, 2020). The production of effective food packaging provides an opportunity to use high-performance, lightweight nanomaterials (Youssef, 2018). The given chapter focuses on providing an overview of nanoparticles and their application to increase the shelf life of food and to assure the safety of foodstuffs.

12.2 NANOPARTICLES

There has been a significant increase in nanotechnology interest in recent years. Nanoparticles are at the center of nanotechnology's applications and advancements. A nanoparticle is a small-sized particle ranging from 1 to 100 nm with a composition that includes carbon, metal, metal oxides, and even organic compounds (Jeevanandam, 2018). Comparing nanoparticles to their larger-scale counterparts, it is clear that they display different physical, chemical, and biological characteristics. As a consequence of the increased area as a function of volume in a chemical process, the reactivity or stability of the chemical process is raised as well (Sudha, 2018). A nanoparticle may have a dimension of zero, in which the length, width, and height are all fixed to a single point, as in nano points; it can also have a dimension with just one parameter, and it can have a dimension with multiple parameters, as in nanospheres. Carbon nanotubes are a good example of two-dimensional graphene; or gold nanoparticles are a good example of three-dimensional graphene, which contains all the properties, including length, breadth, and height. In terms of shape, size, and architecture, nanoparticles are very diverse. They may be cylindrical, spherical, tubular, conical, hollow, spiral, or flat, among other shapes. Its surface may be irregularly shaped or uniformly shaped, with surface variations. Certain nanoparticles are amorphous or crystalline, including scattered or agglomerated multi crystals (Ealias and Saravanakumar, 2017). Various synthesis techniques are being explored at the moment to improve the characteristics and lower the cost of nanoparticles. Several of these technologies are designed specifically to produce nanoparticles with enhanced optical, mechanical, physical, and chemical characteristics for use in the process (Sharma, 2019). As a consequence of advances in instrumentation and application, the performance of nanoparticles has improved. Currently, nanoparticles are being employed in a wide range of applications, spanning from electronics to renewable energy and the aerospace industry. Nanotechnology is essential for ensuring a sustainable environment in the future (Heena, 2019).

12.2.1 Classification of Nanoparticles

Nanoparticles are classified as organic or inorganic, and as carbon.

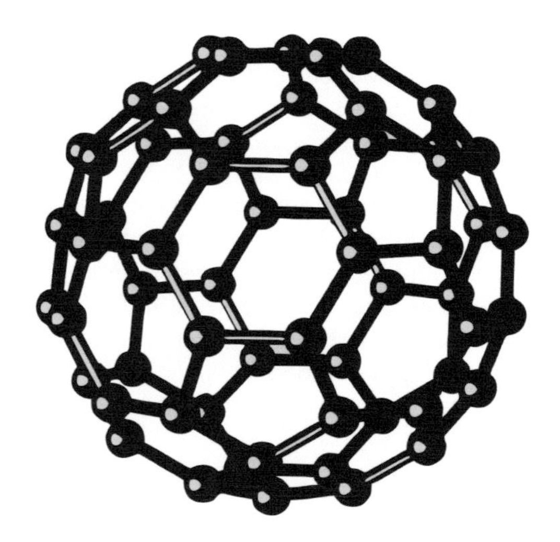

FIGURE 12.1 Structure of organic nanoparticle. (Shin, 2017.)

12.2.1.1 Organic Nanoparticles

Dendrimers, liposomes, micelles, and ferritin are just some of the materials termed "organic particles" or polymers. These nanoparticles damage the environment while, on the other hand, they are not poisonous. Certain particles, such as micelles and liposomes, are hollow on the inside (Figure 12.1); another name for these is nanocapsules. they tend to be sensitive to thermal electromagnetic radiation, through heat and light (Shin, 2017). This quality is taken into account for the production of drugs, considering the stability in their delivery system. The delivery strategy, whether trapped or absorbed, determines the drug's field of administration and effectiveness, in addition to its usual properties such as size, content, surface shape, and so on. Generally, these nanoparticles are employed in biological applications, such as medication delivery systems. They are convenient because they may be injected into selective body areas, a process referred to as selective medication delivery.

12.2.1.2 Inorganic Nanoparticles

Metals and metal oxides used to generate nanoparticles are referred to as "inorganic nanoparticles." These nanoparticles do not include any carbon atoms or molecules.

12.2.1.2.1 Metal-Based

Metal nanoparticles are synthesized in one of two ways: destructively or constructively. The vast majority of metals are available in nanoparticle form (Mourdikoudis, 2018). Aluminum (Al), silver (Ag), lead (Pb), iron (Fe), gold (Au), copper (Cu), cobalt (Co), cadmium (Cd), and zinc (Zn) are the metals most often utilized to produce nanoparticles.

Nanoparticles exhibit a variety of properties, including sizes ranging from 10 to 100 nm, solid and high surface area to volume ratio, charge, cylindrical spherical and amorphous shapes, color, reactivity, and sensitivity to environmental variables such as air, humidity, sunlight, and heat, among others.

12.2.1.2.2 Based on Metal Oxides

Nanoparticles that contain metal oxide are used due to their increased efficiency and reactivity (Shifrina, 2020). Aluminum oxide (Al_2O_3), titanium oxide (TiO_2), silicon dioxide (SiO_2), magnetite (Fe_3O_4), cerium oxide (CeO_2), iron oxide (Fe_2O_3), and zinc oxide (ZnO) are the most often obtained. In comparison to their metallic counterparts, each of these nanoparticles exhibits exceptional characteristics.

12.2.1.3 Carbon-Based

Carbon nanoparticles are well-known and widely used (Xiang, 2019). They are classified as fullerenes, carbon nanotubes (CNTs), graphene, carbon black, and carbon nanofibers, as well as sometimes activated carbon at the nanoscale, as indicated in Figure 12.2.

12.2.1.3.1 The Fullerenes

These are carbon atoms with a spherical shape, and they are generated by carbon atoms bonded by hybridization. Fullerenes are made up of 28 to 1500 atoms. They produce a spherical shape with dimensions ranging from 8.2 to 36 nm for single-layer fullerenes and from 4 to 36 nm for multilayer fullerenes.

12.2.1.3.2 Carbon Nanotubes (CNT)

CNT is graphene nanofilms that have been joined together by carbon atoms, giving them a honeycomb-like appearance. They are wrapped in cylinders with holes to make nanotubes with diameters as tiny as 0.7 nm for a single-layer, 100 nm CNT, and as large as several millimeters for a CNT with multiple layers and extensions. The ends may be left unfilled or sealed with a half molecule of fullerene.

12.2.1.3.3 Carbon Nanofiber

Carbon nanofibers are synthesized using the same graphene nanofibers as CNTs but wrapped in three-dimensional cone shapes instead of regular cylindrical tubes.

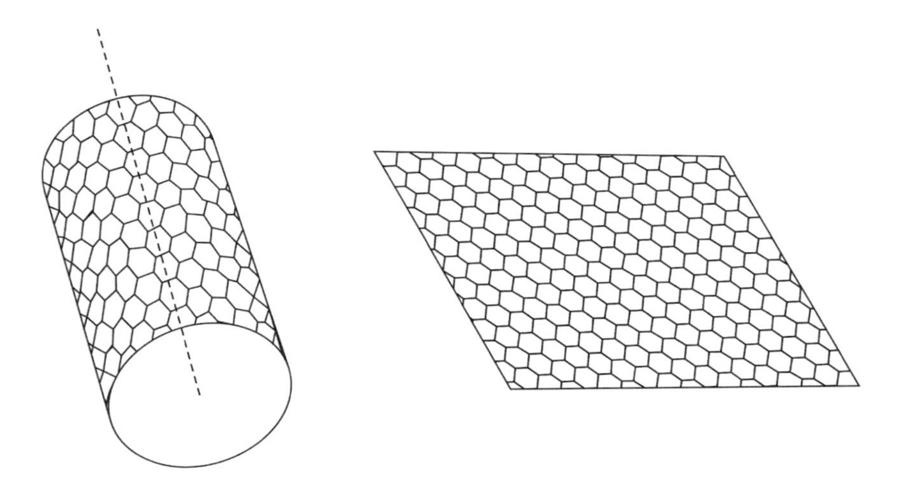

FIGURE 12.2 Structure of carbon nanotube. (Xiang, 2019.)

12.2.2 Nanoparticle Synthesis

Nanoparticles are generally synthesized by different methods classified as ascending or descending methods. Figure 12.3 illustrates the procedure in a simplified manner.

12.2.2.1 Constructive Method

This technique involves the buildup of material at all scales, from atoms to clusters and nanoparticles. Sol-gel, spinning, chemical vapor deposition (CVD), pyrolysis, and biosynthesis are some of the most widely used procedures for the production of nanoparticles.

12.2.2.1.1 Sol-Gel

A colloidal suspension is a solution of colloidal particles suspended in a liquid. The solid macromolecules in gels suspend in a solvent, creating a gelling effect. It is a wet chemical process in which a chemical solution acts as a precursor to the creation of a cohesive system of discrete particles that are then integrated. Typically, metal oxides and chlorides are the pioneers of the sol-gel process (Styskalik, 2017).

Shaking or sonication causes the liquid to dissolve, resulting in a system with a solid and liquid phase. Sedimentation, filtering, and centrifugation are all procedures used to recover nanoparticles. Finally, the dampness is removed by drying (Rane, 2018).

12.2.2.1.2 Centrifugation

Centrifugation can be used to generate nanoparticles because physical parameters such as temperature can be controlled. To prevent chemical reactions from occurring, the reactor is often filled with nitrogen or other inert gases to deplete the inside of oxygen (Zare, 2017).When the liquid, precursors, and water are pushed, the equipment spins at varying rates. This twist results in the molecules fusing, precipitating, collecting, and drying. The flow rate of the liquid, the spinning speed of the disc, the liquid/precursor ratio, the feed location, the surface of the disc, and other factors are taken into consideration and considered to be the parameters that control the properties of the nanoparticles.

FIGURE 12.3 Synthesis process. (Xiang, 2019.)

12.2.2.1.3 *Chemical Vapor Deposition*

This occurs when a thin coating of gaseous reagents forms on a substrate. Accumulation happens when gas molecules are combined at room temperature in a reaction chamber. A chemical reaction occurs when the heated substrate comes into contact with the mixed gas (Cossou, 2019). This reaction results in the formation of a thin layer of product on the surface of the substrate, which is subsequently recovered and used. Its benefit is that it produces nanoparticles that are very pure, homogeneous, complex, and robust. Additionally, they create compounds in the very hazardous gaseous condition (Kulkarni, 2020).

12.2.2.1.4 *Pyrolysis*

Pyrolysis is a process widely practiced in the industry to generate nanoparticles on a large scale. It consists of burning a precursor This must be a liquid or vapor delivered at high pressure into the furnace via small holes where it burns (Hoecker, 2017). Following that, the flue gases are grouped with air to recover the nanoparticles. Certain ovens use lasers and plasma to create high temperatures and speed up the evaporation process (Kuznetsov, 2018).

12.2.2.1.5 *Biosynthesis*

This is an ecological and environmentally friendly approach to synthesizing non-toxic and biodegradable nanoparticles (Issaabadi, 2016). Biosynthesis uses bacteria, fungi, etc., which have unique and enhanced features that make their way into medicinal applications when used to create nanoparticles.

12.2.2.2 Destructive Methods

It consists of degrading thick molecules to a smaller size; that is to say, on a nanometric scale. Among these methods, we could mention the following:

12.2.2.2.1 *Mechanical Milling*

Mechanical milling is used throughout the synthesis process to grind and anneal nanoparticles. During the process, the constituent components are crushed in an inert environment (Vaidya, 2019). Grinding leads to plastic deformation, which results in the particle's shape.

12.2.2.2.2 *Laser Ablation*

This is a well-established method for the production of nanoparticles from a wide range of liquids. When a laser beam is used to irradiate metal that has been immersed in a liquid solution, it condenses a plasma pen, resulting in the formation of nanoparticles.

12.2.3 Nanoparticle Characterization

A nanoparticle's unique properties dictate its potential and use. Therefore, the characterization of the nanoparticles can be realized using various methods summarized in Figure 12.4 (Baimanov, 2019).

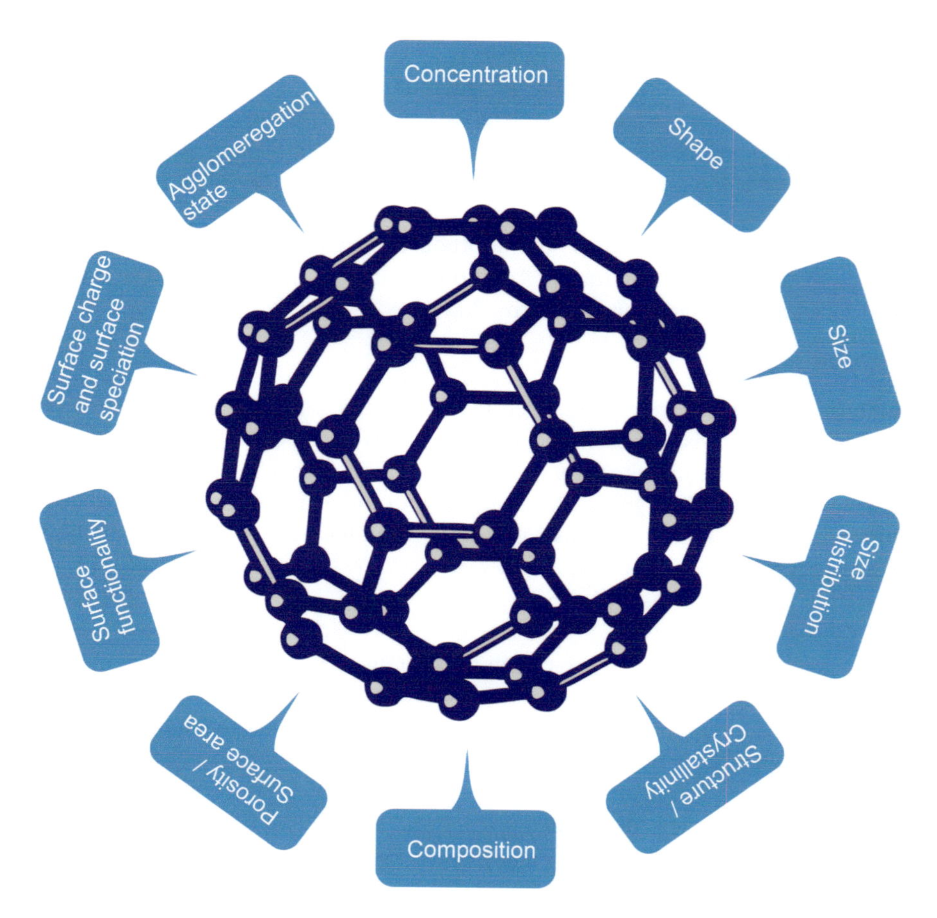

FIGURE 12.4 Nanoparticle characterization. (Baimanov, 2019.)

12.3 APPLICATION OF NANOPARTICLES IN FOODS

The (NPs) are nano-objects with all the external characteristics at the nanoscale; the lengths of the major and minor axes do not differ dramatically from each other. Nanoparticles have the self-assembling ability, and they also have different properties, such as hardness, conductivity, dissolution, permeability rate, and bioavailability. The characteristics of the nanoparticles give rise to new components and other ways of preparing foods with structures and properties that improve their functionalities, resulting in better commercial value in the market (Ojeda et al., 2019).

The most promising nano reinforcements for biodegradable and/or edible films and coatings include polysaccharide nanoparticles, such as starch and chitosan nanocrystals, and cellulose nanofibers (Condés et al., 2016) (see Figure 12.5).

Plant proteins provide the emulsifying properties to nanocomposite systems composed of polysaccharides nanoparticles. These systems prepared with food-grade biopolymers could have a wide range of applications. Thus, they could be used,

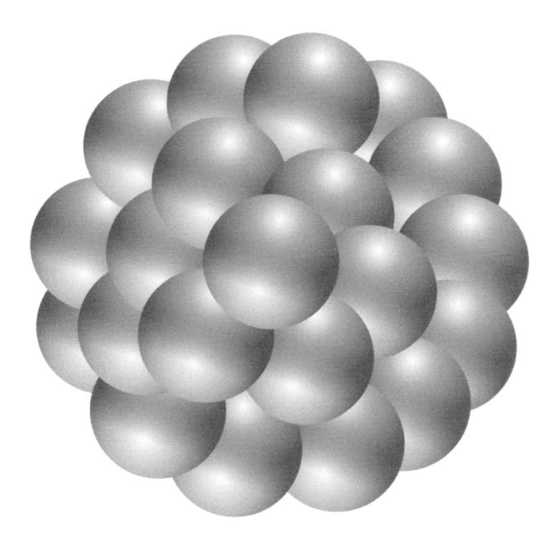

FIGURE 12.5 Inorganic nanoparticles used in the food industry (Ojeda et al., 2019).

for example, to encapsulate active ingredients or to improve the nutritional profile, texture, structure, and the feeling of fullness of various processed foods (Berton-Carabin and Schroën, 2015).

Cellulose nanocrystals can successfully trap selected compounds. In addition, these nanocrystals can generate a more rigid wall material than that obtained from the biopolymers generally used for this purpose (proteins and polysaccharides).

Solid nanoparticles (cellulose nanocrystals (NCC)) derived from cellulose are of great interest and to be evaluated in this application due to their physicochemical properties, abundance, renewability, sustainability, biodegradability, low toxicity, and biocompatibility (Hedjazi and Razavi, 2018).

12.3.1 Nanoparticles in the Food Industry

Nanoparticles (NPs) have always existed on the planet; some examples are smoke and nanoparticles within microorganisms. In some civilizations, nanoparticles were already used for the benefits of their optical and mineral properties. For example, the Egyptian culture used metallic gold nanoparticles as medicinal colloids to safeguard youth and good health while the Chinese civilization used inorganic dyes in porcelain (Berekaa, 2015).

Food manufacturing is a multibillion-dollar sector, and all food firms are searching for methods to increase production efficiency, food safety and security, and resource management. Nanotechnology has emerged as potential aid in improving food production and quality in an industry where competition is fierce and innovation is critical. Nanomaterial applications in food can be found in food processing, food additives, and other materials in contact with food (e.g., packaging). Some applications are already commercialized, and others are still under development (Berekaa, 2015).

12.3.2 Main Applications of Nanoparticles in the Food Industry

12.3.2.1 Application in Agriculture

The application of NPs in agricultural production mainly improves agrochemicals by optimizing synthetic fertilizer, water and phytosanitary products, and protection. However, there are currently only a few products on the market, and most applications are in the development phase (Ávalos et al., 2016).

12.3.2.2 Preparation of Food Additives

Nowadays, industries are investing in nanotechnology studies to obtain healthy, nutritious, and tasty food. Nanoparticles are important in food through the incorporation of food additives. The most used additives are as follow:

12.3.2.2.1 Silicon Oxide (E551)

Silicon oxide is used mainly as a precipitate, with primary dimensions between 30 and 50 nm. Van der Waal forces set these SiO_2 NPs together to shape a body size of 100 nm to 100 μm. Anti-caking additives such as silicon oxide are used in powdered foods and feeds. Recent studies have found that some of this material is nanometric in food size (Pradhan et al., 2015).

12.3.2.2.2 Titanium Oxide (E171)

Among the many applications for titanium dioxide, it is mostly utilized in food items such as milk, cheese, and other dairy products. Silicon dioxide is also used as an anti-caking and drying agent in the manufacturing industry. In addition, it helps absorb water molecules in food, showing hygroscopic applications. As with silicon, the particles aggregate with an average size of 200–300 nm; however, up to 36% of titanium oxide found in food items has particles less than 100 nm in size (Pradhan et al., 2015).

In industry, TiO_2 is used as a bleach for food products; for dairy products, SiO_2 is used as an anti-caking and drying agent; it also helps absorb water molecules in food, with hygroscopic applications.

12.3.2.2.3 Iron Oxide

Iron particles are used to fortify certain foods and different compounds since nutrient deficiencies in human health are common. Metals are also used in industry as food additives, including nano-selenium and colloidal suspensions of metallic particles (copper, gold, silver, platinum, etc.) (Khan et al., 2017).

12.3.2.3 Food Added with Nanoparticles and Adjuvants up to 300 nm

At present, there are many nano encapsulated active particles on the market, such as vitamins and fatty acids, which are used in the industry to process and preserve beverages, meats, cheeses, and other foods. In addition, nanoparticles and particles of several hundred nanometers in size are deliberately added to many foods to improve flow, color, and stability properties during processing or to extend the shelf life of products (Oliveira et al., 2020).

"Edible" NPs are formed from materials such as silicon and ceramics, and other materials such as polymers that react based on temperature or body chemistry. For this reason, work is being done on the creation of ultrasound frequencies to create nanoparticles with aromas, flavors, and colorants for the food industry.

12.3.2.4 Foods Incorporated as Nutritional Additives up to 300 nm

NPs are nutritional additives in food. The "functional foods," also known as "nutraceuticals" (a neologism formed by the terms nutrition and pharmaceutical), are the foods that provide health benefits far superior to those of traditional food products. Functional foods are growing in popularity in national and international markets. NPs are used in the design of functional foods with ingredients that would otherwise be difficult to incorporate due to their low solubility in water (b-carotene), degradation during digestion (curcumin), and volatility (flavors and essential oils). Solid lipid nanoparticles (SLNs), where the particles are spherical and the inner lipid core stabilized by surfactants, fully or partially solidify allowing the encapsulation of bioactive compounds (Pérez Esteve and Gómez Llorente, 2020).

Currently, components such as vitamins, preservatives, and enzymes are added. Although, until very recently, these active ingredients were incorporated into food in the form of nanoparticles, they were also produced in capsules thousands of times smaller (Shelke, 2005).

12.3.2.5 Techniques in Food Processing That Produce Nanoparticles

The discussion of potential health hazards has focused mainly on manufacturing nano-food additives used in food packaging and ignored nanoparticles generated during processing. However, many foods also contain nanoparticles due to the technology used to process them. With the incorporation of food processing technologies that produce nanoparticles in novel foods, the increasing consumption of highly processed foods increases our exposure to nanoparticles in foods, such as the production of salad additives, chocolate syrups, and sweeteners. In addition, in flavored oils and many other processed foods, processing methods generate nanoparticles and nanoscopic emulsions (Sánchez, 2018).

12.3.2.6 Nanoparticles in Packaging and Coating in Food

Silver (Ag) nanoparticles and nanocomposites are widely used in agribusiness for their antimicrobial components in the packaging of fruits and vegetables, herbs, bread, cheese, soups, meats offered under Bags Fresher Longer™ (USA). Likewise, they are added to food trays. They are marketed as Nano Silver Food Container (South Korea), Nano Silver Food (China), and Zeomic (Japan) (Bumbudsanpharoke, 2015). To date, studies show the null or incipient migration of nanocomposites from the container in contact with food. Incorporating NPs TiO_2 to polyethylene, an inhibitory action against *S. aureus* and *E. coli* has been demonstrated, increasing after irradiation with ultraviolet light. Meanwhile, a nanocomposite with Ag-TiO_2 and polyethylene was produced for packaging fresh bread, thus prolonging its shelf life (Cozmuta et al., 2015).

Currently, the incorporation of nanodevices in packaging seeks to improve the useful life of the product. In this way, the exchange of gases, humidity, temperature, flexibility, and mechanical and thermal strength can be controlled. In general, nanocomposites do not cause changes in the density or fluidity of the film or changes in its transparency, and they have the advantage of being recyclable, thereby minimizing environmental pollution. Furthermore, they accept the incorporation of antimicrobial compounds, antioxidants, O_2 absorbers, and water vapor in addition to detecting and providing relevant information about the food, such as freshness and temperature, among others (Ojeda et al., 2019). Ag NPs have significant antimicrobial potential and are among the most scientifically studied and most used for food packaging. They are often used in various forms, such as dispersed particles embedded in containers and liners. These containers allow food to be preserved for longer in optimal consumption conditions (Ávalos et al., 2016).

12.4 NANOPARTICLES IN FOOD PACKAGING

To fulfill customer demand for safe, nutritious, and fresh food while also meeting the need to comply with new food safety rules, the food processing industry is always changing and improving. As a consequence of technological advancements, consumers have grown more demanding, requiring fresh, minimally processed, safe, and ready-to-eat foods that are clearly labeled as to their nutritional content. Food manufacturers, traders, buyers, and food regulatory authorities are all looking for a novel, inexpensive, fast, and consistent method of controlling the quality of packaged food throughout the supply chain, to ensure the safety and authenticity of food. This is because packaging is a critical component of each segment of the food production process (Ling, 2020).

Interventions in food packaging, including nanotechnology, take the following forms: direct inclusion into food products, incorporation into food packaging materials, and usage in the food manufacturing process. Therefore, attachment to new technologies and their applications' sales, and responses to the different nanotechnology applications are defined by consumer behavior (Murthy, 2019).

Research on nanoparticles indicates that consumers tolerate their use as packaging materials and when employed in processing activities, provided they are not directly incorporated into food items (Hoseinnejad, 2018). Richard Feynman defined nanotechnology in 1959, and Norio Taniguchi coined the word "nanotechnology" in 1974. Nanotechnology in polymers includes the design, fabrication, processing, and use of polymeric materials that contain nanoparticles and/or devices (Bailey, 2020). The utilization of these nanoparticles has piqued the interest of academics from a variety of disciplines, including biology, chemistry, engineering, and physics. Due to the positive response worldwide, about three trillion dollars have been projected for its value to the global economy in 2020, resulting in the demand for six million specialists in a variety of connected areas (Basavegowda, 2020). The Institute for Health and Consumer Protection (IHPC) estimates that the market for nanoparticles would reach 20 billion dollars by 2020 (Kalita, 2019). Thus, it is possible to anticipate that nanotechnology will result in the creation of enhanced packaging solutions for the benefit of customers. In comparison to macroscale materials, nanoparticles exhibit unique and enhanced physicochemical features. As a result of their tiny size, nanoparticles exhibit a high correlation between surface volume and activity.

When nanoparticles are combined with suitable polymers, they improve mechanical strength, electrical conductivity, and thermal stability, among other properties.

Thus, nanoparticles are expected to enhance the mechanical properties and barrier characteristics of materials and to enable the development of intelligent packaging systems (Tyagi, 2020). To meet realistic food packaging needs, nanomaterial-optimized polymers will benefit from amplifying the advantages of current polymers while also addressing environmental issues. The packaging design will eliminate any significant interaction between the packaging and food matrices, as well as the impact on consumer health, waste material reduction, increased biodegradability, a protective barrier against gases and light, and CO_2 emissions. The current state of polymeric nanoparticles for food packaging is confined to circumventing regulatory and food safety barriers. By addressing safety and production requirements while maintaining a competitive cost-performance ratio, the new food packaging will be an attractive solution for broad use in food packaging (Omerović, 2021).

Polymer nanoparticles are being used in food packaging to improve the basic characteristics of traditional packaging systems, such as ease of transport and handling; customer friendliness, by preventing leaks or breakages; and protection against microbes by extending the shelf life. Figure 12.6(a) illustrates the nutritional components and preparation recommendations (Dudefoi, 2018).

In the twenty-first century, marketers and customers alike are searching for innovative and enhanced food packaging solutions that provide food safety, quality, and traceability. These requirements necessitate the development of cutting-edge tools and technologies that may be built together to be used in food packaging applications. Current participation and advancements in food packaging must meet certain criteria and provide a justified result that balances the costs of new technology that is economically viable and profitable (Morris, 2017).

The surge of interest in this sector, as seen by the increase in publications since 2000, is the primary drawback associated with permeable food packaging. None of the packaging materials now available provides total resistance to ambient gases, water vapors, and food. Among organic polymeric materials, polypropylene, polystyrene, polyethylene, and polyvinyl chloride remain the industry's preferred options due to their cheap cost, ease of manufacturing, and light weightiness. The fundamental concern, on the other hand, is due to the material's inherent permeability to gases or other small molecules.

Because a single polymer lacks the necessary characteristics for appropriate food packaging, complex multilayer polymers or films are often bonded (Vasile, 2018). This issue may be resolved by developing polymeric nanoparticles for food packaging. Nanomaterials are classed as nanoparticles, nanoclays, and nanoemulsions, each of which has a specific purpose in food (Echiegu, 2017). The uses of polymeric nanoparticles in food packaging fall into three categories: enhanced packaging, intelligent packaging, and active packaging. Because of the nanoparticles in the polymeric matrix, the temperature and humidity parameters may be adjusted with more flexibility in the case of improved packaging applications. Innovative packaging enhances the effectiveness of communication and marketing campaigns. It provides up-to-date information about the true quality of packaged foods and also serves as a false copy of the original. It is possible to achieve preservation and conservation

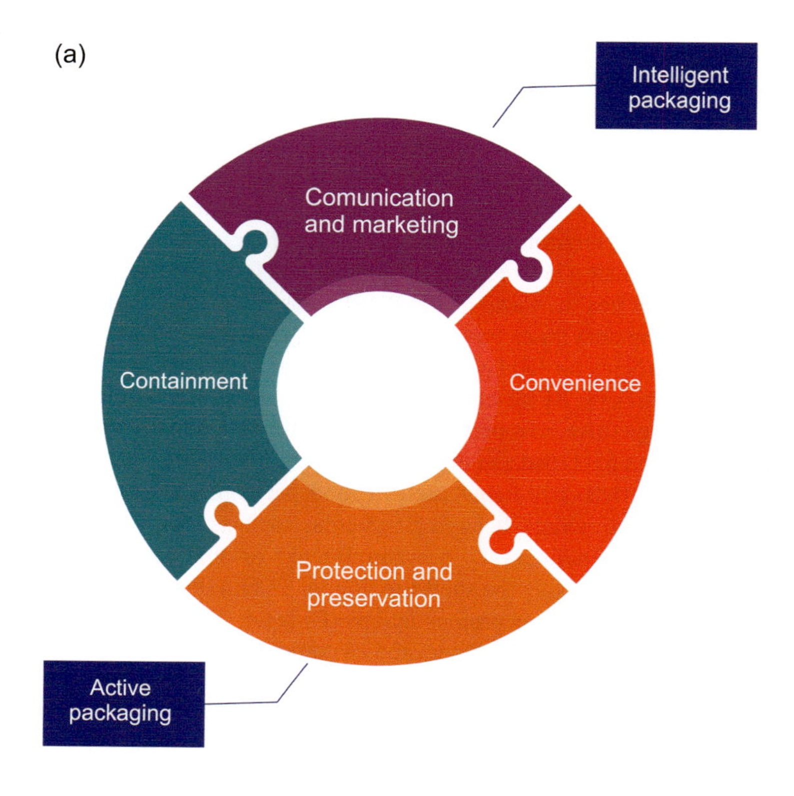

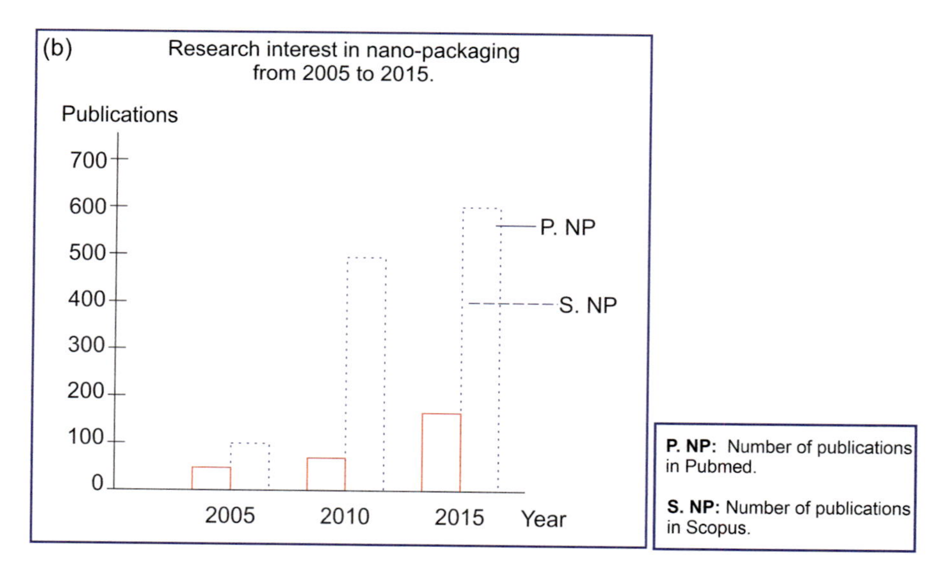

FIGURE 12.6 (a) The concept and mechanism of operation of active and smart packaging. (b) From 2005 to 2015, the worldwide trend in research interest in nano packaging is reflected in PubMed and Scopus papers. (Morris, 2017.)

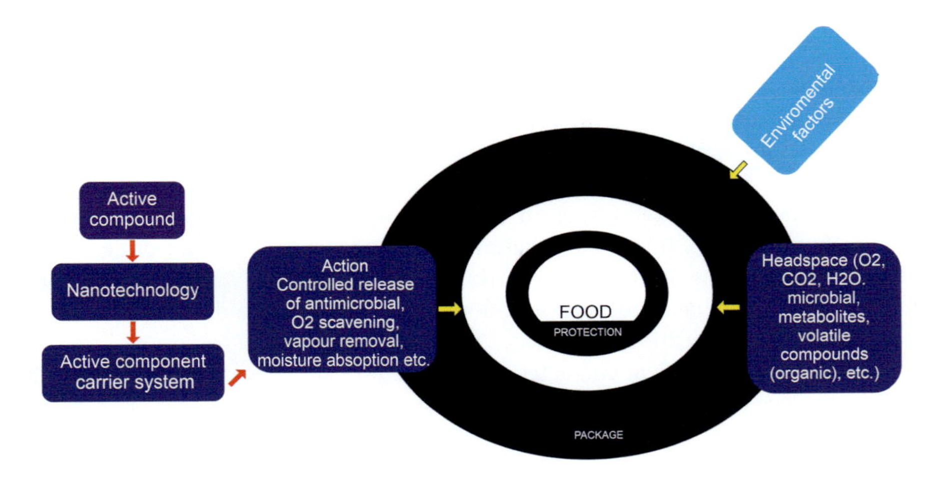

FIGURE 12.7 The use of active packaging and its relationship to nanotechnology. (Patel, 2018.)

via active packaging by activating processes that are triggered by internal and/or external stimuli (Ghoshal, 2018). Active packaging penetrates the component when nanotechnology is used in the food packaging material. Because of the carrier component's ability to react with internal and/or external materials, stimulating activities may be induced, which can help to extend the shelf life of food, improve its quality, and ensure product safety food. In contrast, smart packaging incorporates a sensor based on nanotechnology inside the food container. As seen in Figures 12.7 and 12.8, it may be combined with internal and external environmental elements (Patel, 2018).

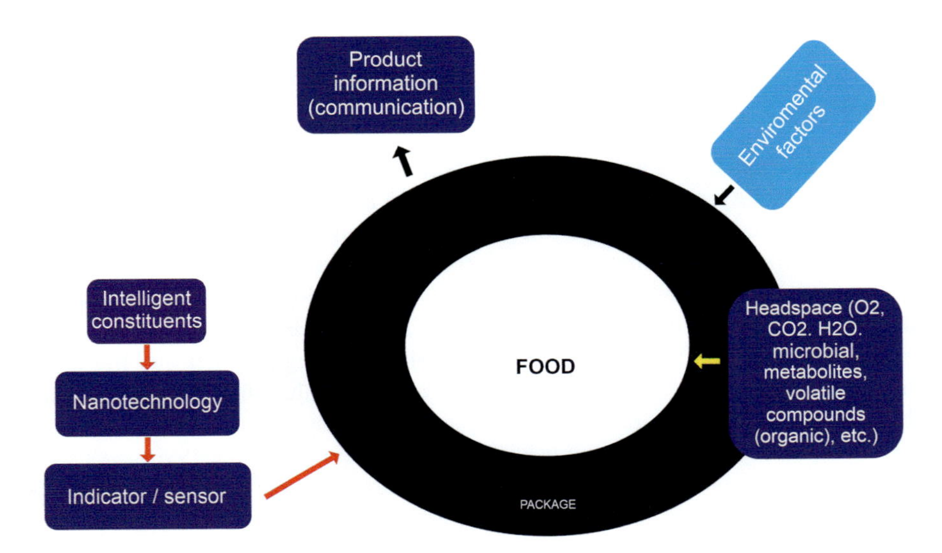

FIGURE 12.8 An illustration of the smart packaging idea and its relationship to nanotechnology is shown. (Patel, 2018.)

12.4.1 Improvement of Packaging Using Nanocomposites

Nanocomposites, which are a combination of conventional food packaging materials and nanoparticles, are gaining traction in the food packaging business. Apart from its obvious antibacterial spectrum, nanocomposites exhibit superior mechanical properties and hardness resistance (Kausar, 2019). Generally, a polymeric matrix produces nanocomposites in either a continuous or discontinuous phase (Senthilkumar, 2018). It is a multiphase material composed of a nano-dimensional phase (discontinuous phase) and a matrix (continuous phase).Nanocomposite contains nanoparticles or nanospheres, nanorods or nanotubes, and nanoplatelets, or nanosheets (Sharma, 2019).

Because the mechanical characteristics of the polymer are improved as a result of the presence of nano-sized phases (elastic stress migrates to the nano-reinforced material), nanocomposites have been acknowledged as the gold standard for increasing the mechanical and barrier characteristics of polymers (Kumar, 2018). Apart from improving mechanical and barrier qualities, nanoparticles provide the packaging system with active or intelligent features (Talegaonkar, 2017). Incorporating nanotechnology into polymer science may offer up new opportunities for enhancing the characteristics of packaging materials as well as their cost-benefit ratios (Yuan, 2019). Polymeric nanocomposites are made up of polymers and inorganic or organic fillers in a variety of morphologies (spheres, flakes, fibers, and particles). In recent years, new packaging materials that are both environmentally friendly and inventive have been created (Ganguly, 2018). The aspect ratio of the filler material in the container (the ratio of its biggest to smallest dimension) is critical.

Numerous nanomaterials are now being investigated for use as fillers, including silica, clay, organoclay, graphene, polysaccharides, nanocrystals, carbon nanotubes, chitosan, and cellulose (Rouf, 2017; Yuan, 2019).

12.4.2 Clay and Silicates

Clay and silicates have garnered researchers' interest as potential nanoparticles due to their availability, cheap cost, and relative simplicity. Due to their two-dimensional structure, layered silicates with a thickness of between 1 nm and a few microns are typically utilized in nanocomposites. The combination of silicates and polymers creates a very effective barrier. Additionally, this combination enhances the diffusion channel of the infiltrating molecule (Dontsova, 2018) as seen in Figure 12.9.

They feature a multilayer structure made of alternating polymer and inorganic layers that are separated by just a few nanometers (Idumah, 2018). Nanocomposites that have been exfoliated are constructed of a range of polymers and have a randomly distributed coating of clay on top of them (Barua, 2019). Since the 1990s, montmorillonite clay has been employed as a nanoparticle in a range of polymers, including polyethylene, nylon, polyvinyl chloride, and starch, to improve their properties. Montmorillonite ($Si_8O_{20}(OH)_4$) is the clay filler that is most often utilized. In this case, the $Al(OH)_3$ film is sandwiched between two silica tetrahedral bilayers, which are

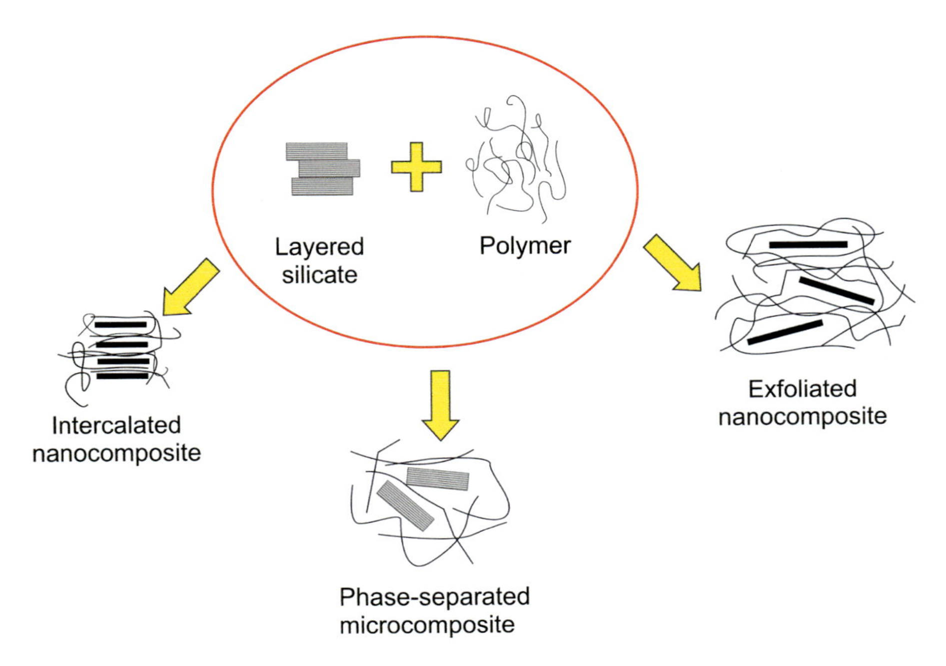

FIGURE 12.9 Types of compounds obtained from the interaction between layered silicate and polymers. (Dontsova, 2018.)

kept together by modest electrostatic forces (Orta, 2020). The presence of exchangeable cations, such as Na^+ and Ca^{2+}, helps to restore equilibrium to the imbalance between negative charges on the surface. Clay layers give gas resistance while also allowing water vapor to pass through.

Additionally, clay aggregates enhance the mechanical resilience of biopolymers. Adding clays to thermoplastic starch, TPS/clay nanocomposites increases the mechanical characteristics of the starch biopolymer while decreasing its water vapor permeability by 5% (w/w) (Raee, 2017). Additionally, it is well established that clay nanoparticles enhance glass transition and thermal degradation temperatures (Orta, 2020). Additionally, nanoclays may be used to create passive and intelligent food nano packaging. Idumah (2018) created a nano packaging system by injecting blueberry extract between the clay silicate intercalated gaps. Blueberries contain anthocyanins, which act as a pH indicator. By adding blueberry extract, these clays might be transformed into active and intelligent nanocomposites.

12.4.3 Cellulose-Based Nanofibers

Cellulose is a naturally occurring polymer that is used to construct long fibrous cells. It is economical, simple to recycle, and eco-friendly. Additionally, the possibility of using cellulose as a support material for a range of nanomaterials is being researched. The addition of cellulose to the nanoparticles enhances their surface area, which explains their increased activity. These additional properties make cellulose nanofibers a promising nanomaterial source (Tsai, 2018). Nanoparticles are classified into two types, microfibrils and whiskers, both of which are made of cellulose. Through hydrogen bonding, cellulose chains are shown as microfibrils (or nanofibers) (Costa, 2017). Each microfibril is created by combining elemental fibrils that include both crystalline and amorphous components. Whiskers, nanocrystals, nanorods, and rod-type cellulose microcrystals are all terms used to describe acid hydrolysis procedures that are used to separate the crystalline components of a substance (Trache, 2017). After hydrolysis, the diameters of the whiskers are determined mostly by the proportion of amorphous areas in the thick fibrils, which vary across species. Microcrystalline cellulose is composed of a large number of cellulose microcrystals clustered together in amorphous regions. With the expansion of polymers, cellulose-based nano reinforcements enhance their heat resistance and modulus (Fardioui, 2018). The incorporation of cellulose nanodiscs and starch improves their thermo-mechanical properties, reduces their water sensitivity, and maintains their biodegradability, while also improving the moisture-barrier qualities of polymeric films. Bano (2017) previously created a cellulose nanocomposite utilizing peanut nanofibers and a solvent mixture of dimethylacetamide/lithium chloride.

12.4.4 Carbon Nanotubes

Single-walled carbon nanotubes are typically one atom thick, but multiple-walled ones are constructed from a series of concentric tubes with a certain aspect ratio and module (Alhans, 2018). Polyethylene naphthalene, polyvinyl alcohol, polyamide, and polypropylene are just a few of the materials that have benefited from the addition of carbon nanotubes to their tensile strength and modulus (Azizi-Lalabadi, 2020). Because carbon nanotubes are capable of directly penetrating microbial cells, they may possess antibacterial properties. Sharma (2019) showed that carbon nanotubes coupled with allyl isothiocyanate were capable of killing *Salmonella choleraesuis* for up to 40 days after storage (Tarfaoui, 2019). Carbon nanotubes inserted into polyethylene sheets used in date packaging have been demonstrated to inhibit fungus growth for up to 90 days. Additionally, single-walled carbon nanotubes were studied in combination with cobalt mesoarylporphyrin complexes to develop a chimera detector that alerts to the presence of amines created during meat decomposition. Despite its widespread usage, issues about its processing and dispersion, as well as its expensive cost, limit its employment in nanocomposites (Baig, 2016).

12.4.5 Starch Nanocrystals

For decades, starch has been intensively studied as a preferred packaging material. Starch offers several benefits, including its abundance, biocompatibility, non-toxicity, affordability, biodegradability, ease of availability, and stability (Martău, 2019), all of which broaden the scope of its potential uses. The combination of inorganic components and synthetic polymers improves the material's resistance to water (Tabasum, 2019). When amorphous areas hydrolyze, native starch granules may be absorbed into elongated hydrolysis at temperatures below gelatinization temperature, permitting the removal of more refractory crystalline sheets. The crystalline starch particles, with a thickness of 6 to 8 nm, enhance the tensile strength and modulus of the pullulan films while decreasing their elongation characteristic (Machado, 2020). According to some, the positive charge boosts its antibacterial activity. Due to the increased surface area of polysaccharides, the antibacterial spectrum of metals incorporated/absorbed increases (Al-Tayyar, 2020).

12.4.6 Active Packaging

In comparison to conventional food packaging, active packaging is a specially engineered method that incorporates antimicrobial or antioxidant chemicals within the packaged food or food item, or makes it permeable to oxygen or water vapor. The combination of active chemicals attached to the polymer increases its effectiveness in terms of quality and shelf life (Shahid, 2021).

12.4.7 Nanoparticles as Part of Food Packaging

At present, there are many investigations and bibliographic reviews related to food packaging with the addition of nanoparticles, as shown in Table 12.1.

TABLE 12.1
Studies of Nanoparticles as a Part of Food Packaging

Qualification	Objective	Results	Bibliography
A thorough assessment of nanomaterials for food packaging applications	This study will synthesize many studies on the use of nanomaterials in food packaging, including zinc oxide, clay, silver, carbon nanotubes, copper, titanium dioxide, and copper oxide.	The use of nanoparticles in food packaging improves the product's physicochemical quality (color, taste, moisture content, weight, bioavailability, and texture). Additionally, it decreases microbial burden as a result of the cell membrane's activity.	Emamhadi et al. (2020).
Recent advancements and problems in nanotechnology's use to food packaging. A critical examination of the literature	The purpose of this review is two-fold. To begin, it will discuss the latest advancements in the use of ENP in food and beverage packaging to give active and intelligent characteristics.	The paper analyzes the present market dynamics for engineered nanoparticles in food and beverage packaging, from idea to current commercial applications.	Enescu et al. (2019).

(Continued)

TABLE 12.1 ***(Continued)***
Studies of Nanoparticles as a Part of Food Packaging

Qualification	Objective	Results	Bibliography
A review of nanoedible films for food packaging	The purpose of this review is to demonstrate that nanoparticles possess unique properties (such as a large surface-to-volume ratio, different optical behavior, and high mechanical strength).	The food and beverage industry is showing great interest in incorporating the benefits of nanotechnology.	Jeevahan and Chandrasekaram (2019).
A quick discussion of the use of encapsulated natural chemicals as antibacterial agents in food packaging.	This study seeks to demonstrate that encapsulation is an intriguing strategy for increasing the physical-chemical and microbiological stability of essential oils and achieving a controlled release.	This study examines contemporary research on nanotechnology and the nano- and microencapsulation of essential oils.	Zanetti et al. (2018).
Advances in food packaging technology – Review	All stakeholders have a common goal: to safeguard food against deterioration caused by environmental and human causes and to guarantee that food is delivered to end consumers in safe and healthy conditions.	The article discusses current advancements in food packaging technology, with an emphasis on the concepts and uses of active and smart packaging, nanotechnology, and antimicrobial packaging.	Nura (2018).
Polyphenol-loaded nanoparticles for use in the food sector	The purpose of this study is to summarize the most recent advances in the use of polyphenol-loaded nanoparticles in food science.	The use of nanoparticles containing polyphenols as active ingredients to enhance the physicochemical and functional qualities of meals is discussed critically.	Milincic et al. (2019).
The migratory potential of nanoparticles in food-contact polymers has been thoroughly investigated.	This article does a critical evaluation of existing research on the migration of nanomaterials that are somewhat conflicting. Additionally, the author discusses the effect of analytical methodologies and experimental design on the findings.	When completely integrated in the polymer, consumers will be protected from food contact polymer nanoparticles.	Stormer et al. (2017).

(Continued)

TABLE 12.1 ***(Continued)***
Studies of Nanoparticles as a Part of Food Packaging

Qualification	Objective	Results	Bibliography
A Review of Cereal Starch Nanoparticles as a Potential Food Additive.	The main objective is to show that raw starch has limited applications due to its inherent disadvantages, such as poor cold-water solubility, the tendency to retrograde, and high viscosity once gelatinized.	Starch nanoparticles may be utilized as food additives due to their wide variety of applications in food, including emulsion stabilization, fat substitution, thickening, and rheological modification.	Kaur et al. (2018).
Characterization, processes, and implications of food crop absorption of synthetic nanoparticles.	The purpose of this review is to describe the in vitro and in vivo characterization of nanoparticles in growth medium and biological matrices, as well as to examine the nanoparticles' absorption patterns.	A thorough knowledge of the effects of nanoparticles on agricultural plants, including their benefits and drawbacks, will aid in optimizing the safe and sustainable employment of nanotechnology in agriculture.	Ma et al. (2018).
The significance and health risks associated with nanoparticles employed in the food sector.	The primary purpose is to raise awareness of the disconnect between expanding research into novel uses for nanoparticles and their safety, which has placed scientists under pressure to find potential side effects.	Numerous researches have been conducted on the use of nanotechnology in food and the impact of nanoparticles on human health. However, there is a void in the literature when it comes to combining these findings.	Naseer et al. (2018).
A review of biopolymer-based polymers integrating silver nanoparticles for usage in food applications as active packaging.	The primary goal is to provide current information on silver nanoparticles used in biopolymer-based packaging materials.	Silver nanoparticles have emerged as an intriguing component of biodegradable biopolymers, owing primarily to their antibacterial capabilities, which enable the production of packaging materials.	Krasniewska et al. (2020).
A survey of advancements in bio-nano composite materials for food packaging.	This article aims to introduce readers to the latest advancements in biopolymer-based food packaging materials such as natural biopolymers, synthetic biopolymers, and biopolymer blends.	Nanocomposite packaging materials based on natural biopolymers have a bright future—a wide range of applications in the food industry, including advanced packaging of functional foods.	Kumar et al. (2017).

12.5 NANOTECHNOLOGY AND FOOD SAFETY

Nanotechnology is a branch of science that is constantly growing in which nanoparticles are used. Nanotechnology offers a variety of applications in food safety, ranging from food processing to contamination analysis (see Figure 12.10) (Krishna et al., 2018). Thus, nanoparticles have been employed to precisely remove pollutants from food, acting as antimicrobials and antioxidants (Kubyshkin et al., 2007).

Recent years have seen the use of nanotechnology in the field of food safety, which has gained greater recognition of nanotechnology among scientists. To address the issue of food safety, detecting systems based on nanotechnology have been developed that differ in terms of their mechanics and designs. They do, however, have a common goal: the early and reliable identification of pathogen residues or other pollutants in drinking water (Krishna et al., 2018).

Nanotechnology has infiltrated various customer products, additives, food preservation, and food packaging and continues to advance in innovations and improvements in food processing and storage to ensure food safety (He et al., 2019).

Food safety and quality concerns must always be considered in their entirety since human life depends on it. Researchers have found several technologies to improve the quality and safety of food, and the participation of nanotechnology in food processing has led to the production of foods with better thermal stability, solubility, environmental impact, and better availability for consumption (Hamad et al., 2017).

The demand for nanoparticle-based materials has increased in different industries. Several of the materials possess essential elements and have also been stable at different temperatures and pressures. Different nanomaterials provide a marked variety in the quality and safety of food and their different health benefits; in that sense, many researchers are generating new techniques and methods (Singh, 2017).

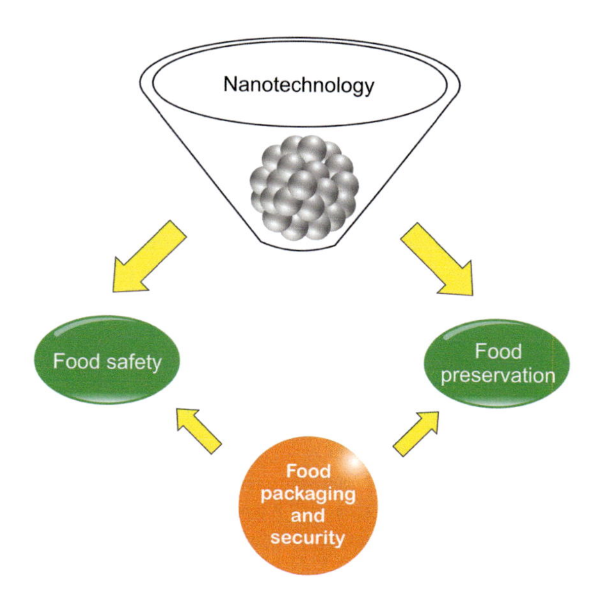

FIGURE 12.10 Nanotechnology and food preservation. (Kubyshkin et al., 2007.)

12.5.1 Nanosensors and Nanofilters

Using nanoparticles for nanosensors serves to detect contaminants and possible pathogens in the food system, which has been another of the possible uses of nanotechnology. Specific nanosensors have been developed; for example, for analyzing food colors, flavors, etc. (Berekaa, 2015).

12.5.2 Nanotechnology in Food Packaging and Safety

Packaging is a vital part of food safety when it comes to processed items. Water vapors, natural chemicals, and ambient gases are nearly impenetrable to practically any kind of packing material. However, it must be remembered that permeability is not always convenient; fruits and vegetables are subjected to cellular respiration. The opposite is the case with carbonated beverages, where it is prudent to cut both oxygen and carbon dioxide flow to prevent decarbonization (Gordon, 2012).

It should be noted that the flow of carbon dioxide or oxygen changes according to the food matrices and the type of container used. Thus, in the complexity of different containers, nanocomposites can be used, including polymers (Abbaspour et al., 2015). With regulatory laws in place, different nanoparticle structures can be used in various food and pharmaceutical industry fields, especially for different investigations based on food science (see Figure 12.11) (Bajpai, 2018).

Current research shows that using nanocomposites in food packaging has increased the capacity of different food packaging to avoid gas segregation, which is a good thing for the environment (Ghanbarzadeh et al., 2014). Nanocapsules have a variety of uses in food packaging, which are outlined in Table 12.2. A recent study on food packaging promotes the use of environmentally friendly biodegradable polymers (Abdollahi et al., 2012). However, like with any new technology, there is worry about these nanoparticles being ingested during food intake. The investigation into how these nanoparticles act in the human body is vital to establishing whether or not they are toxic or have other negative effects on the body (Reddy, 2011). Another point of contention is the polymers' ease of digestion (Klaine et al., 2011). These factors have been thoroughly explored around the globe to identify nanomaterials that coexist harmoniously with the environment and people (Holden, 2016).

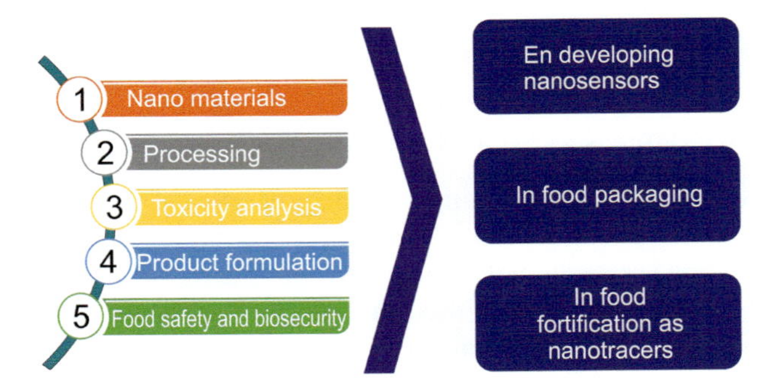

FIGURE 12.11 Steps for the processing and use of nanomaterials in the food sector. (Bajpai, 2018.)

TABLE 12.2
Studies of the Effects of Nanoparticles in Food Packaging

Bibliography	Istiqola and Syafiuddin (2020).	Kubyshkin et al. (2007).	Abdollahi et al. (2012).
Results	To assure their safety prior to broad use, a thorough examination of the toxicological aspects of silver nanoparticles discharged from food packaging must be conducted.	In the first 24 hours after exposure, a 0.02–0.05% nanosilver solution with an alginate stabilizer was observed to inhibit the growth and development of harmful bacteria. The nanosilver solution, at a concentration of 0.0005–0.05%, decreases the viability.	Incorporation of MMT and REO into chitosan improved water gain, water vapor permeability, and chitosan film solubility by more than 50%.
Objective	This publication discusses the use of silver nanoparticles in food packaging technology in detail. The application takes place in food packaging that is controlled by the United States Food and Drug Administration.	Using the Sabouraud agar technique, we investigated the antifungal efficacy of silver nanoparticles in sodium alginate against Candida albicans. Using a luminometer, we evaluated the biocidal effect of silver nanoparticles in solution with sodium alginate.	They used montmorillonite nanoclays (MMT) in relation to chitosan from (1 to 5) and activated three levels of rosemary essential oil (REO) at (0.5%, 1%, 1.5% v / v).
Qualification	A study of silver nanoparticles used in food packaging technology, including their regulation, methods of manufacture, characteristics, migration, and future difficulties.	Silver nanoparticles stabilized in solution with sodium alginate have antimicrobial activity.	A new active bionanocompound film that incorporates rosemary essential oil and nanoclay in chitosan.

12.5.3 Nanocoatings as Smart Packaging for Surfaces

Nanocoatings have the potential to improve the properties of pure polymers as well as the functional characteristics of active and smart packaging in the food packaging industry. Packaging is categorized as "renewed" or "smart," which refers to the kind of innovative material that may be used in a range of applications and be referred to as renewed or smart (Hannon et al., 2015). Furthermore, the European Union has put legal limitations on the use of active and smart packaging materials in a range of commodities, except for titanium nitride encapsulated in plastic bottles, which is permitted (Echegoyen and Nerín, 2013).

Recently, some researchers developed a sophisticated nanocoating film that can detect the presence of contaminants during storage. Additionally, the non-invasive gas content and detection methodology have demonstrated an extraordinary ability to manage the gas content, excess moisture, and oxygen content of a package headspace continuously and efficiently, providing an effective means of verifying the safety quality of food even after the production process (Ducan, 2011). However, the presence of oxygen within the container may jeopardize the shelf life of goods, owing to oxygen's capacity to foster microbial development. Ariyarathna et al. (2017) presented inexpensive biodegradable coatings made of polyvinyl alcohol and chitosan that greatly reduced oxygen and humidity permeability by 25.6 and 10.2%, respectively. In comparison to normal packaging, the films quadrupled the shelf life of cherries. Swaroop and Shukla (2018) produced a food packaging material using a combination of magnesium oxide (MgO) nanostructures and polylactic acid biopolymer, and demonstrated that the material was very effective in preventing the development of bacterial biofilms. Furthermore Foltynowicz et al. (2017) created iron particles with no valence to function as oxygen sensors in food packaging.

12.5.4 Nanotechnology in Food Safety

Food safety is a global public health issue that is escalating at an alarming rate. Food safety's primary goal is to guarantee that food does not hurt the customer during preparation and consumption (Martins et al., 2012). Physical, chemical, and biological contamination of food must be avoided throughout manufacturing, handling, and delivery (Jaimes et al., 2017). Nanotechnology's recent advancements have changed the food sector via its many uses in food processing, safety, and protection. It increased the nutritional content of the nutraceutical, increased its shelf life, and decreased packaging waste (García-Mosquera and Villa-Enciso, 2018). As a result of the fast evolution of recipes and eating habits in the contemporary period, food safety has become a critical concern. Eating-related ailment syndrome (ETAS), poisons, heavy metals, and other pollutants all have the potential to be hazardous to human health if consumed. Using established procedures to detect illnesses and poisons is time-consuming and difficult. Food safety challenges of microbiological contamination, toxin detection, shelf life, and packaging techniques have all been addressed with the help of advancements in nanotechnology (Inbaraj and Chen, 2016). Additional nanomaterials, such as metallic nanoparticles, quantum dots, carbon nanotubes, and other active nanomaterials, may be utilized to make biosensors for the detection of microbes and other food safety testing, among other applications (Dimitrijevic et al., 2015).

12.5.4.1 Nanoparticles for the Detection of Foodborne Pathogens

Nanobiosensors are miniature bioanalytical devices composed of nanoparticles and biological receptors that are used to analyze biological samples (Chandra et al., 2011). The presence of pathogens and decomposing materials in food have been detected using a variety of biosensors (Thakur and Ragavan, 2013). For example, Surface-enhanced Raman Spectroscopy (SERS) is used as a nano biosensor to promptly and accurately diagnose microbial infections (Jarvis and Goodacre, 2004). Additionally, since silver nano colloids increase Raman signals, they are often utilized in SERS to identify bacteria (Baranwal et al., 2016). Similarly, silver nanocolids, graphene oxide (Zuo et al., 2013), magnetic beads (Holzinger et al., 2014), and carbon nanotubes (Yang et al., 2013), as well as plasmonic gold have been used (Yang et al., 2013; Fang et al., 2017). Additionally, nanobarcodes made of synthetic molecular beacons and color-coded probes are used to help in the detection of foodborne infections (Bajpai, 2018).

Direct identification of *E. coli* in particular food samples has also been made feasible by the measurement and detection of the light emitted by the bacteria themselves. This kind of sensor works by forming a bond with a certain protein in the body. It is defined as a bacterium on a silicon chip that coexists with any other *E. Coli* bacteria that have been detected in food samples, according to the manufacturer (Bhattacharya et al., 2007). Researchers at the University of California, San Diego (Chen and Durst, 2006), developed an immunosorbent test based on a matrix for the detection of *E. Coli* and *Salmonella spp.*

L. monocytogenes was also cultured in liposomal G-protein nanovesicles in both pure and mixed medium, and the results were encouraging. Scientists have shown the possibility of using G-protein liposomal nanovesicles in foodborne-pathogen detection as well as their effectiveness as universal immunoassay reagents. DeCory et al. (2005) also developed an immunomagnetic fluorescence test employing beads and immuno liposomes for the quick detection of *E. coli* in aqueous samples. Their results revealed that the use of immunomagnetic beads in combination with sulphorhodamine B encapsulated in immunoliposomes was effective in detecting *E. coli* in water samples in a short period. Apart from that, various researchers have looked at liposome-based techniques for the diagnosis of infection (Shukla et al., 2016). To distinguish proteins and diseases that vibrate at various frequencies depending on their biomass, nanosensors, like nano cantilevers, utilize silicon materials (Jain, 2014).

At the moment, many novel detection systems based on nanoparticles are being developed. For instance (Tominaga, 2018) Palladium nanoparticles were used to construct lateral flow immunological test strips against Klebsiella, which assisted in the binding of the unique shape and identification of mentioned bacteria. A bacterial cell of the *E. coli* type was also seen by employing a field-effect transistor device composed of reduced graphene nanoparticles, as reported by Thakur (2018). *Cronobacter sakazakii*, bacteria that is highly contagious for newborns, has recently been detected in powdered baby formula using an electrochemical detection platform based on gold and graphene oxide nanoparticles. The detection limit for this bacteria was 2×10 CFU/ml (Shukla et al., 2018). Additionally, Shukla et al. (2016) exhibited a fluorescence detection approach for *Cronobacter sp.* by combining immunomagnetic nanoparticles with liposome nanoparticles. When it comes to the gender level, the detection limit is 5.9×103 CFU/mL.

12.5.4.2 Nanoparticles for Detection Against Allergens

To monitor and manage food allergies, nanotechnology has been employed extensively (Pilolli et al., 2013). Certain nanoparticles, on the other hand, have been shown to induce allergic bronchial irritation in people (Yoshida et al., 2011). For example, it has been shown that SiO_2 nanoparticles trigger allergen-specific Th2-like immune responses in female mice when administered in vivo (Yoshida et al., 2011). Swelling, indurations, erythema, skin nodules, and granulomas were seen at the injection site as a consequence of aluminum hydroxide-adjuvant allergy immunotherapy (Vogelbruch et al., 2002). Researchers have looked at the use of polymeric nanoparticles to alleviate this restriction (De Souza et al., 2012), as well as medicines that target toll-like receptors, as alternative adjuvants (Hedayat et al., 2010). Adjuvants such as protamine nanoparticles with cytosine phosphate as a ligand and guanine-oligodeoxynucleotides may be employed. Protamines are roughly 4kDa arginine-rich peptides detected in adult salmon testes. Protamines are biodegradable and are effective in reversing heparin activity during surgical procedures (Schulman and Bijsterveld, 2007) or as an insulin additive (Owens, 2011; Schulman and Bijsterveld, 2007; Gamazo et al., 2014). There has been reported immunotherapy that makes use of nanoparticles to release allergens, hence, decreasing the dose and limiting allergen exposure to immunoglobulin E. Additionally, in recent years, bio-inspired nanoparticles with verified minimum or no adverse effects have been investigated, with the possibility of being employed in the food industry (Prasad et al., 2016). Zhang (2018) utilized a magnetic fluorescence test for identifying food allergies that are based on aptamers and nanoparticles. Brotons-Canto et al. (2018) studied the effectiveness of poly(anhydride) nanoparticles (150 nm) as oral carriers for peanut allergy immunotherapy treatment.

12.5.4.3 Nanoparticles to Prevent Heavy Metal Reduction

Due to the escape of toxic substances from nanomaterials, there is a significant risk of dangerous metals leaking into food and posing significant health risks (Karlsson et al., 2013). Additionally, because of peroxidation and DNA damage, metal and metal oxide nanomaterials, such as Ag, ZnO, and CuO, increase intracellular ROS levels and generate lipids as a consequence of their use (Fukui et al., 2012). The magnetic nanoparticles modified with silica and coated with cationic surfactant operate as submersors for microextraction and are meant to detect levels of Pb, Cd, Co, Ni, Cu, and Mn in environmental samples. Cetylpyridinium bromide is used to synthesize silica surface nanoparticles; when complexed with 8-hydroxyquinoline, it may dissolve metal ions (Karatapanis et al., 2011). Although most nanomaterials have a potential ability for solving contaminants, magnetite nanoparticles have shown that substrates are more attractive and cheap for improving heavy metals (Amin et al., 2014).

12.5.4.4 Nanoparticles for the Inhibition of Biofilm Formation

Densely packed communities of bacteria-like organisms cling to a variety of surfaces and create a robust and difficult-to-penetrate extracellular polymeric family. Biofilm formation includes the primary attachment of bacteria to surfaces followed by proliferation of attached bacterial cells, which leads to the accumulation of multilayer clusters of cells and glycocalix formation (Shakeri et al., 2007). Glycerol monolaurate (GML), a wide spectrum of gram-positive bacteria, including *Bacillus anthracis*, are inhibited in their growth by the antibacterial medication amikacin (Vetter, 2005). GML has been shown to block three separate strains of *Staphylococcus aureus* and MRSA from forming biofilms (Schlievert and Peterson, 2012). Similarly, antimicrobials are included in the nanofibers of sieve membranes to prevent the formation of biofilms (Zhang et al., 2011). Additionally, Shahrokh and Emtiazi (2009) observed that nanosilver samples at low concentrations (0.2 ppm) inhibited bacterial metabolism and, hence, advocated for the adoption of an optimal nanosilver particle dosage for diverse nanomaterials to minimize biofilm development.

12.6 CONCLUSIONS

Nanotechnology for shelf-life extension and food safety is a very up-to-date technology that is increasingly on the rise, bringing rapid progress in this area. This is because nanoparticles have unique chemical, physical, and biological capabilities on a nanoscale that, whether applied directly to food or inside its packaging, enable food preservation. In the past five years, the number of scholarly journals describing research on nanoparticles and their use in food as well as in their containers, packaging, and wrapping, has increased. As a result of this idea, it is possible to conclude that nano encapsulation may be used in a variety of processed foods to increase food preservation. This technology is constantly modifying and improving food to fulfill customer demand for safe, nutritious, and fresh food. Researchers believe that nanoparticles have the potential to enhance the mechanical and barrier qualities of food packaging, hence, extending the shelf life of food while also adding to food safety.

REFERENCES

Abbaspour, A., Norouz-Sarvestani, F., Noori, A., and Soltani, N. 2015. Aptamer-conjugated silver nanoparticles for electro chemical dual-aptamer-based sándwich detection of *staphylococcus aureus*. *Biosensors and Bioelectronic* 68(15): 149–155.

Abdollahi, M., Rezaei, M., and Farzi, G. 2012. A novel active bionanocomposite film incorporating rosemary essential oil and nanoclay into chitosan. *Journal of Food Engineering* 111(2): 343–350.

Alhans, R. 2018. Comparative analysis of single-walled and multi-walled carbon nanotubes for electrochemical sensing of glucose on gold printed circuit boards. *Materials Science and Engineering: C* 90(1): 273–279.

Al-Tayyar, N. 2020. Antimicrobial food packaging based on sustainable bio-based materials for reducing foodborne pathogens: A review. *Food Chemistry* 310(25): 125915.

Amin, M., Alazba, A., and Manzoor, U. 2014. A review of removal of pollutants from water/wastewater using different types of nanomaterials. *Advances in Materials Science and Engineering* 2014(1): 1–10.

Ariyarathna, I., Rajakaruna, R., and Karunaratne, N. 2017. The rise of inorganic nanomaterial implementation in food applications. *Food Control* 77(1): 251–259.

Ávalos, A., Haza, A. I., Mateo, D., and Morales, P. 2016. Nanopartículas de plata: Aplicaciones y riesgos tóxicos para la salud humana y el medio ambiente silver nanoparticles: Applications and toxic risks to human health and environment. *Revista Computense De Ciencias Veterinarias* 7(2): 1–23.

Azizi-Lalabadi, M. 2020. Carbon nanomaterials against pathogens; the antimicrobial activity of carbon nanotubes, graphene/graphene oxide, fullerenes, and their nanocomposites. *Advances in Colloid and Interface Science* 284(2020): 102250.

Bahrami, A. 2020. Efficiency of novel processing technologies for the control of listeria monocytogenes in food products. *Trends in Food Science & Technology* 96(1): 61–78.

Baig, Z. 2016. Recent progress on the dispersion and the strengthening effect of carbon nanotubes and graphene-reinforced metal nanocomposites: A review. *Critical Reviews in Solid State and Materials Sciences* 6(1): 1–46.

Bailey, E. 2020. Dynamics of polymer segments, polymer chains, and nanoparticles in polymer nanocomposite melts: A review. *Progress in Polymer Science* 105(1): 101242.

Baimanov, D. 2019. Understanding the chemical nature of nanoparticle–protein interactions. *Bioconjugate Chemistry* 30(7): 1923–1937.

Bajpai, V. 2018. Prospects of using nanotechnology for food preservation, safety and security. *Journal of Food and Drug Analysis* 3(1): 1–14.

Bano, S. 2017. Studies on cellulose nanocrystals isolated from groundnut shells. *Carbohydrate Polymers* 157(10): 1041–1049.

Baranwal, A., Mahato, K., Srivastava, A., Maurya, P., and Chandra, P. 2016. Phytofabricated metallic nanoparticles and their clinical applications. *RSC Advans* 6(1): 105996–106010.

Barua, S. 2019. Silicon-based nanomaterials and their polymer nanocomposites. *Nanomaterials and Polymer Nanocomposites* 2019(1): 261–305.

Basavegowda, N. 2020. Bimetallic and trimetallic nanoparticles for active food packaging applications: A review. *Food and Bioprocess Technology* 13(1): 30–44.

Berekaa, M. 2015. Nanotechnology in food industry: Advances in food processing. *Packaging and Food Safety* 4(5): 345–357.

Berton-Carabin, C., and Schroën, K. 2015. Pickering Emulsions for food applications: Background, trends, and challenges. *Annual Review of Food Science and Technology* 6(1): 263–297.

Bhattacharya, S., Jang, J., Yang, L., Akin, D., and Bashir, R. 2007. Biomems and nanotechnology-based approaches for rapid detection of biological entities. *Journal of Rapid Methods y Automation in Microbiology* 15(1): 1–32.

Biswas, et al. 2019. Nano silica-carbon-silver ternary hybrid induced antimicrobial composite films for food packaging application. *Food Packaging and Shelf Life* 19 (1):104–113.

Brotons-Canto, A., Gamazo, C., Martin-Arbella, N., Abdulkarim, M., Matías, J., Gumbleton, M., and Irache, J. 2018. Evaluation of nanoparticles as oral vehicles for immunotherapy against experimental peanut allergy. *International Journal of Biological Macromolecules* 110(15): 328–335.

Bumbudsanpharoke, N. 2015. Nano-food packaging: An overview of market, migration research, and safety regulations. *Journal of Food Science* 80(1): 910–923.

Chandra, P., Noh, H., Won, M., and Shim, Y. 2011. Detection of daunomycin using phosphatidylserine and aptamer co-immobilized on Au nanoparticles deposited conducting polymer. *Biosensors and Bioelectronics* 26(11): 4442–4449.

Chen, C. and Durst, R. 2006. Simultaneous detection of *Escherichia coli* O157:H7, Salmonella spp. and Listeria monocytogenes with an array-based immunosorbent assay using universal protein G-liposomal nanovesicles. *Talanta* 69 (1):232–238.

Condés, M. C., Echeverría, I., Añón, M. C., and Mauri, A. N. 2016. Nanocompounds as formulating aids. In *Edible Films and Coatings*. New York: Pocket Books.

Cossou, B. 2019. Synthesis and optimization of low-pressure chemical vapor deposition-silicon nitride coatings deposited from $SiHCl_3$ and NH_3. *Thin Solid Films* 681 (1): 47–57.

Costa, A. 2017. Production of bacterial cellulose by Gluconacetobacter hansenii using corn steep liquor as nutrient sources. *Frontiers in Microbiology* 8(1): 132–145.

Cozmuta, M., Peter, A., Cozmuta, M., Nicula, C., Crisan, L., and Baia, L. 2015. Active packaging system based on Ag/TiO2 nanocomposite used for extending the shelf life of bread. *Packaging Technology and Science* 28(1): 271–284.

De Souza, J., Esparza, I., Ferrer, M., Sanz, M., Iracha, M., and Gamazo, C. 2012. Nanoparticulate adjuvants and delivery systems for allergen immunotherapy. *Journal of Biomedicine and Biotechnology* 2012(1): 13–15.

DeCory, T., Durst, R., Zimmerman, S., Garringer, L., Paluca, G., DeCory, H., and Montagna, R. 2005. Development of an immunomagnetic bead-immunoliposome fluorescence assay for rapid detection of *Escherichia coli* O157:H7 in aqueous samples and comparison of the assay with a standard microbiological method. *Applied and Environmental Microbiology* 10(1): 1856–1864.

Dimitrijevic, M., Karabasil, N., Boskovic, M., Teodorovic, V., Vasilev, D., Djordjevic, V., Kilibarba, N., and Cobanovic, N. 2015. Safety aspects of nanotechnology applications in food packaging. *Procedia Food Science* 5(1): 57–60.

Dontsova, T. 2018. Directional control of the structural adsorption properties of clays by magnetite modification. *Hindawi Journal of Nanomaterials* 2018(9): 65–73.

Ducan, T. 2011. Applications of nanotechnology in food packaging and food safety: Barrier materials, antimicrobials and sensors. *Colloid and Interface Science* 363(3): 1–24.

Dudefoi, W. 2018. Nanoscience and nanotechnologies for biobased materials, packaging and food applications: New opportunities and concerns. *Innovative Food Science & Emerging Technologies* 46(1): 107–121.

Ealias, A., and Saravanakumar, M. 2017. A review on the classification, characterisation, synthesis of nanoparticles and their application, *IOP conference series. Materials Science and Engineering* 263(2017): 032019.

Echegoyen, Y., and Nerín, C. 2013. Nanoparticle release from nano-silver antimicrobial food containers. *Food Chemical Toxicology* 62(1): 16–22.

Echiegu, E. 2017. Nanotechnology applications in the food industry. *Nanotechnology* 2017(1): 153–171.

El Sheikha, A. F., Levin, R. E., and Xu, J. 2018. *Molecular Techniques in Food Biology: Safety, Biotechnology, Authenticity and Traceability*, John Wiley & Sons, USA.

Emamhadi, M., Sarafraz, M., Akbari, M., Thai, V., Fakhri, Y., Linh, Y., and Khaneghah, M. 2020. Nanomateriales para aplicaciones de envasado de alimentos: Una revisión sistemática. *Toxicología Alimentaria y Química* 146(1): 111825.

Enescu, D., Cerqueira, M., Fucinos, P., and Pastrana, L. 2019. Avances y desafíos recientes en las aplicaciones de la nanotecnología en el envasado de alimentos. Una revisión de la literatura. *Toxicología Alimentaria y Química* 10(12): 110814.

Fang, Z., Zhao, Y., Warner, R., and Johnson 2017. Active and intelligent packaging in meat industry. *Trends in Food Science y Technology* 61(1): 60–71.

Fardioui, M. 2018. Bio-active nanocomposite films based on nanocrystalline cellulose reinforced styrylquinoxalin-grafted-chitosan: Antibacterial and mechanical properties. *International Journal of Biological Macromolecules* 114(15): 733–740.

Foltynowicz, Z., Bardensshtein, A., Sangerlaub, S., Antvorskov, H., and Kozak, W. 2017. Nanoscale, zero valent iron particles for application as oxygen scavenger in food packaging. *Food Packaging* 11(1): 74–83.

Fukui, H., Horie, M., Endoh, S., and Kato, H. 2012. Association of zinc ion release and oxidative stress induced by intratracheal instillation of ZnO nanoparticles to rat lung. *Chemico-Biological Interactions* 198(1): 1–3.

Gamazo, C., Gastaminza, G., Ferrer, M., Sanz, M., and Irache, J. 2014. Nanoparticle based-immunotherapy against allergy. *Immunotherapy* 6(7): 885–897.

Ganguly, S. 2018. Polymer nanocomposites for electromagnetic interference shielding: A review. *Journal of Nanoscience and Nanotechnology* 18(11): 7641–7669.

García-Mosquera, J., and Villa-Enciso, E. 2018. Análisis de las principales tendencias de innovación en nanotecnología de alimentos: Una aproximación a su estudio a partir de vigilancia tecnológica. *Revista CEA* 4(8): 95–115.

Garofalo, E. 2018. Tuning of co-extrusion processing conditions and film layout to optimize the performances of PA/PE multilayer nanocomposite films for food packaging. *Polymer Composites* 2017(1): 24–32.

Ghanbarzadeh, B., Amir, S., and Almasi, H. 2014. Nano-structured materials utilized in biopolymer based plastics for food packaging applications. *Food Science and Nutrition* 2(4): 37–41.

Ghoshal, G. 2018. Recent trends in active, smart, and intelligent packaging for food products. *Food Packaging and Preservation* 2018(1): 343–374.

Gordon, R. 2012. *Food Packaging, Principles and Practice.* Third Edition, Editorial Taylor & Francis Group, New York.

Hamad, A., Han, J., Kim, B., and Rather, I. 2017. El entrelazamiento de la nanotecnología con la industria alimentaria. *Saudi Journal of Biological Sciences* 25(1): 1–194.

Hannon, J., Kerry, J., Cruz-Romero, M., Morris, M., and Cummins, E. 2015. Advances and challenges for the use of engineered nanoparticles in food contact materials. *Food Science e Technology* 10(5): 1–20.

He, X., Deng, H., and Hwang, H. 2019. The current application of nanotechnology in food and agriculture. *Journal of Food and Drug Analysis* 27(2019): 1–21.

Hedayat, M., Tajeda, K., and Rezaei, N. 2010. Prophylactic and therapeutic implications of toll-like receptor ligands. *Medicinal Research Reviews* 32(2): 294–325.

Hedjazi, S., and Razavi, S. H. 2018. A comparison of Canthaxanthine Pickering emulsions, stabilized with cellulose nanocrystals of different origins. *International Journal of Biological Macromolecules* 106: 489–497.

Heena, S. 2019. Green nanotechnology for the environment and sustainable development. *Green Materials for Wastewater Treatment* 38(1): 13–46.

Hoecker, C. 2017. The influence of carbon source and catalyst nanoparticles on CVD synthesis of CNT aerogel. *Chemical Engineering Journal* 314(15): 388–395.

Hoelzer, K. 2018. Emerging needs and opportunities in foodborne disease detection and prevention: From tools to people. *Food Microbiology* 75(1): 65–71.

Holden, P. 2016. Considerations of environmentally relevant test conditions for improved evaluation of ecological hazards of engineered nanomaterials. *Environmental Science & Technology* 3(4): 1–70.

Holzinger, M., Le, A., and Cosnier, S. 2014. Nanomaterials for biosensing applications: A review. *Frontiers in Chemistry* 2(1): 63–65.

Hoseinnejad, M. 2018. Inorganic and metal nanoparticles and their antimicrobial activity in food packaging applications. *Critical Reviews in Microbiology* 44(2): 161–181.

Idumah, C. 2018. Recently emerging trends in polymer nanocomposites packaging materials. *Polymer-Plastics Technology and Engineering* 2018(10): 1054–1109.

Inbaraj, B., and Chen, B. 2016. Nanomaterial-based sensors for detection of foodborne bacterial pathogens and toxins as well as pork adulteration in meat products. *Journal of Food and Drug Analysis* 24(1): 15–28.

Issaabadi, Z. 2016. Green synthesis of the copper nanoparticles supported on bentonite and investigation of its catalytic activity. *Journal of Cleaner Production* 142(4): 3584–3591.

Istiqola, A., and Syafiuddin, A. 2020. Una revisión de las nanopartículas de plata en las tecnologías de envasado de alimentos: Regulación, métodos, propiedades, migración y desafíos futuros. *Journal of the Chinese Chemical Society* 2020: 1–15.

Jafari, S. 2018. Neural networks modeling of Aspergillus flavus growth in tomato paste containing microencapsulated olive leaf extract. *Journal of Food Safety* 2017(10): 17–20.

Jaimes, J., Rios, I., and Severiche, C. 2017. Nanotecnologia y sus aplicaciones en la industria de alimentos - Nanotechnology and its applications in the food industry. *Revista De La Asociación Colombiana De Ciencia y Tecnología* 25(1): 11–15.

Jain, K., 2014. Nanodiagnostics: Application of nanotechnology in molecular diagnostics. *Expert Review of Molecular Diagnostics* 153–161.

Jarvis, R., and Goodacre, R. 2004. Discrimination of bacteria using surface-enhanced Raman spectroscopy. *Analytical Chemiustry* 76(1): 40–47.

Jeevahan, J., and Chandrasekaram, M. 2019. Películas nanocomestibles para envasado de alimentos: Una revisión. *Journal of Materials Science* 54(1): 12290–12318.

Jeevanandam, J. 2018. Review on nanoparticles and nanostructured materials: History, sources, toxicity and regulations. *Journal of Nanotechnology* 9(1): 1050–1074.

Kalita, D. 2019. Chapter 11 - The impact of nanotechnology on food. *Nanomaterials Applications for Environmental Matrices* 2019(10): 369–379.

Karatapanis, A., Fiamegos, Y., and Stalikas, C. 2011. Silica-modified magnetic nanoparticles functionalized with cetylpyridinium bromide for the preconcentration of metals after complexation with 8-hydroxyquinoline. *Talanta* 84(3): 834–839.

Karimi, N. 2020. Thymol, cardamom and lactobacillus plantarum nanoparticles as a functional candy with high protection against Streptococcus mutans and tooth decay. *Microbial Pathogenesis* 148(1): 104–481.

Karlsson, H., Cronholm, P., Hedberg, Y., Tornberg, M., De Battice, L., Svedhem, S., and Odnevall, I. 2013. Cell membrane damage and protein interaction induced by copper containing nanoparticles—Importance of the metal release process. *Toxicology* 313(1): 59–69.

Kaur, J., Savita, G., and Jeet, K. 2018. Nanopartículas de almidón de cereal: Un posible aditivo alimentario: Una revisión. *Critical Reviews in Food Science and Nutrition* 58: 1097–1107.

Kausar, A. 2019. A review of high performance polymer nanocomposites for packaging applications in electronics and food industries. *Journal of Plastic Film and Sheeting* 36(1): 2020.

Khan, I., Saeed, K., and Khan, I. 2017. Nanoparticles: Properties, applications and toxicities. *Arabian Journal of Chemistry* 12(7): 908–931.

Khaneghah, A. 2018. Antimicrobial agents and packaging systems in antimicrobial active food packaging: An overview of approaches and interactions. *Food and Bioproducts Processing* 111(1): 1–19.

Klaine, S., Koelmans, A., Carley, S., Handy, R., Kapustka, L., Nowack, B., and Kammer, F. 2011. Paradigms to assess the environmental impact of manufactured nanomaterials. *Environmental Toxicology and Chemistry* 31(1): 1–10.

Krasniewska, K., Galus, S., and Gniewosz, M. 2020. Materiales a base de biopolímeros que contienen nanopartículas de plata como envasado activo para aplicaciones alimentarias: Una revisión. *International Journal of Molecular Sciences* 21(3): 698.

Krishna, V., Wu, K., Su, D., Cheeran, M., Wang, J., and Perez, A. 2018. Nanotechnology: Review of concepts and potential application of sensing platforms in food safety. *Food Mircrobiiology* 20(17): 22–33.

Kubyshkin, A., Chegodar, D., Katsev, A., Petrosyan, A., Krivorutchenko, Y., and Postnikowa, O. 2007. Antimicrobial effects of silver nanoparticles stabilized in solution by sodium alginate. *Biochemistry and Molecular Biology Journal* 2(2): 13.

Kulkarni, M. 2020. Microfluidic devices for synthesizing nanomaterials—a review. *Nano Express* 1(3): 032004.

Kumar, S. 2018. Recent advances and remaining challenges for polymeric nanocomposites in healthcare applications. *Progress in Polymer Science* 80(1): 1–38.

Kumar, N., Kaur, P., and Bhatia, S. 2017. Avances en los materiales bio-nanocompuestos para el envasado de alimentos: Una revisión. *Nutrition & Food Science* 47(4): 591–606.

Kuznetsov, G. 2018. Unsteady temperature fields of evaporating water droplets exposed to conductive, convective and radiative heating. *Applied Thermal Engineering* 131(25): 340–355.

Ling, E. 2020. Integrity of food supply chain: Going beyond food safety and food quality. *International Journal of Productivity and Quality Management* 29(2): 2020.

Ma, C., White, J., Zhai, J., Zhao, Q., and Xing, B. 2018. Absorción de nanopartículas de ingeniería por los cultivos alimentarios: Caracterización, mecanismos e implicaciones. *Annual Review of Food Science and Technology* 2018(9): 129–153.

Machado, L. 2020. Pea protein isolate nanocomposite films for packaging applications: Effect of starch nanocrystals on the structural, morphological, thermal, mechanical and barrier properties. *Emirates Journal of Food & Agriculture* 32(7): 40–49.

Martău, G. 2019. The use of chitosan, alginate, and pectin in the biomedical and food sector—Biocompatibility, bioadhesiveness and biodegradability. *Polymers* 11(11): 1837.

Martins, A., Benelmekki, M., Teixeira, V., and Coutinho, P. 2012. Platinum nanoparticles as pH sensor for intelligent packaging. *Journal of Nano Research* 18(1): 18–19.

Milincic, D., Popovic, D., Levic, S., Kostic, A., Tesic, Z., Nedovic, V., and Pesic, M. 2019. Aplicación de las nanopartículas cargadas de polifenoles en la industria alimentaria. *Nanomaterials* 9(1): 16–29.

Mohajerani, A. 2019. Nanoparticles in construction materials and other applications, and implications of nanoparticle use. *Materials* 12(1): 30–52.

Morris, M. 2017. Development of active, nanoparticle, antimicrobial technologies for muscle-based packaging applications. *Meat Science* 132(1): 163–178.

Mourdikoudis, S. 2018. Characterization techniques for nanoparticles: Comparison and complementarity upon studying nanoparticle properties. *Royal Society of Chemistry* 10(1): 12871–12934.

Murthy, C. 2019. Metal oxide nanoparticles and their applications in nanotechnology. *SN Applied Sciences* 2(607): 2019.

Naseer, B., Srivastava, G., Qadri, O., Faridi, S., Islam, R., and Younis, K. 2018. Importancia y peligros para la salud de las nanopartículas utilizadas en la industria alimentaria. *Nanotechnology Reviews* 23(3): 22–54.

Nura, A. 2018. Avances en la tecnología de envasado de alimentos – Revisión. *Jpht* 6(4): 55–64.

Ojeda, G., Arias Gorman, A., and Sgroppo, S. 2019. La nanotecnología y su aplicación en alimentos. Mundo nano. *Revista Interdisciplinaria En Nanociencias Y Nanotecnología* 12(23): 1e–14e.

Oliveira, M. G. L., da Rocha, P. G. L., Lemos, P. V. F., de Jesus Assis, D., de Almeida Júnior, A. R., and da Silva, J. B. A. 2020. Avaliação de nanopartículas de amido como aditivo a lubrificantes. *Brazilian Applied Science Review* 4(5): 3190–3201.

Omerović, R. 2021. Antimicrobial nanoparticles and biodegradable polymer composites for active food packaging applications. *Comprehensive Reviews in Food Science and Food Safety* 20(3): 2428–2454.

Orta, M. 2020. Biopolymer-clay nanocomposites as novel and ecofriendly adsorbents for environmental remediation. *Applied Clay Science* 198(15): 105838.

Owens, D. 2011. Insulin preparations with prolonged effect. *Diabetes Technology & Therapeutics* 13(1): 32–33.

Pandey, V. 2020. Antimicrobial biodegradable chitosan-based composite nano-layers for food packaging. *International Journal of Biological Macromolecules* 157(15): 212–219.

Patel, A. 2018. Chapter 1 - Application of nanotechnology in the food industry: Present status and future prospects. *Impact of Nanoscience in the Food Industry* 2018(1): 1–27.

Pérez Esteve, E., and Gómez Llorente, H. 2020. Sistemas nanoestructurados basados en lípidos para la encapsulación de compuestos bioactivos. *Physical Review* 47(10): 777–780.

Pilolli, R., Monaci, L., and Visconti, A. 2013. Advances in biosensor development based on integrating nanotechnology and applied to food-allergen management. *TrAC Trends in Analytical Chemistry* 47(1): 12–26.

Pradhan, N., Singh, S., Ojha, N., Shrivastava, A., Barla, A., Rai, V., and Bose, S. 2015. Facets of nanotechnology as seen in food processing, packaging, and preservation industry. *BioMed Research International* 34(3): 34–56.

Prasad, A., Mahato, K., Chandra, P., Srivastava, A., Shrikrishna, N., and Kumar, P. 2016. Bioinspired composite materials: Applications in diagnostics and therapeutics. *Journal of Molecular and Engineering Materials. Synthesis of Inorganic Nanomaterials* 4(1): 21–22.

Raee, E. 2017. Biodegradable polypropylene/thermoplastic starch nanocomposites incorporating halloysite nanotubes. *Journal of Applied Polymer Science* 135(4): 45740.

Rane, A. 2018. Chapter 5 - Methods for synthesis of nanoparticles and fabrication of nanocomposites. *Synthesis of Inorganic Nanomaterials* 10(8): 121–139.

Rasouli, M.R., and Tabrizian, M. 2019. An ultra-rapid acoustic micromixer for synthesis of organic nanoparticles. *Lab on a Chip* 19(19): 3316–3325.

Reddy, B. 2011. Advances in diverse industrial applications of nanocomposites. *Web of Science*. New York: Pocket Books.

Rouf, T. 2017. Chapter 8 - Natural biopolymer-based nanocomposite films for packaging applications. *Bionanocomposites for Packaging Applications* 2018(10): 83–97.

Schlievert, P., and Peterson, M. 2012. Glycerol monolaurate antibacterial activity in broth and Biofilm cultures. *Plos One* 28(1): 1–5.

Schulman, S., and Bijsterveld, N. 2007. Anticoagulants and their reversal. *Transfusion Medicine Reviews* 21(1): 37–48.

Senthilkumar, P. 2018. Nanocomposites: Recent trends and engineering applications. *Nano Hybrids and Composites* 20(1): 65–80.

Shahid, R. 2021. Role of active food packaging developed from microencapsulated bioactive ingredients in quality and shelf-life enhancement: A review. *Journal of American Science* 17(2): 2021.

Shahrokh, S., and Emtiazi, G. 2009. Toxicity and unusual biological behavior of nanosilver on Gram positive and negative bacteria assayed by microtiter-plate. *Journal of Biological Sciences* 1(3): 28–31.

Shakeri, S., Kermanshahi, R., Momeni, M., and Emtiazi, G. 2007. Assessment of biofilm cell removal and killing and biocide efficacy using the microtiter plate test. *The Journal of Bioadhesion and Biofilm Research* 2013(1): 37–41.

Sharma, R. 2020. Antimicrobial bio-nanocomposites and their potential applications in food packaging. *Food Control* 112(1): 107086.

Sharma, S. 2019. Nanotechnology in cancer therapy: An overview and perspectives (review). *International Journal of Pharmaceutical Chemistry and Analysis* 6(4): 110–114.

Shelke, K. 2005. Hidden ingredients take cover in a capsule. *The Journal of Bioadhesion and Biofilm Research* 34(1): 65–86.

Shifrina, Z. 2020. Role of polymer structures in catalysis by transition metal and metal oxide nanoparticle composites. *Chemical Review* 120(2): 1350–1396.

Shin, T. 2017. Synergism of nanomaterials with physical stimuli for biology and medicine. *Accounts of Chemical Research* 50(3): 567–572.

Shukla, S., Haldorai, Y., Bajpai, V., and Rengaraj, A. 2018. Electrochemical coupled immunosensing platform based on graphene oxide/gold nanocomposite for sensitive detection of Cronobacter sakazakii in powdered infant formula. *Biosensors and Bioelectronics* 109(30): 139–149.

Shukla, S., Lee, G., Song, X., Park, S., and Kim, M. 2016. Immunoliposome-based immunomagnetic concentration and separation assay for rapid detection of Cronobacter sakazakii. *Biosensors and Bioelectronics* 77(15): 986–994.

Singh, T. 2017. Application of nanotechnology in food science: Perception and overview. *Frontiers in Microbiology* 2017(10): 01–15.

Stormer, A., Bott, J., Kemmer, D., and Franz, R. 2017. Revisión crítica del potencial de migración de nanopartículas en plásticos en contacto con alimentos. *Tendencias En Ciencia y Tecnología De Los Alimentos* 63(1): 39–50.

Styskalik, A. 2017. The power of non-hydrolytic sol-gel chemistry: A review. *Catalysts* 7(6): 168.

Sudha, P. 2018. Nanomaterials history, classification, unique properties, production and market. *Emerging Applications of Nanoparticles and Architecture Nanostructures* 1(2018): 341–384.

Swaroop, C., and Shukla, M. 2018. Nano-magnesium oxide reinforced polylactic acid biofilms for food packaging applications. *Biological Macromolecules* 113(1): 729–736.

Tabasum, S. 2019. A review on blending of corn starch with natural and synthetic polymers, and inorganic nanoparticles with mathematical modeling. *International Journal of Biological Macromolecules* 122(1): 969–996.

Talegaonkar, S. 2017. Bionanocomposites: Smart biodegradable packaging material for food preservation. *Food Packaging* 2017(1): 79–110.

Tarfaoui, M. 2019. Self-heating and deicing epoxy/glass fiber based carbon nanotubes buckypaper composite. *Journal of Materials Science* 54(2019): 1351–1362.

Thakur, M. 2018. Rapid detection of single E. coli bacteria using a graphene-based field-effect transistor device. *Biosensors and Bioelectronic* 110(1): 16–22.

Thakur, M., and Ragavan, K. 2013. Biosensors in food processing. *Journal of Food Science and Technology* 50(4): 625–641.

Tominaga, T. 2018. Rapid detection of Klebsiella pneumoniae, Klebsiella oxytoca, Raoultella ornithinolytica and other related bacteria in food by lateral-flow test strip immunoassays. *Journal of Microbiological Methods* 147(1): 43–49.

Trache, D. 2017. Recent progress in cellulose nanocrystals: Sources and production. *Nanoscale* 2017(9): 1763–1786.

Tsai, Y. 2018. Drug release and antioxidant/antibacterial activities of silymarin-zein nanoparticle/bacterial cellulose nanofiber composite films. *Carbohydrate Polymers* 180(15): 286–296.

Tyagi, P. 2020. Nanotechnology foods and nanotechnology food packaging. *Journal of Critical Reviews* 7(7): 2020.

Vaidya, M. 2019. High-entropy alloys by mechanical alloying: A review. *Journal of Materials Research* 34(5): 2019.

Vasile, C. 2018. Polymeric nanocomposites and nanocoatings for food packaging: A review. *Materials* 11(10): 1834.

Vetter, S. S. 2005. Glycerol monolaurate inhibits virulence factor production in Bacillus anthracis. *Antimicrobial Agents and Chemotherapy* 49(4): 1302–1305.

Vogelbruch, M., Nuss, B., Korner, M., Kapp, A., Kiehl, P., and Bohn, W. 2002. Aluminium-induced granulomas after inaccurate intradermal hyposensitization injections of aluminium-adsorbed depot preparations. *Allergy* 55(1): 883–887.

Xiang, Z. 2019. Enhanced electromagnetic wave absorption of nanoporous Fe_3O_4 @ carbon composites derived from metal-organic frameworks. *Carbon* 142(1): 20–31.

Yang, M., Peng, Z., Ning, Y., Chen, Y., Zhou, Q., and Deng, L. 2013. Highly specific and cost-efficient detection of Salmonella Paratyphi a combining aptamers with single-walled. *Carbon Nanotubes* 13(1): 6865–6881.

Yoshida, T., Yoshioka, Y., Fujimura, M., Yamashita, K., Higashisaka, K., Morishita, Y., Kayamuro, H., Nabeshi, H., Nagano, K., Abe, Y. A., Kamada, H., Tsunosa, S., Itoh, N., Yoshikawa, T., and Tsutsumi, Y. 2011. Promotion of allergic immune responses by intranasally-administrated nanosilica particles in mice. *Nanoescale Research Letters* 6(1): 195–107.

Youssef, A. 2018. Bionanocomposites materials for food packaging applications: Concepts and future outlook. *Carbohydrate Polymers* 193(1): 19–27.

Yuan, S. 2019. Polymeric composites for powder-based additive manufacturing: Materials and applications. *Progress in Polymer Science* 91(1): 141–168.

Zanetti, M., Carniel, T., Dalcanton, F., Anjos, R., and Riella, G. 2018. Uso de compuestos naturales encapsulados como aditivos antimicrobianos en el envasado de alimentos: Una breve reseña. *Tendencias En Ciencia y Tecnología De Los Alimentos* 81(1): 51–60.

Zare, E. 2017. Simple biosynthesis of zinc oxide nanoparticles using nature's source, and it's in vitro bio-activity. *Journal of Molecular Structure* 1146(1): 96–103.

Zhang, Y. 2018. Magnetic-assisted aptamer-based fluorescent assay for allergen detection in food matrix. *Sensors and Actuators B: Chemical* 263(15): 43–49.

Zhang, L., Luo, J., Menkhaus, T., Varadaju, H., Sun, Y., and Fong, H. 2011. Antimicrobial nano-fibrous membranes developed from electrospun polyacrylonitrile nanofibers. *Journal of Membrane Science* 369(1): 1–2.

Zuo, P., Li, X., Dominguez, D., and Ye, B. 2013. A PDMS/paper/glass hybrid microfluidic biochipintegrated with aptamer-functionalized graphene oxidenano-biosensors for one-step multiplexed pathogendetection. *Lab on a Chip* 13(1): 3921–3928.

Part III

Nanotechnology and Food Safety

13 Potential Risks, Health Safety Features, and Public Acceptance of Nanoparticles in Packaging

Monika Hans[1], *Rosy Bansal*[2], *Gulzar Ahmad Nayik*[3], *Ioannis K. Karabagias*[4], and *Mohammad Javed Ansari*[5]
[1]Department of Food Science & Technology, Padma Shri Padma Sachdev Govt. PG College for Women Gandhi Nagar, J&K, India
[2]Department of Food Processing & Engineering, GSSDGS Khalsa College Patiala, Punjab, India
[3]Department of Food Science & Technology, Govt. Degree College Shopian, J&K, India
[4]Department of Food Science & Technology, School of Agricultural Sciences, University of Patras, Agrinio, Greece
[5]Department of Botany, Hindu College Moradabad, (Mahatma Jyotiba Phule Rohilkhand University), Bareilly, UP, India

CONTENTS

13.1 INTRODUCTION

Nanotechnology is a technique of using particles at a scale of one billionth of a meter, and materials of less than 100 nm are known as nanomaterials. Thus, nanotechnology states that components/materials, assemblies, and manufacturing arrangements having size ranging from 1–100 nm are nanoparticles (Moraru et al., 2003).

DOI: 10.1201/9781003207641-16

These nanoparticles, also referred to as "magic bullets" have various applications in foods, medicine, and textile. The nano-techniques can be practiced at any stage of processing of foods, in storage and transportation and for value-added foods. The beneficial effects of nanoparticles in food packaging have brought a dramatic revolution in the food sector across the world. There has been a massive development in the study of the use of nanoparticles and of its hostile effects. The safety concerns of nanoparticles need to be researched more as these can threaten our environment and have harmful impact on the atmosphere (Maynard, 2006). Nanoparticles can pose a threat to human health if these get entry into the digestive tract, respiratory tract, or skin, and knowledge related to the safety of nanoparticles is inadequate and a lot more data is needed. Despite all the discussions and possible health issues, nanoparticles are now used in food packaging and people are worried about its safety (Sharma et al., 2017).

Nanoparticles act as a carrier of micronutrients, and pharmaceutical companies are producing nano-capsules to transmit vitamins directly to the blood using encapsulation techniques. Nanoparticles are used to enrich food and beverages and as supplements of foods without changing the food composition and taste (Thies, 2012). Nano-sense technology in food packaging is beneficial for tracking the internal and external characteristics of foods and its components in the food supply chain. The use of silicate particles reduces the oxygen migration into the packaging and also the moisture leakage, thus, maintaining the food freshness(Thakur and Gupta, 2016). Sensors detect the growth of bacteria like *E. coli.* in packaged foods. Nanotechnology also helps in reducing the waste generated through packaging.

Nano-packaging is a smart packaging and has applications in extending the shelf life of foods. It includes sensors and nano-antimicrobial agents which help in increasing the stored quality of foods so that supermarkets can keep food safe for longer before selling it. In future, it is feasible to build packages with enhanced mechanical barrier and thermal qualities by incorporating appropriate nanoparticles. Bacteria and germs will not be able to invade nanostructured materials, which is an issue for food safety. If a meal has deteriorated and can no longer be consumed, the nanosensors included in the package may inform the consumer.

13.2 APPLICATIONS OF NANONTECHNIQUES IN FOOD PACKAGING INDUSTRY

Recently, research and development in the field of nanotechnology has gained popularity and has led to the development of nano-products with varied properties. The technology can be applied not only in the field of foods but it also has a bright future for agriculture, pharmaceuticals, and bio- nanotechnology, as well as pest control and the production of safer pesticides (Figure 13.1). The nano-scale products have improved flavour and texture. It can also be applied as a carrier for nutrients and it can aid in enhancing the stored quality of items with the use of nano-packaging (Chung et al., 2017).

Nano-packaging is a smart packaging technique which uses active components, like antimicrobials, and it also can detect freshness of food with the help of nano-sensor particles which can monitor the changes in food packaging. Plastics and biodegradable materials were previously utilised in food packaging but due to their mechanical

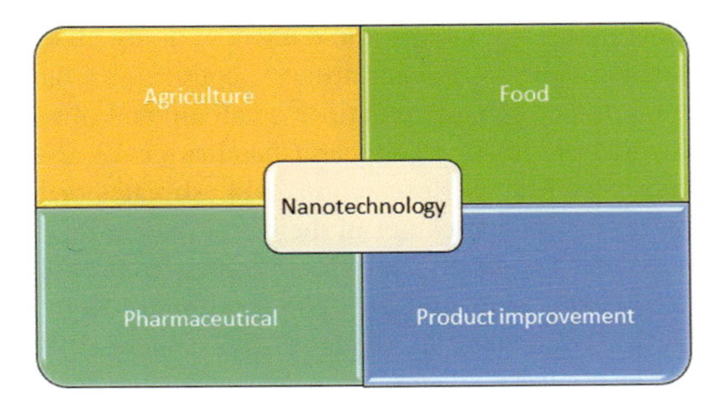

FIGURE 13.1 Applications of nanotechnology in different sectors.

strength limitations and the permeation of oxygen and moisture, nano-packaging with unique properties has replaced the conventional packaging materials. Nanomaterials have gained popularity over biopolymers due to their superior mechanical, thermal, and gas barrier qualities (Neethirajan and Jayas, 2011). Nanomaterials of food grade origin are classified as organic and inorganic having both natural and engineered nanoparticles (ENP) (Figure 13.2). The inorganic nanoparticles include titanium dioxide, iron oxide, silver, and silicon dioxide and these particles can be found in both crystalline and amorphous forms and have different characteristic depending upon their processing conditions (McClements and Xiao, 2017). The nano-packaging materials include zinc oxide coated silicate, titanium dioxide, montmorillonite, and kaolinite. These materials possess heat resistance as well as O_2 scavenging and anti-microbial properties.

Cellulose nano-fibres, micro-fibrils, or cellulose nano-whiskers are all terms for nano-scaled cellulose. It is made up of alternating crystalline and amorphous strings, in most cases. The crystalline area is generated by cellulose's strong hydrogen bond network, which makes it a relatively stable polymer. Cellulose chains have a relatively high axial stiffness, which is a crucial quality for a composite filler, due to the intricate

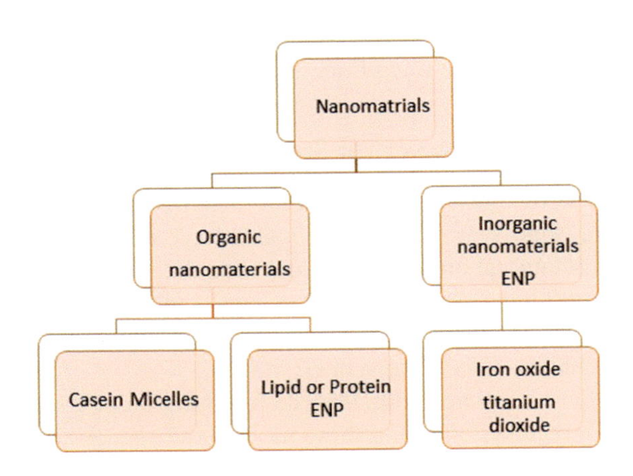

FIGURE 13.2 Classification of food grade nanoparticles.

system of hydrogen bonding. As a result, these are commonly used as reinforcement fibre additives since their qualities and performance considerably outperform conventional fillers. Because of the high aspect ratio, a small amount of nano-cellulose can increase the efficacy and rigidity of a polymer (Chaudhary et al., 2020).

A report by Persistence Market Research in 2014 estimates that nanotechnology is being adopted by up to 400 companies in the world (Neethiraja and Jayas, 2011). The nanoparticles that exist naturally, like casein protein in milk, have shown to be 300–400 nm in size; pectin in plants has a polymer chain length of 100–400 nm. They can be engineered from material sources also (Zhang et al., 2008).

Nanomaterials with higher packaging barriers have been found to retain food quality during transportation, extend the freshness of fruits and vegetables during storage, and protect meat and poultry from harmful bacteria (Chaudhary et al., 2020). These materials represent an alternative option of additives to increase the polymeric qualities of food packaging material due to their remarkable barrier, mechanical, and heat resistance properties.

Nano-packaging of foods can be of three types:

i. **Active packaging:** This packaging allows good protection of packed materials. Materials like nano-copper, nanosilver, nano-titanium dioxide, nano-magnesium oxide, and carbon nanotubes offer antimicrobial properties. Currently, the use of silver nanoparticles is of increasing popularity due to its antibacterial properties.
ii. **Intelligent packaging:** This packaging is based on sensors which can sense the chemical changes of foods and can detect the development of pathogens in foods or the gases liberated due to the spoilage of foods. The smart packaging also traces the food safety and colour changes.
iii. **Improved packaging:** The polymer materials are amalgamated with nanomaterials to improve the permeation properties, temperature, and moisture resistance.

Newer advancements in packaging based on nanoparticles provide better freshness, seal ability, and "self-spoken" products, for sensing the quality and expiration of products packaged. The nanopackaging is resistant to wear and tear, and can identify the microflora of contents.

13.2.1 Nano-Sensors Enhance the Shelf Life of Foods

Food preservation is the prevention of food spoilage by slowing the pace of microbiological growth. Nano-sensors make it possible to detect physical, chemical, and microbiological contamination in food in real time. Food packaging can be incorporated with low-cost nano-sensors to detect changes in food quality during storage and transit. These sensors use visual, optical, or electrical outputs to communicate and record these changes (Neethirajan et al., 2009).

Nano-sensors have been created in Canada for detection of insects or pests present inside grain storage systems, and these have the advantages of low power consumption, light weight, and ease of installation. The electronic tongue may disclose a visible colour shift as the package environment changes. The electronic tongue from

Kraft foods, which is used in smart food and beverage packaging, contains a large number of nanoparticles that are sensitive to changes in fresh food during storage. When compared to the human tongue, such devices are demonstrated to be far more sensitive to varied tastes. The electronic nose, which is constructed of nanowires and has gas sensors, can detect and identify many types of odours in food items. An electronic sniffer was used to detect quality changes in grain samples as a result of fungal contamination.

Nanofabricated glucose biosensors and liposome nanoparticles have been utilised to detect and measure glucose and allergenic proteins in food (Ruengruglikit, 2004). It has been reported that a glucose-sensitive enzyme with gold nanoparticles can successfully detect the amount of glucose in beverages, while nano-immunosensors have been used for detection of aflatoxin B1 to a range of 0.01 ng/ml. The microfluidic nano-sensor, a silicon chip, is a fast instrument that can detect microbial contamination in small sample quantities. Polychromix (Wilmington, MA, USA) developed a digital transformer that estimates trans-fat levels in food using a nanoelectromechanical system. These quality-control devices use transducers to detect biochemical signals created by any type of adulteration in food and during storage, and these respond at different frequencies. These gadgets are small, inexpensive, and simple to use (Abdullah et al., 2011).

Carbon nanotube–based nano-sensors have also shown superiority over traditional diagnosis methods, like high-performance liquid chromatography. The advantages of a carbon nanotube–based nano-sensor are rapid and high throughput detection, simplicity and cost effectiveness, reduced power requirements, easy recycling, and the absence of external chemicals or labels. Multi-walled carbon nanotubes and centred biosensors to detect bacteria, harmful proteins, and breakdown products in food and beverages have also been developed. Electronic noses and tongues for tasting and smelling odours and flavours have been built into packages to change colour and warn consumers when a food is about to expire or has been contaminated by germs. Nestlé, British Airways, and MonoPrix Supermarkets use chemical nano-sensors to detect colour change in real-world applications (Kuswandi, 2016).

13.2.2 Nanocomposites

Nanocomposites are made up of a continuous and discontinuous polymer matrix and can be multiphasic. These are a blend of conventional and modern nanoparticles, and they are more advanced food packaging materials, possessing antimicrobial properties as well as greater mechanical strength and toughness of material. The polymers mixed with organic or inorganic fillers with specific geometries, like fibre, spheres, flakes, and particulates are known as polymer nano- composites and these comprise novel packaging materials. The proportion of the biggest to the smallest dimensions of packaging filler material is noteworthy, and a greater ratio results in more surface area and higher strengthening properties. Many nanomaterials, like silica, clay, organoclay, graphene, chitosan, carbon nanotubes, nanocrystals, and colloidal copper or titanium are being used as fillers (Thakur and Gupta, 2016). Nanocomposite technology and its contact with food have been given approval by United States Food and Drug Administration (USFDA).

The sorts of composites produced are determined by the nature of the components used (layered silicate, organic cation, and polymer matrix) and the method of synthesis. When the polymer chain is unable to intercalate into the silicate layer, micro-composites are generated, resulting in phase separated polymer/clay composites. The interlayer distance is extended when a polymer chain is inserted between clay layers, yet the layers still have a well-defined spatial relationship. When all of the clay layers have been removed and dispersed throughout the polymeric matrix, exfoliated nano-composites are created. When compared to beginning polymers and microscale composites, these bio-nanocomposites have significantly better mechanical, heat tolerant, barrier, physical and chemical properties as a result of nanometre-size particles generated through dispersion. Because of the existence of layers of clay that might slow down the molecular transit, making the diffusive path more convoluted, these show a tremendous promise in offering superior barrier properties (Bharadwaj et al., 2002).

13.3 POTENTIAL RISKS AND FOOD SAFETY ASPECTS OF NANO-PACKAGING

Despite the valuable features that nanomaterial-based improved food processing may bring to traditional food processing, its availability in the current market is growing slowly. Due to the potential migration of nano-particles from food processing into foodstuffs and the potentially dangerous effects on consumer health, which can result from ingestion and exposure to nano-particles, its widespread commercialisation faces a significant hurdle. The health risks linked with eating food that contains nano-scale chemicals transmitted from packaging is still unknown. Smaller particles are more easily absorbed and disseminated further into the organs, where these may cause harm to cells and tissues by creating reactive oxygen species (ROS) or toxicity; that is, direct or indirect (Alfei et al., 2020).

Nano techniques can be involved in each and every segment of the food industry, like processing and packaging of food, food safety, and food security, because of the fact that human cells and food components of nano-scale level can interact easily at cellular level. The food supplements of nano-scale level can deliver nutrients directly to the cells and are acceptable, but yet, due to ethical concerns and lack of knowledge, the use of nano-packaging is not recognised among consumers (Bambang, 2016).

Foodborne illnesses are a major health concern, especially the infections caused by *Salmonella, E. coli, L. monocytogenes, and Campylobacter sps. Clostridium perfringens* accounts for nearly 20 million cases of infection per year across the world (Scharff, 2010). The food must be processed, handled, and distributed in a manner that it should be free from contamination, and nano-packaging is imperative with its numerous applications in all areas of food preservation as it extends the shelf life of food, enhances food safety, and decreases packaging waste, thus maintaining food freshness and flavours. Besides knowledge of the advantages and possible uses of nanoparticles in food packaging, there is little understanding for the concerns of toxic effects of these particles on regular consumption or intake which can lead to adverse health problems (Siegrist et al., 2007).

Nanoscience-based pathogen detection technologies are gaining hold in food analysis due to their specificity, rapidity, reproducibility, and high performance while requiring minimal measurement time. The binding of nanoparticles to antibodies of the target microbe produces easily detectable optical and electrical signals, making microbiological detection of bacteria, viruses, and poisons much easier. Nanotechnology separates the intended pathogen from a complicated dietary food medium using magnetic separation assays and antigen–antibody interaction, and then identifies it using near or mid-infrared spectroscopy. *Listeria monocytogenes* may be separated from contaminated milk using magnetic iron oxide nanoparticles. *Brucella* antibodies are detected in contaminated blood serum of cows using a similar method (Naseer et al., 2018).

The nanoparticles must be tested before their release in the market in any form for human consumption (Sozer and Kokini, 2009).The health evaluation of these engineered particles having assorted materials and diverse coatings on the surface is a complicated process as authenticated diagnostic methods are required for their description (Aschberger et al., 2011). The nanoparticles should be assessed prior to their release in the market due to their possibility of entering the human body on exposure to nanoparticles. The risk assessment is based on many parameters, like shape, size, structure, functionality, and concentration (Savolainen et al., 2010).

Recent research has shown that silver nanoparticles migrate into food in amounts that are much lower than the allowed range, although there is a shortage of safety evaluation data in this area. The National Packaging Products Quality Supervision and Inspection Centre found that as time and temperature increased, silver diffusion from fresh nanosilver bags rose in all food solutions, implying that nanosilver particles were released via dual sorption diffusion and embedding. In addition to inorganic silver metal diffusion, the modified cellulose nanocrystals migration trends have been examined, with cellulose nanocrystal migration levels in isooctane being higher in 10% ethanol but still being within the European Union's legal limits (Neethirajan et al., 2009). Researchers are primarily interested in knowing how far these particles travel into food, as well as what happens when nanoparticles are eaten, absorbed by various organs, and then digested until ejected from the body. Silver, TiO_2, tin, and CNT, according to certain studies, reach the gastrointestinal tract through the bloodstream. Physical characteristics such as size, surface charge, mass, crystal structure, surface porosity, chemical composition, and state of agglomeration govern the harmful effects of nanoparticles on target organs, such as the spleen, liver, kidney, and brain. When nano-sized hydrophilic and positively charged particles enter the bloodstream, blood circulation improves considerably (Naseer et al., 2018).

Food serves a variety of functions in the body, and the characteristics of the food play a vital role in those functions. A functional element in a food may be specific to it, such as vitamin B12 in beef. The primary purpose of such food processing is to reduce the amount of valuable components that are lost. Food, by its characteristics, is a vast reservoir of metabolic reactions, and the addition of a highly reactive species of nanoparticles to food could spark a variety of reactions. Nanoparticle interactions with such constituents have yet to be investigated.

The risks of breathing, ingesting, and skin absorption of nanoparticles with unknown toxicity need the development of trustworthy analytical instruments to undertake a safety risk evaluation of nano-food products. Safety techniques to categorise and quantify nanoparticles in food matrices are being developed for future regulatory testing on the distribution and migration of modified particles in food stuffs prior to their inclusion in foods. This will considerably increase consumer trust in nanoparticles-based items, ensuring quality control and shelf protection (Bouwmeester et al., 2009).

13.4 NANOPARTICLE EXPOSURE AND ITS IMPACT ON HUMAN HEALTH

The exposure of nanoparticles is mainly through ingestion but these can gain entry via inhalation, or through penetration can be through skin also. Many medical devices like injections or implants also allow their entry into human body. The foods packaged in nano-packaging materials is in direct contact with nanoparticles and these can leach from package to food and, on ingestion, can result in their distribution in the intestine, liver, and spleen, and thus, will be in circulation in the entire body (Aschberger et al., 2011). It is still not clear whether the nanoparticles can penetrate the skin of humans as no data is available yet. It has been researched and shown that titanium dioxide and magnesium oxide are found in systemic circulation in rats and in the olfactory bundle under the forebrain, respectively, on inhalation (Nurkiewicz et al., 2008). The studies indicate that circulation is more likely if the surface of nanoparticles has a positive charge. Moreover, the particles affect the circulation and can result in blood clots and lead to cardiovascular problems (Pekkanen et al., 2002). The nanoparticles can also be passed on to a foetus and can interact with metabolism of the fetus (Maynard, 2006). Nanoparticles possess more surface area, which makes them more reactive chemically, and thus, cause more biological reactions, which can be a disaster for human health and the environment. It has been reported that their size distinguishes nanoparticles from other particles, and the smaller the size is, the more is the surface area there is. The nanoparticles are absorbed in a varied absorption route, crossing all biological barriers to settle in body tissues. These also can initiate immune responses. Thus, the size of nanoparticles should surely be considered when assessing toxicity levels and the exposure and absorption in tissue (Borm and Kreyling, 2004).

All of the possible uses of nanotechnology in food have been identified by food industry stakeholders and research scientists, yet the value of any developing technology is highly dependent on its expense as well as public views of its hazards and benefits. Nanotechnology has significant benefits for agriculture and the food business, but its significance in the food industry is still limited and uncertain, despite its rapid adoption in other industrial sectors, including health, biotechnology, information technology, and physical sciences. The absence of sufficient returns in comparison to the substantial initial investment, a lack of regulatory framework, and a range of public attitudes all contribute to the restricted market coverage of nanoagrotech products (Naseer et al., 2018).

Nanoparticles may endanger all organs; however, long-term exposure studies are confined to unknown implications, which must be investigated further. Nanoparticles can enter the body through a variety of routes, including dermal contact, ingestion, and inhalation. People have already expressed worries about pulmonary inflammation disorders and vascular illness as a result of long-term exposure to carbon nanoparticles. TiO_2 nanoparticles have the potential to penetrate the skin's dermis layers, interacting with the immune system through lymph nodes and causing oxidative damage to the skin by creating hydroxyl radicals. Given their small size, particulates go deeper into the alveolar region than do larger particles, resulting in an acute and chronic toxicity when inhaled. Once within the body, the particles can pass the blood-brain barrier, resulting in pulmonary granuloma, oxidative damage, and pneumonia. It is impossible to make a comment on the toxicity of all nano-sized particles unless numerous research trials have been completed (Savolainen et al., 2010). When determining toxicity, it's crucial to consider particle size and mass. In one experiment, it was discovered that 20–30% of polystyrene nanoparticles with diameters of 50 and 100 nm were absorbed by the intestinal mucosa, with smaller particles penetrating faster (Chau et al., 2007).

This generally means that regulatory guidelines should be adjusted by the abundance of secure and efficient risk-assessment tools, that do not currently exist for nano-packaging, as, if nano-foods are to be applied sequentially in our food cycle, the advantages of nanotech-packaging should always be supported by greater public clarity of the risks of such foods in order to build consumer trust and acceptance of nano-packaged foods. This will improve the performance of nanomaterials in the food preservation and packaging industries (McClements and Xiao, 2017).

Thus, to ensure the safe and successful implementation of nano-packaging applications, three regulations must be met:

i. Food safety,
ii. Health safety,
iii. Environmental safety.

These are required to ensure that society benefits from novel nano-packaging applications while still maintaining high levels of health, safety, and environmental protection. Furthermore, these are required to ensure the safe use of these nano-packaging, with a particular emphasis on improving knowledge of their eventual toxicological effects, migration potential, and levels of exposure for both workers and consumers, and with a particular emphasis on the effects of the selected nanomaterials on human health following chronic exposure. Furthermore, businesses must not only assure product quality while conforming to regulatory regulations, but customers must also be engaged by offering clear information about the balance of benefits and potential dangers as well as environmental protection. If all regulations are followed, the successful inclusion of nanomaterials into food packaging will play a significant part in making the world's food supply healthier, safer, tastier, and more nutritious while also being environmentally friendly (Johnston et al., 2008).

The identification of novel detection methods, on the other hand, will provide a justifiable evaluation on the use of these nanoparticles, which may be viewed as a guarantee for humans to enjoy high-tech products safely in the food sector.

Regulatory organisations should issue guidelines on the criteria to be used in assessing the safety of food packaging, as well as the usage of nanomaterials with unique properties and functions. Novel methods, approaches, and standardised test procedures to study the effects of nanomaterials upon ingestion, or the potential interaction of nanomaterial-based food contact materials with food components, are urgently needed for the evaluation of potential hazards relating to human exposure to nanoparticles (Kuswandi, 2016).

13.5 CONCLUSION

Nano-packaging offers a wide range of benefits to the food industry, especially to meat and meat products, pertaining to active and intelligent packaging. There must be sufficient knowledge of nanoparticles before implementing them in nano-packaging as health risks are involved if not used properly. Moreover, a regulatory framework with effective governance needs to be implemented. The progress of nano-packaging is based on close associations between nanoparticles inventors, hazard evaluators, supervisors, and investigators. Further advanced research is required for nano-packaging as it has been researched that many nanoparticles produce harmful effects (Primožič et al., 2021).

Despite the nanoparticle's miraculous properties, it may have certain negative impacts on human health. The scientific community is under a lot of pressure to be transparent about the safety of nanomaterials used in food because of rising health consciousness, knowledge, and widespread Internet access. Various studies on the security of nanoparticles have also been carried out, with some nanoparticles proving to be hazardous and having negative effects on humans. Apart from knowledge of their individual toxicity, much more research appears to be required to confirm the toxicity of nanoparticles used in food. The interactions of nanoparticles with food systems, which may affect the digestibility of food elements, must be clarified.

The food service sector is the industry most affected by nano-packaging techniques as it can affect the food nature and composition. Food packaging is developed with inert materials for more hygienic concerns and this helps in increasing storage life but also the distribution of food becomes easier. The active and inert particles responsible for antimicrobial features of nano-packaging include zinc oxide, manganese oxide, and silver nanoparticles. Some other materials, like gold nanostructures, C-tubes, and QD (quantum tubes) are used as effective devices for detecting microbes in food packaging. The unique characteristics of nano-packaging material, although a benefit to industrialists, may be the cause of hazardous side effects to the human body and deleterious to environment. The nano-toxicology should deliver clear strategies and technical know-how to decrease the risks posed by them. Intensive research is required for determining the exposure level of the nanoparticles in consumers as well as in workers directly dealing with them. A safe and beneficial nano-packaging material can be developed with a specific study of various properties, like size, surface area, chemistry, and composition, etc. Therefore, a thorough knowledge of nano-packaging materials is required by food industries to derive more benefits from them. Despite this, nanotechnology-derived food packaging is projected to become more readily available to consumers worldwide in the near future.

REFERENCES

Abdullah, A. H., Adom, A. H., Ahmad, M. N., Saad, M. A., Tan, E. S., Fikri, N. A., and Zakaria, A. 2011. Electronic nose system for Ganoderma detection. *Sensor Letters* 9(1): 353–358.

Alfei, S., Marengo, B., and Zuccari, G. 2020. Nanotechnology application in food packaging: A plethora of opportunities versus pending risks assessment and public concerns. *Food Research International* 137: 109664.

Aschberger, K., Micheletti, C., Sokull-Klüttgen, B., and Christensen, F. M. 2011. Analysis of currently available data for characterising the risk of engineered nanomaterials to the environment and human health—Lessons learned from four case studies. *Environment International* 37(6): 1143–1156.

Bharadwaj, R. K., Mehrabi, A. R., Hamilton, C., Trujillo, C., Murga, M., Fan, R., and Thompson, A. K. 2002. Structure–property relationships in cross-linked polyester–clay nanocomposites. *Polymer* 43(13): 3699–3705.

Borm, P. J., and Kreyling, W. 2004. Toxicological hazards of inhaled nanoparticles—Potential implications for drug delivery. *Journal of Nanoscience and Nanotechnology* 4(5): 521–531.

Bouwmeester, H., Dekkers, S., Noordam, M. Y., Hagens, W. I., Bulder, A. S., De Heer, C., and Sips, A. J. 2009. Review of health safety aspects of nanotechnologies in food production. *Regulatory Toxicology and Pharmacology* 53(1): 52–62.

Chau, C. F., Wu, S. H., and Yen, G. C. 2007. The development of regulations for food nanotechnology. *Trends in Food Science & Technology* 18(5): 269–280.

Chaudhary, P., Fatima, F., and Kumar, A. 2020. Relevance of nanomaterials in food packaging and its advanced future prospects. *Journal of Inorganic and Organometallic Polymers and Materials* 30(12): 5180–5192.

Chaudhary, P., Fatima, F., and Kumar, A. 2020. Relevance of nanomaterials in food packaging and its advanced future prospects. *Journal of Inorganic and Organometallic Polymers and Materials* 30: 5180–5192. https://doi.org/10.1007/s10904-020-01674-8

Chung, I. M., Rajakumar, G., Gomathi, T., Park, S. K., Kim, S. H., and Thiruvengadam, M. 2017. Nanotechnology for human food: Advances and perspective. *Frontiers in Life Science* 10(1): 63–72.

Johnston, J. H., Borrmann, T., Rankin, D., Cairns, M., Grindrod, J. E., and Mcfarlane, A. 2008. Nano-structured composite calcium silicate and some novel applications. *Current Applied Physics* 8(3-4): 504–507.

Kuswandi, B. 2016. Nanotechnology in food packaging. In: Ranjan S., Dasgupta N., and Lichtfouse E. (eds.), *Nanoscience in Food and Agriculture 1. Sustainable Agriculture Reviews*, vol 20. Springer, Cham. https://doi.org/10.1007/978-3-319-39303-2_6

Maynard, A. D. 2006. Nanotechnology: Assessing the risks. *Nano Today* 1(2): 22–33.

McClements, D. J., and Xiao, H. 2017. Is nano safe in foods? Establishing the factors impacting the gastrointestinal fate and toxicity of organic and inorganic food-grade nanoparticles. *npj Science of Food* 1(1): 1–13.

Moraru, C. I., Panchapakesan, C. P., Huang, Q., Takhistov, P., Liu, S., and Kokini, J. L. 2003. Nanotechnology: A new frontier in food science understanding the special properties of materials of nanometer size will allow food scientists to design new, healthier, tastier, and safer foods. *Nanotechnology* 57(12): 24–29.

Naseer, B., Srivastava, G., Qadri, O. S., Faridi, S. A., Islam, R. U., and Younis, K. 2018. Importance and health hazards of nanoparticles used in the food industry. *Nanotechnology Reviews* 7(6): 623–641.

Neethirajan, S., and Jayas, D. S. 2011. Nanotechnology for the food and bioprocessing industries. *Food and Bioprocess Technology* 4(1): 39–47.

Neethirajan, S., Gordon, R., and Wang, L. 2009. Potential of silica bodies (phytoliths) for nanotechnology. *Trends in Biotechnology* 27(8): 461–467.

Nurkiewicz, T. R., Porter, D. W., Hubbs, A. F., Cumpston, J. L., Chen, B. T., Frazer, D. G., and Castranova, V. 2008. Nanoparticle inhalation augments particle-dependent systemic microvascular dysfunction. *Particle and Fibre Toxicology* 5(1): 1–12.

Pekkanen, J., Peters, A., Hoek, G., Tiittanen, P., Brunekreef, B., de Hartog, J., and Vanninen, E. 2002. Particulate air pollution and risk of ST-segment depression during repeated submaximal exercise tests among subjects with coronary heart disease: The exposure and risk assessment for fine and ultrafine particles in ambient air (ULTRA) study. *Circulation* 106(8): 933–938.

Primožič, M, Knez, Ž, and Leitgeb, M. 2021. (Bio)nanotechnology in food science-food packaging. *Nanomaterials (Basel)* 11(2): 292. https://doi.org/10.3390/nano11020292. PMID: 33499415; PMCID: PMC7911006.

Ruengruglikit, C., Kim, H. C., Miller, R. D., and Huang, Q. R. 2004. Fabrication of nanoporous oligonucleotide microarray for pathogen detection and identification. *Abstracts of Papers of the American Chemical Society* 227: U464–U464.

Savolainen, K., Alenius, H., Norppa, H., Pylkkänen, L., Tuomi, T., and Kasper, G. 2010. Risk assessment of engineered nanomaterials and nanotechnologies—a review. *Toxicology* 269(2-3): 92–104.

Scharff, R. L. 2010. Health-related costs from foodborne illness in the United States. CiteSeerx

Sharma, C., Dhiman, R., Rokana, N., and Panwar, H. 2017. Nanotechnology: An untapped Resource for food packaging. *Frontiers in Microbiology* 8: 1735. https://doi.org/10.3389/fmicb.2017.01735

Sozer, N., and Kokini, J. L. 2009. Nanotechnology and its applications in the food sector. *Trends in Biotechnology* 27(2): 82–89.

Siegrist, M., Cousin, M. E., Kastenholz, H., and Wiek, A. 2007. Public acceptance of nanotechnology foods and food packaging: The influence of affect and trust. *Appetite* 49(2): 459–466.

Thakur, V. K., and Gupta, R. K. 2016. Recent progress on ferroelectric polymer-based nanocomposites for high energy density capacitors: Synthesis, dielectric properties, and future aspects. *Chemical Reviews* 116(7): 4260–4317.

Thies, C. 2012. Nanocapsules as delivery systems in the food, beverage and nutraceutical industries.In Huang, Q (Ed.), *Nanotechnology in the Food, Beverage and Nutraceutical Industries*, (pp. 208–256), Woodhead Publishing, Sawston, UK.

Zhang, L., Chen, F., An, H., Yang, H., Sun, X., Guo, X., and Li, L. 2008. Physicochemical properties, firmness, and nanostructures of sodium carbonate-soluble pectin of 2 Chinese cherry cultivars at 2 ripening stages. *Journal of Food Science* 73(6): N17–N22.

14 Toxicity, Government Regulations, and the Future of Nanotechnology in Food Packaging

Syeda Saniya Zahra[1], *Muhammad Younas Khan*[2], *and Gulzar Ahmad Nayik*[3]
[1]Department Pharmacognosy, Shifa Tameer-e-Millat University, Islamabad, Pakistan
[2]Department Pharmacognosy, Islamia University of Bahawalpur, Punjab, Pakistan
[3]Department of Food Science & Technology, Govt. Degree College Shopian, J&K, India

CONTENTS

DOI: 10.1201/9781003207641-17

14.1 INTRODUCTION

The benefits of food nanotechnology span, but are not limited to, food processing, preservation, and packaging. The nanoscale size range gives the nanomaterials a unique property of increased surface area and, thus, mass transfer rates to help enhance penetration, biological activity, and quantum attributes (Avella et al., 2007). Richard Feynman is known to have introduced the world to the concept of nanotechnology in an annual meeting of the American Physical Society held in 1959 (Khademhosseini and Lager, 2006). The size range in nanomaterials lies between 1 and 100 nm, which helps devise systems and devices with entirely new properties (Roco, 2003). Various innovations in food technology have taken place which provide alternative ways to improve and expand food products in terms of their quality and ease of availability and economical at the same time. The various options from the field of nanotechnology are being efficiently utilised in order to minimize contamination in food by the introduction of nanomaterial-based antimicrobial sensors, packaging substances, preservatives, and flavouring agents (Bajpai et al., 2012). Therefore, the field of nanotechnology has found its application in providing innovative alternatives towards a better quality product in nutraceuticals, bioactives, and functional foods (Samal, 2017)

14.2 SYNTHESIS OF NANOMATERIALS

There are, mainly, three methods to synthesize various types of nanomaterials: physical, chemical, and biological methods, as shown in Figure 14.1.

14.2.1 Physical Method

The nanomaterial is synthesized by utilizing mechanical processes. These involve high ball milling method, wire discharge method, pulse laser ablation, physical vapour deposition, and mechanical chemical synthesis methods (Satyanarayana and Reddy, 2018).

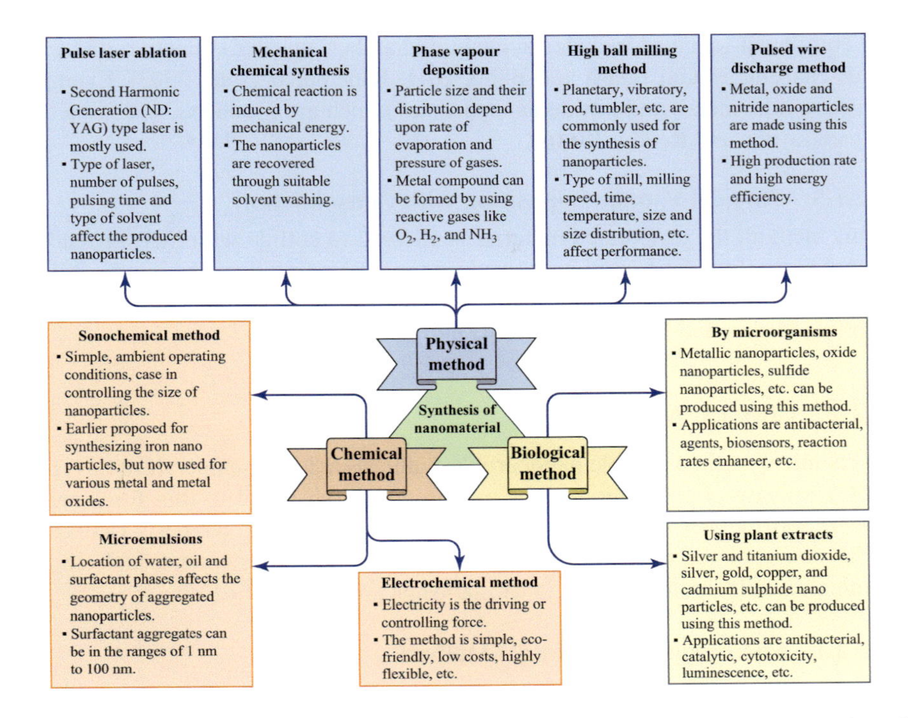

FIGURE 14.1 Various methods and techniques for the synthesis of nanomaterials. (Adopted from Sahoo et al., 2021.)

14.2.1.1 Pulse Laser Ablation Method

The technique involves generating the nanoparticles from laser vapourised material via nucleation and agglomeration processes (Satyanarayana and Reddy, 2018); for example, nanocomposites made from Cu-CS (copper-chitosan) (Sportelli et al., 2015) and silver (Ag) nanoparticles (Sportelli et al., 2018).

14.2.1.2 High Ball Milling Method

In this method, material is finely ground in mills, such as rod, planetary, vibratory, tumbler, etc., which make use of steel or tungsten carbide balls to execute the process. The examples are nanomaterials made up from Cr, Co, W, Al-Fe, Ni-Ti, and Ag-Fe (Satyanarayana and Reddy, 2018).

14.2.1.3 Mechanical Chemical Synthesis Method

The technique involves a composite powder which is produced by providing heat or mechanical force over a mixture containing oxides, metals, and chlorides, followed by suspension in stable salt matrix, and then washing to recover the ultrafine particles (Satyanarayana and Reddy, 2018).

14.2.1.4 Pulsed Wire Discharge Method

The pulsed wire discharge method involves a current passing through a metallic wire to evaporate it. The vapours formed in that process are cooled, resulting in nanoparticles.

The equipment involved includes a vacuum chamber, powder collection filter, and the discharging circuit (Dash and Balto, 2011; Kinemuchi et al., 2003). Examples include, oxide, metal, and nitride nanoparticles which are produced by this method (Satyanarayana and Reddy, 2018).

14.2.1.5 Physical Vapour Deposition with Consolidation

In this method, the evaporated material is allowed to collide with inert or reactive gases, followed by condensation to produce the desired product (Kulkarni, 2015a). For example, this is how aluminium film is developed to be used as a food packaging material (Mishra and Ningthoujam, 2017).

14.2.2 Chemical Method

The chemical methods enable nanoparticle synthesis at lower temperature (less than 350°C), an ease of conversion of the end-product from liquid to thin films or dry powder, and the integration of metals, such as iron, during synthesis (Kulkarni, 2015b). Some of the methods such as microemulsion, sonochemical, and electrochemical, are elaborated on here (Satyanarayana and Reddy, 2018).

14.2.2.1 Microemulsion/Colloidal Method

A thermodynamically stable dispersion of nanoparticle range is made by mixing one immiscible fluid into another; for example, oil in water, water in oil, or water in supercritical carbon dioxide to give O/W, W/O, or W/SC-CO_2 microemulsions, respectively, by using a suitable surfactant (Ghorbani, 2014).

14.2.2.2 Sonochemical Method

In this method, both ultrasonic pulses (20 kHz to 10 MHz) and electrolytes are utilised for the production of nanoparticles. The ultrasound radiations produce acoustic cavitations which are further utilised for synthesizing nanoparticles (Satyanarayana and Reddy, 2018). For example, silver nanoparticles are synthesized using this method (Yakoot and Salem, 2016).

14.2.2.3 Electrochemical Method

In this technique, the nanoparticles are produced at the interface between the electrode and electrolyte when the electric current is passed through the system (Satyanarayana and Reddy, 2018). For example, silver nanoparticles are produced by using glassy carbon and silver metal electrodes dipped in silver nitrate solution (0.01 mM) (Nakamura et al., 2019).

14.2.3 Biological Method

Because they provide the benefits of non-toxicity, ease of scale up, robustness, distinct morphology, and environment friendly processes, biological methods have been adopted. The nanoparticle synthesis involving these methods can be done by using microbes (such as fungi, bacteria, yeasts, etc.), and plant extracts (Singh, 2016).

14.2.3.1 Nanoparticle Synthesis by Microorganism

Microorganisms pick the targeted metal ions from the environment and integrate them to form the nanoparticle either at the exterior or interior of the microbial cell (Kulkarni, 2015c). As an example, *Aeromonas hydrophila* have been used to produce zinc oxide (ZnO) nanoparticles (Iravani, 2014).

14.2.3.2 Nanoparticle Synthesis by Plant Extracts

The nanomaterial can be prepared from any part of the plant (Naseem and Farrukh, 2015). The metal is reduced to form the nanoparticle by combining with proteins, vitamins, amino acids, flavonoids, polyphenols, alkaloids, terpenoids, and polysaccharides (Duan et al., 2015). As an example, the silver nanoparticles, produced by combining biomolecules, exhibited antibacterial activity and, therefore, used in making food packaging material (Ahmed et al., 2016).

14.3 NANOTOXICOLOGY OF FOOD PACKAGING

Due to the prodigious and remarkable growth and impact of nanotechnology in almost every field, and particularly in food processing and packaging materials, comprehensive and thorough consumer safety studies need to be conducted on a fast-track basis (Chaudhry and Castle, 2011). There have been reports of nanoparticles entering the body via oral, nasal, and cutaneous routes (Maisanaba et al., 2015). Once they are bioavailable, the nanoparticles (e.g., titanium and silver NPs) would reach various deposition sites in the body (Carrero-Sánchez et al., 2006; Kim et al., 2008; Rhim et al., 2013). From there they may be transported to and fro from liver and spleen (Dimitrijevic et al., 2015). NPs may migrate from nanopackaging and are ingested along with food, thus gaining access to the gastrointestinal tract (He and Hwang, 2016). The extent of migration depends upon particle size, concentration, molecular weight, solubility, diffusivity, temperature, polymer viscosity and structure, pH value, type of mechanical stress, and contact period with the food product. The metallic NPs showed a direct and inverse relation with temperature and pH, respectively, in enhancing the migration rate (Huang et al., 2015). Aschberger et al. (2011) found the toxicity signs with higher doses of silver or titanium dioxide NPs ingested orally. Some NP materials become genotoxic to epidermal cells which otherwise are non-toxic in bulk form, such as ZnO (Sharma et al., 2009). This may be due to the nano-range diameter. Smaller NPs are established as being more harmful than the larger ones (Chithrani et al., 2006). This may be attributed to the higher surface area of nanoparticles, more easily allowing the biological molecules to interact and produce any adverse effect. However, the health risk and safety issues related to inhalation and skin penetration for personnel engaged in manufacturing and for consumers are still under investigation (Youssef, 2013). Cationic NPs have been shown to be more toxic than anionic or neutral ones (Love et al., 2012), which may be in line with high affinity for the plasma membrane which is oppositely charged. Moreover, the cationic NP results in cytotoxicity due to lysosomal damage (Nel et al., 2009), therefore, there is information about passing across the blood brain and placental barriers (Dimitrijevic et al., 2015). There are studies for silica NPs (Avella et al., 2005) silver and zinc oxide NPs (Panea et al., 2014), which have safety limits set by the European Commission (EC). Further research is required for NPs to be assumed as totally safe. The toxicity of some NPs are explained below.

14.3.1 Silicon Oxide Nanoparticles

Minimal cytotoxicity was observed in intestinal epithelial cells (Moos et al., 2011). The least cellular toxicity was demonstrated with fluorescent SiO_2 NPs co-incubated with Caco-2 cells (Schübbe et al., 2012). It was seen that no toxicity was exhibited at 200 μg/ml. In contrast, silicon oxide NPs induced a considerably higher level of allergen specific IgE and IgG antibodies when exposed to ovalbumin (OVA) intranasally (Yoshida et al., 2011). It was observed that at higher dosages, as much as 200 μg/ml, the adverse effects by SiO_2 NPs were more prominent as compared to lower concentrations.

14.3.2 Titanium Dioxide Nanoparticles

Titanium dioxide NPs, on inhalation, have been reported to cause oxidative stress priming towards cell cycle arrest, DNA damage, and mitochondrial abnormalities (Shi et al., 2013). At 100 μg/ml, the cytotoxicity of P25 TiO_2 was observed on Caco-2, SW480, and epithelial cell line (Chalew and Schwab, 2013). Contrary to this, a few studies suggested that TiO_2 NP had the ability to cause the anatomical changes to the normal microvilli structure resulting in altered nutrient absorption (McCracken et al., 2016).

14.3.3 Zinc Oxide Nanoparticles

Zinc oxide NPs penetrate the cells and trigger the formation of free radicals which destabilise the lysosomes and mitochondria, thus, leading to cytotoxicity (Vandebriel and De Jong, 2012). Similarly, cytotoxicity was confirmed by ZnO NPs in a dose-dependent manner via LDH and MTT bioassays in Caco-2 cells (Kang et al., 2013).

14.3.4 Silver Nanoparticles

Silver nanoparticles were seen to cause oxidative stress to the cells due to the generation of free radicals (Pradhan et al., 2015). In another study, human lung fibroblasts and glioblastoma cells were exposed to Ag NPs (6–20 nm in size) which caused the cell cycle arrest in G2/M phase and enhanced the mitochondrial injury and DNA damage (AshaRani et al., 2009). Similarly, decreased cell viability was witnessed when Caco-2 cells were incubated with Ag NPs (90 nm in size) at 10 μg/ml (Song et al., 2014). Moreover, Kumar (2015) demonstrated that silver NPs adversely affect the cytoskeleton of the cell due to the depolarization of α-tubulin. Figure 14.2 explains the health hazards due to the accidental exposure to metal NPs by industry workers and consumers.

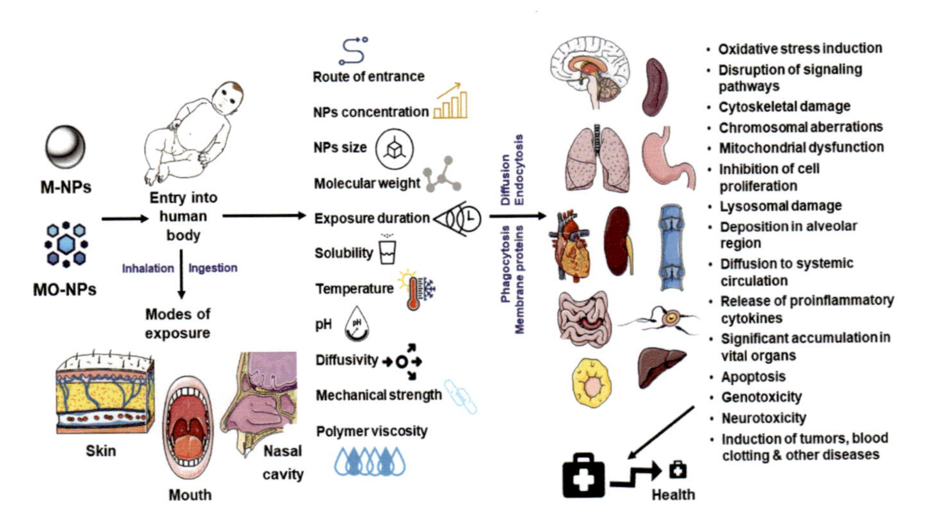

FIGURE 14.2 Effects of metal NPs exposure and its implications on living cells. (Adopted from Kumar et al., 2021.)

14.3.5 Carbon Nanoparticles

Carbon NPs, such as carbon nanotubes, when tested in already sensitized mice with OVA egg allergen, enhanced the allergen-specific IgE levels and, thus, resulted in alveolar inflammation. Also, pre-existing inflammation in mice (with allergic asthma) was exacerbated to airway fibrosis due to inhalation of multiwalled CTs (He and Hwang, 2016).

All of the above mentioned problems of toxicity come into play as the packaging integrity is compromised. However, the chances of contamination still persist due to accidental damage to the packaging material while in transport or from any technical flaw in manufacturing. Therefore, the manufacturers should ensure that the nanoparticle incorporation within the coatings or films is immaculate and free of fragile spots (Störmer et al., 2017). Moreover, the toxicity of NPs depends on the physicochemical properties and the amount present in the packaging. The various forms of analysis would involve biological assays using in vitro and in vivo models. The systematic kinetic studies involving absorption, disposition, and elimination would be helpful in determining the safe and toxic levels, and thus the risk could be assessed (He and Hwang, 2016).

14.4 EFFECT OF FOOD NANOPACKAGING ON ORGAN SYSTEMS

The nanoparticle-based products have rapidly gained so much utility in various aspects of life, and therefore, a thorough understanding of their effects on living system is required (Bouwmeester et al., 2009). Nanotoxicology, a sub-discipline of nanotechnology, deals with the estimation of effects on living organisms produced by consumption of various nanomaterials (Oberdörster et al., 2005; Smart et al., 2006). Generally, the impact of nanoparticles depends upon properties such as chemical composition, mass, particle size, and nanoparticles aggregation (Oberdörster et al., 2005). Toxicity of nanoparticles depends on bioavailability of nanoparticles at various sites, distribution, and deposition in the human body (Oberdörster et al., 2005).

For evaluation of nanotoxicity, certain parameters are important, such as environmental and biological exposure, transportation, and transformation of nanomaterials (Dreher, 2004). Causes of toxicity include the electronic, optical, and magnetic properties of NPs (Nel et al., 2006) and the oxidative and catalytic reactions (Nel et al., 2006). Nanoparticles may have several toxic effects in the body as well as potential and unpredictable risks because of their increased contact and use in food packaging. Their contact with cells and their components cause production of free radicals. Oxidative stress induced by nanoparticles enhances inflammation, modified mitochondrial function, and redox signalling (Vega-Villa et al., 2008).

Nanoparticles are transported to the bone marrow, spleen, colon, brain, lungs, GI tract, and lymphatic system (Hagens et al., 2007). NPs may also be distributed to the lungs, heart, liver, and kidneys when inhaled (Oberdörster et al., 2005). There may be accumulation of nanoparticles in the liver after first-pass metabolism (Oberdörster et al., 2005). Moreover, genotoxicity, carcinogenesis, and teratogenicity may occur (Bouwmeester et al., 2009).

14.4.1 Acute Toxicity

Different NPs have been investigated in rodents for selenium, copper, zinc, titanium dioxide cute, subacute, and subchronic toxicity upon oral exposure. It was found that the nanoparticles exhibit acute toxicity at high doses (Bouwmeester et al., 2009).

14.4.2 Respiratory Tract Toxicity

Inhaled, ultrafine NPs can move from lungs to other organs of the body via the blood stream (Azarmi et al., 2008). It was seen that ultrafine carbon particles were found to be more toxic to the pulmonary organs than the larger carbon particles in rats (Warheit et al., 2004). The toxicity involved multifocal granulomas and production of pro-inflammatory markers, such as interleukin-1β; tumour necrosis factor-α reduced macrophage phagocytotic abilities, leading towards inhibited cell proliferation; reduced phagocytic capability; and decreased cell viability (Xia et al., 2006). Moreover, they induce lung lesions, (Smart et al., 2006), acute inflammation and granuloma formation, alveolar wall thickening, and diffuse interstitial fibrosis (Smart et al., 2006).

14.4.3 Neurotoxicity

There has been evidence of nanoparticles crossing the blood-brain barrier (BBB) (Silva, 2007) but the extent of toxicity has not been measured yet.

14.4.4 Reproductive Toxicity

There is a notion that nanoparticles may lead to toxicity of the embryo as the NPs can cross the placenta (Fujimoto et al., 2005).

14.4.5 Mutagenicity

NPs interact with cell components, such as intracellular proteins, DNA, and organelles, which enhances toxicity (Bouwmeester et al., 2009).

14.4.6 Cardiovascular System Toxicity

Exposure of NPs causes purple lump, phlebitis, edema, blood clot, (Laverman et al., 2001), hemolysis platelet aggregation, and vascular thrombosis (Fornaguera et al., 2015).

14.4.7 Hepatic System Toxicity

NPs indirectly and quantitatively induced hepatic functional changes, like liver fibrosis, and alterations in serum biomarkers, such as alanine aminotransferase, aspartate aminotransferase, γ-glutamyl transferase, and alkaline phosphatase (Yamagishi et al., 2013).

14.4.8 Renal System Toxicity

NPs caused different pathologic changes in renal variables, such as albumin, total proteins, urine protein, urine albumin, hematuria, glomerular filtration rate, and creatinine ratio (Gandhi et al., 2013; Petrica et al., 2015).

14.4.9 Gastrointestinal System Toxicity

The polymeric nanoparticles resulted in acid-base imbalances and digestive malfunction (Lin et al., 2014). Copper nanoparticles caused pathological changes to spleen, liver, and kidneys (Curtis et al., 2006).

14.4.10 Skin Irritation

The NPs were proven to cause skin irritation, as confirmed by skin tests (Huczko and Lange, 2001). Nanoparticles lead to a loss in cell viability, increased oxidative stress, pro-inflammatory cytokine (IL-8), and alteration in cellular structure (Smart et al., 2006).

14.4.11 Hypersensitivity or Allergy

NPs may absorb biomolecules due to the actively charged surfaces while passing through the gastrointestinal tract, which may result in a food allergy (e.g., coatings). For example, lectins produce gastrointestinal irritation and inflammatory responses which may be cytotoxic (Bouwmeester et al., 2009).

14.5 REGULATORY ASPECTS OF NANOMATERIALS

Nanopackaging provides not only a smarter way of storing and transporting the food products, but it may also present with a constant threat of exposure of food items to the packaging material. The extent of the risk involved may largely depend on the nature of nanomaterial and the degree of ingestion, inhalation, and/or penetration through skin (Cushen et al., 2012; He and Hwang, 2016; Yang et al., 2010). Therefore, bioaccumulation of nanoparticle-based products might pose substantial safety concerns (Cushen et al., 2012; Jovanovic, 2015). Henceforth, the suitable risk assessment may be of utmost importance in toxicity management (He and Hwang, 2016). The silver nanoparticle–based packaging material may contaminate food (Echegoyen and Nerín, 2013) but the toxicity studies are limited. However, there is evidence of their deposition in organs, such as the stomach, small intestine, liver, spleen, and kidneys, in animals (McClements and Xiao, 2017). There are several reports of toxicity issues involving renal, hepatic, and alveolar injuries due to a single orally administered dose of ZnO nanoparticles (Esmaeillou et al., 2013). Moreover, the exploitation of titanium oxide and its metabolites may cause potential human and environmental danger (Yang et al., 2014).

Recently, there has been a huge surge amongst the policy makers, regulatory agencies, and other stakeholders in relation to the legitimate use of nanoscale-based particles. It is quite vital that further studies be carried out for the formulation of proper legislation. One of the hurdles may be that each nanomaterial may behave differently depending upon the processing conditions (He et al., 2018). Due to the increasing applications in the food industry, it has become imperative that the permissible limits for food products be set. This is possible with vigorous measures through research and development. A global-level platform for sharing the experience and valuable knowledge input from academics, industrialists, and consumers would assist in the research and development of food packaging, as shown in Table 14.1.

14.6 POTENTIAL FOR COMMERCIALIZATION AND FUTURE PROSPECTS

Despite the ever-increasing utility of nanoparticles in various fields, including food packaging, extensive research is required for improvement;

- Considerable research needs to be carried out on the novel and smart packaging ideas for the development of antigen specific biomarkers. These could be harnessed at commercial level to rapidly detect the food spoilage as caused by various microbes, such as mycotoxins, bacteria, and viruses (Cho et al., 2008).
- As the nanocomposites are carbon neutral molecules, ways could be created for commercial-level use in food packaging. Also, the nanosilica which provides modified barrier properties and finds its use as a surface coating material could also be worked upon to scale up to a commercial scale (Chaudhry and Castle, 2011).
- The film packaging could be supplemented with nanosensors which could detect generation of various gases due to food spoilage (Chaudhry and Castle, 2011; Das et al., 2011), thus benefiting the food producers and retailers.
- Similarly, carbon nanotubes could also be utilised for detecting the microorganisms growing in food products (Tully et al., 2006). This would help launch the food packaging systems with added benefits.

TABLE 14.1
Regulations on Global Use of Nanomaterials (NMs) in Food Nanopackaging

S. no.	Country	Regulation on Nanomaterials
1	EU	A provision in guideline EU 10/2011 clearly specifies NMs as additional substances in plastics. The added substances included titanium nitride, coated and uncoated ZnO, carbon dark, and SiO_2, and silanes and size range equal or greater than 100 nm. The definitions as to the characterization of substances to be used as NMs and then the risk assessment would be executed from time to time.
2	USA	The regulatory bodies, namely, the Occupational Safety and Health Administration (OSHA), the Environmental Protection Agency (EPA), the Food and Drug Administration (FDA), and the Consumer Product Safety Commission (CPSC), have been involved in the regulation of NMs which have been revised from time to time, and the last version (2011) by the title, "NT applications in the agricultural, feed and food sector" is applicable today.
3	Japan	The safety standards for food and agriculture in Japan were revised by Ministry of Health, Labour and Welfare (MHLW) in the 190th conference held in 2019. However, there was little/or no coverage of NMs.
4	Australia and New Zealand	The regulation of hazardous risks linked with the NMs used in food or packaging were guarded under Food Standards Australia New Zealand (FSANZ) Act. In 2008, the Application Handbook was revised. It provides guidelines, with practical approaches to the application of food nanotechnology (NT), particularly the measurement of risks associated with it. A research group, Scientific Nanotechnology Advisory Group (SNAG), was put into execution by FSANZ, constituting the experts from nanotechnology, nanosafety, and nanotoxicology to formulate, recommend, and improve the safety guidelines related to the applications of NMs.
5	India	The Nano Mission, a program of the Department of Science and Technology (DST) watches the likely health and environmental risks involved with NT. In 2016, a guideline was originated by DST along with other regulatory agencies, such as the ISO, the US National Institute for Occupational Safety and Health (NIOSH), OECD, and OSHA, to assure the safe use and disposal of NMs in research and various industries, including food processing and packaging.
6	China	In 2003, the work of formulating safety regulations of NMs was taken up by the National Center for Nanoscience and Technology (NCNST). The Commission on NT Normalisation joined hands with NCNST in 2005 and started to oversee and control the manufacturing and safety regulations. However, there still lies the grave need to frame proper legislation specific to the NMs related to food and agriculture.

Source: Adapted from Hossain et al. (2021).

Further research could be carried out to explore the possible utilisation of nanofibers in food packaging systems and the interactions with food particles. Owing to the above points, nanotechnology-derived food packaging systems would revolutionize the food industry by imparting superb qualities in terms of shelf-life extension and additive features to the food products.

14.7 CONCLUSION

The use of nanomaterial for food packaging finds a vast variety of application, and every day the utilisation is escalating. The reason for this is the tremendous benefits that it imparts to the food packaging, that range beyond enhancing the product shelf life to the inculcation of nutritional substances in food products. It has also helped in preventing microbial contamination. Nanotechnology has gained so much importance in the packaging industry that it is even more required now, to overcome the prevailing challenges of cost and safety. It requires finding a real packaging solution after the restraints of environment safety and economic viability are solved. The advancement would have a great impact on the quality of food, benefiting both consumers and producers. However, further work is needed due to the risk of migration of nanostructured materials into the food products and the possible impact on human health and environment. Therefore, owing to the recent developments in the nanoparticle-based food packaging it is very crucial to reset processing and production cycles and consumers' food habits, thereby paving the way to sustainable agricultural growth.

REFERENCES

Ahmed, S., Ahmad, M., Swami, B. L., and Ikram, S. 2016. A review on plants extract mediated synthesis of silver nanoparticles for antimicrobial applications: A green expertise. *Journal of Advanced Research* 7(1): 17–28.

Aschberger, K., Micheletti, C., Sokull-Klüttgen, B., and Christensen, F. M. 2011. Analysis of currently available data for characterising the risk of engineered nanomaterials to the environment and human health—Lessons learned from four case studies. *Environment International* 37(6): 1143–1156.

AshaRani, P. V., Low Kah Mun, G., Hande, M. P., and Valiyaveettil, S. 2009. Cytotoxicity and genotoxicity of silver nanoparticles in human cells. *ACS Nano* 3(2): 279–290.

Avella, M., Bruno, G., Errico, M. E., Gentile, G., Piciocchi, N., Sorrentino, A., and Volpe, M. G. 2007. Innovative packaging for minimally processed fruits. *Packaging Technology and Science: An International Journal* 20(5): 325–335.

Avella, M., De Vlieger, J. J., Errico, M. E., Fischer, S., Vacca, P., and Volpe, M. G. 2005. Biodegradable starch/clay nanocomposite films for food packaging applications. *Food Chemistry* 93: 467–474.

Azarmi, S., Roa, W. H., and Löbenberg, R. 2008. Targeted delivery of nanoparticles for the treatment of lung diseases. *Advanced Drug Delivery Reviews* 60(8): 863–75.

Bajpai, S. K., Chand, N., and Chaurasia, V. 2012. Nano zinc oxide-loaded calcium alginate films with potential antibacterial properties. *Food and Bioprocess Technology* 5(5): 1871–1881.

Bouwmeester, H., Dekkers, S., Noordam, M. Y., Hagens, W. I., Bulder, A. S., Heer, C. D., and Ten Voorde, S. E. C. G., et al. 2009. Review of health safety aspects of nanotechnologies in food production. *Regulatory Toxicology and Pharmacology* 53(1): 52–62.

Carrero-Sánchez, J. C., Elias, A. L., Mancilla, R., Arrellin, G., Terrones, H., and Laclette, J. P. 2006. Biocompatibility and toxicological studies of carbon nanotubes doped with nitrogen. *Nano Letters* 6(8): 1609–1616.

Chalew, T. E. A., and Schwab, K. J. 2013. Toxicity of commercially available engineered nanoparticles to Caco-2 and SW480 human intestinal epithelial cells. *Cell Biology and Toxicology* 29(2): 101–116.

Chaudhry, Q., and Castle, L. 2011. Food applications of nanotechnologies: An overview of opportunities and challenges for developing countries. *Trends in Food Science & Technology* 22(11): 595–603.

Chithrani, B. D., Ghazani, A. A., and Chan, W. C. 2006. Determining the size and shape dependence of gold nanoparticle uptake into mammalian cells. *Nano Letters* 6(4): 662–668.

Cho, Y. J., Kim, C. J., Kim, N. A. M. S. O. O., Kim, C. T., and Park, B. 2008. Some cases in applications of nanotechnology to food and agricultural systems. *Biochip Journal* 2(3): 183–185.

Curtis, J., Greenberg, M., Kester, J., Phillips, S., and Krieger, G. 2006. Nanotechnology and nanotoxicology. *Toxicological Reviews* 25(4): 245–260.

Cushen, M., Kerry, J., Morris, M., Cruz-Romero, M., and Cummins, E. 2012. Nanotechnologies in the food industry–Recent developments, risks and regulation. *Trends in Food Science & Technology* 24(1): 30–46.

Das, S., Jagan, L., Isiah, R., Rajesh, B., Backianathan, S., and Subhashini, J. 2011. Nanotechnology in oncology: Characterization and in vitro release kinetics of cisplatin-loaded albumin nanoparticles: Implications in anticancer drug delivery. *Indian Journal of Pharmacology* 43(4): 409.

Dash, P. K., and Balto, Y. 2011. Generation of nano-copper particles through wire explosion method and its characterization. *Research Journal of Nanoscience and Nanotechnology* 1(1): 25–33.

Dimitrijevic, M., Karabasil, N., Boskovic, M., Teodorovic, V., Vasilev, D., Djordjevic, V., Kilibarda, N., and Cobanovic, N. 2015. Safety aspects of nanotechnology applications in food packaging. *Procedia Food Science* 5: 57–60.

Dreher, K. L. 2004. Health and environmental impact of nanotechnology: Toxicological assessment of manufactured nanoparticles. *Toxicological Sciences* 77(1): 3–5.

Duan, H., Wang, D., and Li, Y. 2015. Green chemistry for nanoparticle synthesis. *Chemical Society Reviews* 44(16): 5778–5792.

Echegoyen, Y., and Nerín, C. 2013. Nanoparticle release from nano-silver antimicrobial food containers. *Food and Chemical Toxicology* 62: 16–22.

Esmaeillou, M., Moharamnejad, M., Hsankhani, R., Tehrani, A. A., and Maadi, H. 2013. Toxicity of ZnO nanoparticles in healthy adult mice. *Environmental Toxicology and Pharmacology* 35(1): 67–71.

Fornaguera, C., Calderó, G., Mitjans, M., Vinardell, M. P., Solans, C., and Vauthier, C. 2015. Interactions of PLGA nanoparticles with blood components: Protein adsorption, coagulation, activation of the complement system and hemolysis studies. *Nanoscale* 7(14): 6045–6058.

Fujimoto, A., Tsukue, N., Watanabe, M., Sugawara, I., Yanagisawa, R., Takano, H., Yoshida, S., and Takeda, K. 2005. Diesel exhaust affects immunological action in the placentas of mice. *Environmental Toxicology: An International Journal* 20(4): 431–440.

Gandhi, S., Srinivasan, B. P., and Akarte, A. S. 2013. An experimental assessment of toxic potential of nanoparticle preparation of heavy metals in streptozotocin induced diabetes. *Experimental and Toxicologic Pathology* 65(7-8): 1127–1135.

Ghorbani, H. R. 2014. A review of methods for synthesis of al nanoparticles. *Oriental Journal of Chemistry* 30(4): 1941–1949.

Hagens, W. I., Oomen, A. G., de Jong, W. H., Cassee, F. R., and Sips, A. J. 2007. What do we (need to) know about the kinetic properties of nanoparticles in the body? *Regulatory Toxicology and Pharmacology* 49(3): 217–229.

He, X., and Hwang, H. M. 2016. Nanotechnology in food science: Functionality, applicability, and safety assessment. *Journal of Food and Drug Analysis* 24(4): 671–681.

He, X., Fu, P., Aker, W. G., and Hwang, H. M. 2018. Toxicity of engineered nanomaterials mediated by nano–bio–eco interactions. *Journal of Environmental Science and Health, Part C* 36(1): 21–42.

Hossain, A., Skalicky, M., Brestic, M., Mahari, S., Kerry, R. G., Maitra, S., Sarkar, S., Saha, S., Bhadra, P., Popov, M., and Islam, M. 2021. Application of nanomaterials to ensure quality and nutritional safety of food. *Journal of Nanomaterials* 2021. doi: 10.1155/2021/9336082

Huang, J. Y., Li, X., and Zhou, W. 2015. Safety assessment of nanocomposite for food packaging application. *Trends in Food Science & Technology* 45(2): 187–199.

Huczko, A., and Lange, H. 2001. Carbon nanotubes: Experimental evidence for a null risk of skin irritation and allergy. *Fullerene Science and Technology* 9(2): 247–250.

Iravani, S. 2014. Bacteria in nanoparticle synthesis: Current status and future prospects. *International Scholarly Research Notices*. doi: 10.1155/2014/359316

Jovanovic, B. 2015. Critical review of public health regulations of titanium dioxide, a human food additive. *Integrated Environmental Assessment and Management* 11: 10–20.

Kang, T., Guan, R., Chen, X., Song, Y., Jiang, H., and Zhao, J. 2013. In vitro toxicity of different-sized ZnO nanoparticles in Caco-2 cells. *Nanoscale Research Letters* 8(1): 1–8.

Khademhosseini, A., and Langer, R. 2006. Drug delivery and tissue engineering. *Chemical Engineering Progress* 102(2): 38–42.

Kim, Y. S., Kim, J. S., Cho, H. S., Rha, D. S., Kim, J. M., Park, J. D., Choi, B. S., Lim, R., Chang, H. K., Chung, Y. H., and Kwon, I. H. 2008. Twenty-eight-day oral toxicity, genotoxicity, and gender-related tissue distribution of silver nanoparticles in Sprague-Dawley rats. *Inhalation Toxicology* 20(6): 575–583.

Kinemuchi, Y., Murai, K., Sangurai, C., Cho, C. H., Suematsu, H., Jiang, W., and Yatsui, K. 2003. Nanosize powders of aluminum nitride synthesized by pulsed wire discharge. *Journal of the American Ceramic Society* 86(3): 420–424.

Kulkarni, S. K. 2015a. Synthesis of nanomaterials—I (physical methods). In Kulkarni, S. K. (Ed.), *Nanotechnology: Principles and Practices* (pp. 55–76). Cham: Springer. doi: 10.1007/978-3-319-09171-6

Kulkarni, S. K. 2015b. Synthesis of nanomaterials—II (chemical methods). In Kulkarni, S. K. (Ed.), *Nanotechnology: Principles and Practices* (pp. 77–109). Cham: Springer. doi: 10.1007/978-3-319-09171-6

Kulkarni, S. K. 2015c. Synthesis of nanomaterials—III (biological methods). In Kulkarni, S. K. (Ed.), *Nanotechnology: Principles and Practices* (pp. 111–123). Cham: Springer. doi: 10.1007/978-3-319-09171-6

Kumar, A., Choudhary, A., Kaur, H., Mehta, S., and Husen, A. 2021. Metal-based nanoparticles, sensors, and their multifaceted application in food packaging. *Journal of Nanobiotechnology* 19(1): 1–25.

Kumar, L. Y. 2015. Role and adverse effects of nanomaterials in food technology. *Journal of Toxicology and Health* 2(2): 1–11.

Laverman, P., Dams, E. T. M., Storm, G., Hafmans, T. G., Croes, H. J., Oyen, W. J., Corstens, F. H., and Boerman, O. C. 2001. Microscopic localization of PEG-liposomes in a rat model of focal infection. *Journal of Controlled Release* 75(3): 347–355.

Lin, S., Wang, X., Ji, Z., Chang, C. H., Dong, Y., Meng, H., Liao, Y. P., Wang, M., Song, T. B., Kohan, S., and Xia, T. 2014. Aspect ratio plays a role in the hazard potential of CeO2 nanoparticles in mouse lung and zebrafish gastrointestinal tract. *ACS Nano* 8(5): 4450–4464.

Love, S. A., Maurer-Jones, M. A., Thompson, J. W., Lin, Y. S., and Haynes, C. L. 2012. Assessing nanoparticle toxicity. *Annual Review of Analytical Chemistry* 5: 181–205.

Maisanaba, S., Pichardo, S., Puerto, M., Gutierrez-Praena, D., Camean, A. M., and Jos, A. 2015. Toxicological evaluation of clay minerals and derived nanocomposites: A review. *Environmental Research* 138: 233–254.

McClements, D. J., and Xiao, H. 2017. Is nano safe in foods? Establishing the factors impacting the gastrointestinal fate and toxicity of organic and inorganic food-grade nanoparticles. *Npj Science of Food* 1(1): 1–13.

McCracken, C., Dutta, P. K., and Waldman, W. J. 2016. Critical assessment of toxicological effects of ingested nanoparticles. *Environmental Science: Nano* 3(2): 256–282.

Mishra, R., and Ningthoujam, R. S. 2017. High-temperature ceramics. In Tyagi, A. K., and Banerjee, S. (Eds.), *Materials Under Extreme Conditions* (pp. 377–409). Amsterdam, the Netherlands: Elsevier.

Moos, P. J., Olszewski, K., Honeggar, M., Cassidy, P., Leachman, S., and Woessner, D., et al. 2011. Responses of human cells to ZnO nanoparticles: A Gene transcription study. *Metallomics* 3: 1199–1211.

Nakamura, S., Sato, M., Sato, Y., Ando, N., Takayama, T., Fujita, M., and Ishihara, M. 2019. Synthesis and application of silver nanoparticles (Ag NPs) for the prevention of infection in healthcare workers. *International Journal of Molecular Sciences* 20(15): 3620. doi: 10.3390/ijms20153620

Naseem, T., and Farrukh, M. A. 2015. Antibacterial activity of green synthesis of iron nanoparticles using Lawsonia inermis and Gardenia jasminoides leaves extract. *Journal of Chemistry* 912342: 1–7.

Nel, A. E., Mädler, L., Velegol, D., Xia, T., Hoek, E. M., Somasundaran, P., Klaessig, F., Castranova, V., and Thompson, M. 2009. Understanding biophysicochemical interactions at the nano–bio interface. *Nature Materials* 8(7): 543–557.

Nel, A., Xia, T., Mädler, L., and Li, N. 2006. Toxic potential of materials at the nanolevel. *Science* 311(5761): 622–27.

Oberdörster, G., Maynard, A., Donaldson, K., Castranova, V., Fitzpatrick, J., Ausman, K., Carter, J., Karn, B., Kreyling, W., Lai, D., and Olin, S. 2005a. Principles for characterizing the potential human health effects from exposure to nanomaterials: Elements of a screening strategy. *Particle and Fibre Toxicology* 2(1): 1–35.

Panea, B., Ripoll, G., González, J., Fernández-Cuello, Á, and Albertí, P. 2014. Effect of nanocomposite packaging containing different proportions of ZnO and Ag on chicken breast meat quality. *Journal of Food Engineering* 123: 104–112.

Petrica, L., Vlad, A., Gluhovschi, G., Zamfir, A., Popescu, C., Gadalean, F., Dumitrascu, V., Vlad, D., Popescu, R., Velciov, S., and Gluhovschi, C. 2015. Glycated peptides are associated with proximal tubule dysfunction in type 2 diabetes mellitus. *International Journal of Clinical and Experimental Medicine* 8(2): 2516.

Pradhan, N., Singh, S., Ojha, N., Shrivastava, A., Barla, A., Rai, V., and Bose, S. 2015. Facets of nanotechnology as seen in food processing, packaging, and preservation industry. *BioMed Research International* 2015: 17. doi: 10.1155/2015/365672

Rhim, J. W., Park, H. M., and Ha, C. S. 2013. Bio-nanocomposites for food packaging applications. *Progress in Polymer Science* 38(10-11): 1629–1652.

Roco, M. C. 2003. Nanotechnology: Convergence with modern biology and medicine. *Current Opinion in Biotechnology* 14(3): 337–46.

Sahoo, M., Vishwakarma, S., Panigrahi, C., and Kumar, J. 2021. Nanotechnology: Current applications and future scope in food. *Food Frontiers* 2(1): 3–22.

Samal, D. 2017. Use of nanotechnology in food industry: A review. *International Journal of Environment, Agriculture and Biotechnology* 2(4): 238902.

Satyanarayana, T., and Reddy, S. S. 2018. A review on chemical and physical synthesis methods of nanomaterials. *International Journal for Research in Applied Science & Engineering Technology* 6: 2885–2889.

Schübbe, S., Schumann, C., Cavelius, C., Koch, M., Müller, T., and Kraegeloh, A. 2012. Size-dependent localization and quantitative evaluation of the intracellular migration of silica nanoparticles in Caco-2 cells. *Chemistry of Materials* 24(5): 914–923.

Sharma, V., Shukla, R. K., Saxena, N., Parmar, D., Das, M., and Dhawan, A. 2009. DNA damaging potential of zinc oxide nanoparticles in human epidermal cells. *Toxicology Letters* 185(3): 211–218.

Shi, H., Magaye, R., Castranova, V., and Zhao, J. 2013. Titanium dioxide nanoparticles: A review of current toxicological data. *Particle and Fibre Toxicology* 10(1): 1–33.

Silva, G. A. 2007. Nanotechnology approaches for drug and small molecule delivery across the blood brain barrier. *Surgical Neurology* 67(2): 113–116.

Singh, H. 2016. Nanotechnology applications in functional foods; Opportunities and challenges. *Preventive Nutrition and Food Science* 21(1): 1–8.

Smart, S. K., Cassady, A. I., Lu, G. Q., and Martin, D. J. 2006. The biocompatibility of carbon nanotubes. *Carbon* 44(6): 1034–47.

Song, Y., Guan, R., Lyu, F., Kang, T., Wu, Y., and Chen, X. 2014. In vitro cytotoxicity of silver nanoparticles and zinc oxide nanoparticles to human epithelial colorectal adenocarcinoma (Caco-2) cells. *Mutation Research/Fundamental and Molecular Mechanisms of Mutagenesis* 769: 113–118.

Sportelli, M. C., Ancona, A., Picca, R. A., Trapani, A., Volpe, A., Trapani, G., and Cioffi, N. 2015. Laser ablation synthesis in solution of nanoantimicrobials for food packaging applications. *MRS Online Proceedings Library (OPL)* 1804: 37–42.

Sportelli, M. C., Izzi, M., Volpe, A., Clemente, M., Picca, R. A., Ancona, A., Lugarà, P. M., Palazzo, G., and Cioffi, N. 2018. The pros and cons of the use of laser ablation synthesis for the production of silver nano-antimicrobials. *Antibiotics* 7(3): 67.

Störmer, A., Bott, J., Kemmer, D., and Franz, R. 2017. Critical review of the migration potential of nanoparticles in food contact plastics. *Trends in Food Science & Technology* 39–50.

Tully, E., Hearty, S., Leonard, P., and O'Kennedy, R. 2006. The development of rapid fluorescence-based immunoassays, using quantum dot-labelled antibodies for the detection of listeria monocytogenes cell surface proteins. *International Journal of Biological Macromolecules* 39(1-3): 127–134.

Vandebriel, R. J., and De Jong, W. H., 2012. A review of mammalian toxicity of ZnO nanoparticles. *Nanotechnology, Science and Applications* 5 (61). doi: 10.2147/NSA.S23932

Vega-Villa, K. R., Takemoto, J. K., Yáñez, J. A., Remsberg, C. M., Forrest, M. L., and Davies, N. M. 2008. Clinical toxicities of nanocarrier systems. *Advanced Drug Delivery Reviews* 60(8): 929–938.

Warheit, D. B., Laurence, B. R., Reed, K. L., Roach, D. H., Reynolds, G. A., and Webb, T. R. 2004. Comparative pulmonary toxicity assessment of single-wall carbon nanotubes in rats. *Toxicological Sciences* 77(1): 117–125.

Xia, T., Kovochich, M., Brant, J., Hotze, M., Sempf, J., Oberley, T., Sioutas, C., Yeh, J. I., Wiesner, M. R., and Nel, A. E. 2006. Comparison of the abilities of ambient and manufactured nanoparticles to induce cellular toxicity according to an oxidative stress paradigm. *Nano Letters* 6(8): 1794–1807.

Yakoot, S. M., and Salem, N. A. 2016. A sonochemical-assisted simple and green synthesis of silver nanoparticles and its use in cosmetics. *International Journal of Pharmacology* 12(5): 572–575.

Yamagishi, Y., Watari, A., Hayata, Y., Li, X., Kondoh, M., Tsutsumi, Y., and Yagi, K. 2013. Hepatotoxicity of sub-nanosized platinum particles in mice. *Die Pharmazie-An International Journal of Pharmaceutical Sciences* 68(3): 178–182.

Yang, F. M., Li, H. M., Li, F., Xin, Z. H., Zhao, L. Y., Zheng, Y. H., and Hu, Q. H. 2010. Effect of nano-packing on preservation quality of fresh strawberry (*Fragaria ananassa Duch. cv Fengxiang*) during storage at 4°C. *Journal of Food Science* 75(3): C236–C240.

Yang, Y., Doudrick, K., Bi, X., Hristovski, K., Herckes, P., Westerhoff, P., and Kaegi, R. 2014. Characterization of food-grade titanium dioxide: The presence of nanosized particles. *Environmental Science & Technology* 48(11): 6391–6400.

Yoshida, T., Yoshioka, Y., Fujimura, M., Yamashita, K., Higashisaka, K., Morishita, Y., Kayamuro, H., Nabeshi, H., Nagano, K., Abe, Y., and Kamada, H. 2011. Promotion of allergic immune responses by intranasally-administrated nanosilica particles in mice. *Nanoscale Research Letters* 6(1): 1–6.

Youssef, A. M. 2013. Polymer nanocomposites as a new trend for packaging applications. *Polymer-Plastics Technology and Engineering* 52(7): 635–660.

15 Environmental Toxicity of Emerging Micro and Nanoplastics

A Lesson Learned from Nanomaterials

Mansoor Ahmad Bhat, Kadir Gedik, and Eftade O. Gaga
Department of Environmental Engineering, Faculty of Engineering, Eskişehir Technical University, Eskişehir, Turkey

CONTENTS

15.1 INTRODUCTION

The growing concern and study of plastic pollution has engrossed scientists and researchers working on the environmental problems and risks to human health and safety of nanomaterials. Although numerous of the approaches established in nano setting, human health, and safety research relate to particulate plastics, the nanometric size of nanoscale plastics have significant implications for examining the analytical challenges and environmental consequences of these incidental nanomaterials. Plastic pollution has risen to the forefront of environmental study. Nanoplastics have been recognized as new pollutants of concern in recent years (Bianco and Passananti, 2020; Koelmans, 2019; Lambert and Wagner, 2016a) and there have

DOI: 10.1201/9781003207641-18

been many publications on microplastic contamination (Dris et al., 2017; Free et al., 2014; Xiong et al., 2018; Zhang et al., 2020a). They are one of the least researched kinds of plastic waste yet one of the most dangerous. Due to their ubiquitousness, light weight, low cost, resilience, and adaptability, plastics have been a material of choice for decades. Polyamide, polystyrene, polypropylene, polyethylene, polyethylene terephthalate, and polyvinyl chloride are the most widely produced and utilized plastic polymers. With population expansion globally, the consumption and usage of plastic items will continue to grow. Statistically, worldwide plastics output has grown from 1.5 million to over 359 million tonnes between 1950 and 2018 (Wang et al., 2021). It is projected that overall plastic output would reach 34,000 million tonnes by 2050 (Zhang et al., 2021). Sadly, energy recovery and recycling procedures recover only a tiny part of polymers. Owing to the lower recycling efficacy, most plastics end up in the environment via several paths, like wind, sewage, erosion, and water currents, etc. Even though the presence of microplastics and bigger plastic particles in the marine environment has gotten a lot of notice, colloidal plastic particles may constitute only a fraction of emitted plastic that goes unaccounted for in marine flow models (Eriksen et al., 2014; Geyer et al., 2017; Jambeck et al., 2015).

Plastic contamination on land and in freshwater (Horton et al., 2017) is predicted to be several times larger than the 4.8 to 12.7 million tonnes (Jambeck et al., 2015) of plastic emitted yearly in the ocean, and still less is known about the quantities of colloidal polymeric particles in these environmental settings. Nanoplastics are the colloids' smaller nanoscale parts, and they're more likely to be formed by accident as larger plastic waste breaks down. Heat, ultra violet degradation, mechanical wear, and, in several situations, biological action (Dawson et al., 2018) lead to a comparatively quick disintegration of polymeric waste down to the nano range, even though total disintegration of bigger plastic debris can take decades or hundreds of years. The degradation mechanisms and the rate of degradation of microplastics to nanoplastics have not yet been defined due to technical limitations. However, the breakdown processes will undoubtedly continue, and microplastics will continue to produce nanoplastics. It is a long term and long lasting environmental process. Because there are no practical methods available for determining, quantifying, and assessing nanoplastics in the different environmental sectors, the quantity of nanoplastics in the environmental settings is unknown. Nanoplastics can be widely dispersed in the different ecosystems and might be easily consumed by living organisms owing to their nano size range. Micro and nanoplastics are produced by different human activities, like plastic, industrial, construction, and agricultural waste, and enter the different compartments of the environment (atmosphere, soil, and water) where the plastic particles can either degrade or accumulate in the environment and have harmful impact on plants, humans, and animals. All these activities will be discussed in the following sections.

Micro and nanoplastics research are still in their early stages in contrast to other nanomaterials, like carbon- and metal-based nanomaterials. The knowledge of how accidental nanoplastics interact with the environment is still developing. Nanoplastics in environmental samples are mainly unquantified due to methodological challenges (Nguyen et al., 2019). Nanoplastics have been generally considered an extension of microplastics due to their comparable composition and origin as well as the nature of the research communities engaged. Nanoplastics, on the other hand, differ from microplastics in terms of transport characteristics, interactions with natural colloids and light, bioavailability, analytical difficulties, possible harmful effects, and additive leaching times due to their size-dependent features (Gigault et al., 2021). Incidentally generated nanoplastics in the environment, unlike engineered nanomaterials which can incorporate polymer compositions, are simply detritus from the environmental fragmentation of bigger plastic items.

This chapter draws on current information from the area of environmental toxicology of nanomaterials to review the potential environmental toxicological consequences of micro and nanoplastics. The chapter's primary focus will be on sources and emissions of micro and nanoplastics. We discuss the relationship between nanomaterials and plastic particles. Furthermore, we will compare nanoplastics with microplastics and nanomaterials and discuss their expected behavior in the environment. One of the critical challenges in the environmental toxicology of micro and nanoplastics has been the application of existing test methods. This mainly involves test amendments and the exceptional concentration on physical-chemical characterization and the statistical interpretation. The environmental impacts of nanomaterials and the biological effects of micro and nanoplastics on air, soil, and water will be discussed. Last, we summarize the lessons learned from nanomaterials and make recommendations on how these might be applied to micro and nanoplastic ecotoxicity testing guidelines.

15.2 SOURCES AND EMISSIONS OF MICRO AND NANOPLASTICS

The possible sources and entry paths for micro and nanoplastics are quite similar to those of engineered nanomaterials (Pessoni et al., 2019). As the name implies, engineered nanomaterials are purposefully developed and manufactured for precise applications, procedures, or items. Production can occur by synthesizing via the bottom-up approach or comminuting more extensive resources in the top-down approach. This is analogous to the manufacture of primary micro and nanoplastics, such as microbeads intended for cosmetic goods or plastic pellets used as a feed source in plastics manufacturing. Primary nanoplastics, depending on the definitions used, would fall within the concept of engineered nanomaterials. In 2012, an estimated 4000 tonnes of primary microplastic beads were used in cosmetics in Europe (Gouin et al., 2015). Approximately 1500 tonnes per year of microplastics from personal care products flow out to wastewater treatment plants and enter the global aquatic environment (Sun et al., 2020). Based on the consumption of personal care products and microplastics levels, the global emission of personal care products derived from microplastics reaches up to 1.2×10^4 tonnes/year. The emission of personal care products derived from microplastics in the last 50 years reached 3.0×10^5 tonnes (Sun et al., 2020). Although the European Union responded quickly in 2014 when scientists raised concerns about the impact of microbeads, requiring cosmetics containing microbeads to have the European Union Ecolabel, there has been no European-wide ban in place, but it is expected to take effect in 2022 (Anagnosti et al., 2021). Inorganic nanoparticles, like titanium dioxide, zinc oxide, and silica, are the most common inorganic nanoparticles found in cosmetics. Nanoscale plastics (nanoplastics) are used in cosmetics on purpose, primarily as nanocapsules (Fytianos et al., 2020; Pastrana et al., 2018; Raj et al., 2012).

Although primary microplastics represent the smallest proportion of the projected total environmental microplastics burden (Lassen et al., 2015), that portion may certainly be concentrated and decreased. However, unmanaged processes, like abrasion and decomposition of larger plastic goods and pieces, are secondary sources of anthropogenic origin and are significant causes of micro and nanoplastic pollution (Wagner and Lambert, 2018b). This includes unmanaged plastic trash, which is either thrown directly into the environment or poorly assembled and discharged into landfills before reaching different parts of the environment by wind and surface runoff water (Duis and Coors, 2016). Manufacturing abrasion procedures (e.g., air blast), synthetic paint, and vehicle tires also contribute to the production of micro and nanoplastics (Eraslan et al., 2021; Lassen et al., 2015). Another major source can be synthetic fabrics (Bhat et al., 2021): micro and nanoplastics are released when fabric is washed, entering the wash water and also the atmosphere during drying. Compared to engineered nanomaterials, the comparative relevance of secondary sources is distinct to micro and nanoplastics in that designed nanomaterials are created by managed engineering procedures and are not formed from bulk materials in the environment. Engineered nanomaterials emission is thus connected to certain goods or manufacturing applications and is, hence, analogous to primary microplastics. Risk management and regulatory alternatives are affected by the differences between engineered/industrially generated primary particles and accidentally created secondary particles. The danger of engineered nanomaterials can be kept at an acceptable limit by regulating their manufacture and by upstream regulation. Guidelines addressing requirements for air pollutants from several combustion procedures can assist in minimizing the discharge of human-made but accidentally generated nanomaterials. Upstream regulation of micro and nanoplastics has the potential to reduce primary microplastic emissions to the environment. Examples are the US Microbead-Free Waters Act 2015, (US Congress, 2015) which prohibited plastic microbeads in cosmetics that are rinsed, including in toothpaste, and the UK's prohibition on cosmetic microbeads in 2017 (UK Department for Environment, Food & Rural Affairs, 2016). For secondary microplastics, however, it is essential to take broad action regarding plastics manufacturing, usage, and waste management to reduce their environmental occurrences. Taxation or prohibition of plastic bags and bottle deposit refund schemes are examples of regulatory initiatives to reduce the general environmental impact of plastic. The production of microplastics is dependent on the intrinsic characteristics of the plastic and the environmental circumstances present once it has entered the environment (Wagner and Lambert, 2018a), making it difficult to limit by regulatory measures.

15.3 NANOMATERIALS VERSUS PLASTIC PARTICLES

There is a distinct difference in the chemical makeup of engineered nanomaterials and of micro/nanoplastics. In general, engineered nanoparticles may be created from any solid material. Higher-volume engineered nanomaterials are often composed of metals or metal oxides (e.g., titanium dioxide, ceric dioxide, and silver) or carbon (e.g., carbon nanotubes) (Hendren et al., 2011); organic nanomaterials, on the other hand, are also produced from polymers, monomers, and lipids (Lacour, 2012). However, the micro and nanoplastics include synthetic polymers made from polymerization and cover various materials, including polyethylene, polypropylene, polystyrene, and polyvinyl chloride (Andrady, 2011). The porosity, density, and matter of non-polymeric additives within synthetic polymers vary. Additives might represent up to 50% of the total plastic mass and may consist of organic and inorganic elements (Oehlmann et al., 2009). As a result, whereas micro and nanoplastics are made up of particular synthetic polymers (such as polypropylene, polystyrene, etc.), there are numerous differences between them because of additive mixtures and ratios. These additives might change the characteristics of the substance, causing it to act differently in the setting and have various environmental impacts. The same applies to nanomaterials that are manufactured: the characteristics alter with the various crystalline forms and surface coatings of manufactured nanoparticles, each with a particular chemical configuration (e.g., titanium dioxide). Similarly, engineered nanoparticles may be formed from a diversity of ingredients and material combinations. There is ongoing debate regarding "sameness" in engineered nanomaterials: When are two particles regarded as the same, and when are they so dissimilar that they cannot be deemed the same? This has effects on regulatory categorization and comparsion (Organisation for Economic Cooperation and Development (OECD), 2016). For instance, if there is statistical data on the toxicity of several nanomaterials, can this information be utilized to evaluate the security of a comparable nanomaterial? What characteristics should be comparable between these two particles: shape, size, and surface chemistry? And when is "similar" sufficiently comparable to "similar"? If the legislative framework demands regulating data reagrding the environmental safety of particles, this topic will be important for micro and nanoplastics. Under European law, polymers are presently excluded from REACH (Registration, Evaluation, Authorization and Restriction of Chemicals) registration (European Chemicals Agency (ECHA), 2012). This might change in the future, though, as the sameness debate is equally applicable in relation to primary micro and nanoplastics. Sameness is also essential for the categorization of particles in the environment when it comes to secondary microplastics, just as it is for comparison of experimental behavior and the impacts of micro and nanoplastic particles between various scientific research.

The immediate and long-term dangers caused by chemicals in the environment have been more evident over the last half-century, making environmental pollution a global social problem (Bhat and Gaga, 2022a; 2022b). Insecticides, persistent organic pollutants, medicines, trace metals, and endocrine-disrupting compounds, plus the effects of elemental combinations, have kept the attention of researchers, lawmakers, and the general public. Particles, a novel class of chemical compounds, have risen to prominence in the recent decade. With the increased usage of nanotechnology and the manufacturing of nanomaterials, the community, scientific, and governing consideration on their possible effects on the different environmental components and human health has shifted, resulting in establishment of a new scientific discipline called nanomaterial ecotoxicology. The issues connect not just to designed nanomaterials but also to human-made nanomaterials created accidentally, like ultrafine particles produced by combustion processes. Likewise, tiny plastic particles are becoming increasingly visible in the atmosphere due to industrial manufacturing, anthropogenic activities, and poor waste methods (Rist and Hartmann, 2018). This plastic waste is found in the environment as microplastics. Still, submicron range polymeric particles (i.e., nanoplastics) are also predicted to develop in the environment due to the ongoing disintegration of bigger plastic fragments (Andrady, 2011; Avio et al., 2015). Plastics of relevance in toxicological risk assessment are categorized as nanoplastics <1 μm as well as <100 nm, and microplastics <5 mm, depending on their size (Arthur et al., 2009; Browne et al., 2007; Frias and Nash, 2019; Koelmans et al., 2015). In contrast, engineered nanomaterials are more clearly characterized by having a size range of 1–100 nm (European Commission, 2011). Nanoparticles are a subset of three-dimensional nanomaterials within this spectrum. The word "nanomaterials" is used throughout; however, nanoparticles are used in several places to emphasize the material's particulate character.

Throughout non-natural weathering of bigger plastic particles, nanoparticle tracking research showed the production of nanoplastics in a size down to 30 nm (Lambert and Wagner, 2016b). This is significant evidence that this mechanism can also occur in the physical world. As particles become greater environmental contaminants it is necessary to have a more profound knowledge of their behavior in the environment and their potential harm to species. Particle environmental toxicity testing marks a change away from testing soluble compounds and requires a review of current test protocols and methodologies, including standardized approaches. On the one hand, ecotoxicological testing of particles may be the same whether the particles are designed for nanomaterials or plastic particles (Syberg et al., 2015). On the other hand, it is critical to know where the parallels between the two types end, to avoid unnecessary methods, the deployment of ineffective procedural techniques, and the creation of useless information. Nanoplastics and microplastics encompass a broad spectrum of particle sizes. Microplastics with a diameter of less than a few microns and the submicron-sized nanoplastic particles mentioned below are more likely to have similarities with manufactured nanomaterials. There are also significant variations in their element characteristics, sources, and associated procedural problems that are explained in additional detail below. Particles of comparable size are more likely to share aspects such as behavior, fate, and consequences.

15.4 COMPARISON OF NANOPLASTICS WITH MICROPLASTICS AND NANOMATERIALS

A rise in microplastic publications has corresponded with a modification in terminology and the entry of new scientific groups to tackle plastic waste issues as an increasing environmental problem. Polystyrene spheres, for instance, are increasingly characterized as micro or nanoplastics in recent academic literature rather than merely as particles or nanoparticles (Gigault et al., 2021). The designation of polystyrene spheres as micro or nanoplastics offers novel names for similar research done several decades ago with a different objective (Gigault et al., 2021; Petosa et al., 2010). The reframing of the microplastic issue as one that involves nanoplastics might indicate a retooling of the nano environmental health and safety community for approaches to the plastic debris problem. Indeed, a substantial body of information from studies of engineered nanomaterials behavior in indoor environments, consumer items, and natural and complex environmental settings may be used for nanoplastics research. During the last two decades, experimental advancements have permitted us to describe, detect, and compute the environmental destiny and impacts of engineered nanomaterials in aquatic media, such as fullerenes (Chen et al., 2008), silver nanoparticles (Benn and Westerhoff, 2008), and titanium dioxide nanoparticles (Wang et al., 2012). Engineered nanomaterial research has shown us that correctly defining words early in the research process and developing rigorous taxonomies allows data to be shared across communities. Therefore, micro and nanoplastic terminology and their relationships with engineered nanomaterials must be articulated. While every plastic is polymerized, not every polymer is plastic, and nanoplastic is not synonymous with nanopolymers (Gangadoo et al., 2020). Plastics usually comprise materials consisting of polymers and additives. The word nanoplastic has traditionally been used concerning size alone, but there is some ambiguity as to the actual size cutoff between nano and microplastic. The most frequent size cutoffs are 100 nm and 1000 nm (Gigault et al., 2018; Hartmann et al., 2019; Rist and Hartmann, 2018).

Although nanoplastics have many of the same characteristics as engineered nanomaterials, nanoplastics in the environment generate special concerns and problems. To begin with, compared to engineered nanomaterials, environmental nanoplastics, which are primarily accidental in origin, have a far more significant exposure potential. In addition, as a class of pollutants, environmental nanoplastics are far more diverse than engineered nanomaterials (Gigault et al., 2021). Engineered nanomaterials are designed to meet the necessary standards, usually with a consistent configuration for a specific matter. Separating and classifying engineered nanomaterials by these requirements and their resultant characteristics in complicated media are assisted by optical, magnetic, and conductor properties. The scientific community is now extensively using standardized methodologies suited to engineered nanomaterials. Note that engineered nanomaterials consisting of polymer types can be termed plastics. For example, in the engineered nanomaterial community, spherical and monodispersal polystyrene nanoparticles are utilized as reference/model tools, like field flow fractionation, chromatographic segregation of size, static light diffusion, and additional approaches detailed elsewhere for the calibration or models (Fytianos et al., 2020). These plastic spheres are easily trackable during the analysis due to their excellent homogeneity but do not represent the diversity of accidental nanoplastics in the environment (Sander et al., 2019). Environmental nanoplastics are unintentionally produced. As a result of diverse source supplies, fragmentation paths, and ecological exposure, they vary considerably in form, size, adsorbed pollutants, polydispersity, composition, surface characteristics, and additives (Rochman et al., 2019). The resultant chemical and physical variability of nanoplastics can change their reactivity and likely disturb interactions with organisms and natural colloids. Furthermore to altering substances' chemical characteristics, like polarity and crystallinity, which substantially impact adsorption, weathering by ultraviolet radiation promotes the disintegration of plastic into micro and nanoplastics (Holmes et al., 2014; Liu et al., 2020).

Some researchers create model matter meant to be operated as nanoplastic substitutions (Balakrishnan et al., 2019; Mitrano et al., 2019; Pessoni et al., 2019; Sander et al., 2019); however, there are no standard materials that can adequately mimic or track ambient nanoplastics. Numerous factors impact the creation of environmental nanoplastics. As the majority of these procedures are yet poorly understood, nanoplastic substitutions employed in the laboratory may contain idealizations that are incomplete to the environment or that cannot be generalized to others (Koelmans, 2019; Wagner and Reemtsma, 2019). Since cupric oxide nanoparticles might have distinct effects from titanium dioxide particles, interpreting experimental data should thus not be overlooked. The varied effects that different kinds of nanoplastic can have should not be discounted. A comparison of the findings from studies employing nanoplastic proxies may be made using field-collected experiments and their succeeding fractionation with a top-down method, followed by their classification (Gigault et al., 2021).

15.5 ENVIRONMENTAL IMPACTS OF MICRO AND NANOPLASTICS CONCERNING NANOMATERIALS

Laboratory tests are necessary to assess the environmental jeopardy presented by micro and nanoplastics and to analyze the impacts of particulates under defined conditions. Researchers found that microplastics exist extensively in human feces (Schwabl et al., 2019), foodstuffs (Toussaint et al., 2019), home dust (Zhang et al., 2020b), and several environmental settings (Petersen and Hubbart, 2021), as microplastics are becoming increasingly frequent in the environment through a range of pathways containing mainly engineered sources (polymeric beads), transport (rubber tires), and anthropogenic activities (cosmetics and synthetic materials). Environmental impacts of micro and nanoplastics on air, water, and soil will be discussed because the natural environment is a whole.

15.5.1 AIR

The primary source of airborne microplastics is synthetic textiles. Synthetic fibers are extensively used worldwide due to their excellent abrasion resistance, tenacity, and exquisite feel and touch. The release of tiny fibers is aided by the drying, washing, and wearing of clothing and other fiber items. Micro and nanoplastics in the atmosphere are likely to pollute aquatic and terrestrial ecosystems. With the help of wind, micro and nanoplastics in the atmosphere might move far away and accumulate in the aquatic ecosystems. On the other hand, it is unclear to what degree microplastics from water are the cause of microplastics in the atmosphere. Owing to the comparable forms of microplastics in the aquatic environment and atmosphere, Cai et al. (2017) hypothesized that some microplastic particles in the aquatic environment might derive from atmospheric deposition. Atmospheric microplastics can also accumulate on the ground, where they can be transferred into aquatic environments by rainwater and surface runoff or re-suspended into the air by wind (Dris et al., 2015). Rain and snow were also seen as factors for the deposition of microplastics in remote places (Bergmann et al., 2019; Klein and Fischer, 2019; Wright et al.,

2019). A dynamic cycle is established for micro and nanoplastics in the environment through their interchange across aquatic, terrestrial, and air ecosystems. Air microplastics may travel through the respiratory system and build up in infants' and adults' lungs, which Pauly et al. (1998) proved.

Studies have shown that microplastics are widespread in the air (Abbasi et al., 2019; Allen et al., 2019; Klein and Fischer, 2019; Liu et al., 2019; Vianello et al., 2019). There is, therefore, no question that people are vulnerable to contamination by micro and nanoplastics. Many individuals are subjected to ambient microplastics because they spend maximum time indoors where microplastics are far more polluting than outside (Chen et al., 2020). The high concentration of indoor micro and nanoplastics might be due to the numerous sources of micro and nanoplastics in synthetic textiles and household products (Bhat et al., 2021; Eraslan et al., 2021) and to the diverse dispersion means indoors, like weather situations, room partitions, airflow, and ventilation rate. Vianello et al. (2019) utilized a mannequin that imitates individual breathing to assess human exposure to ambient microplastics inside. They discovered that breath samples had microplastics consisting primarily of polyester with concentrations ranging from 1.7 to 16.2 particles m^{-3}. Research conducted by Gasperi et al. (2015) has shown that concentrations of indoor microplastics in one office and several private homes ranged from 3 to 15 particles m^{-3}. This gradient of indoor concentrations was according to the sampling height (30, 125, and 250 cm) suggesting that the microplastics are mainly resuspended from the floor probably due to human activity and movement. Microplastics inhaled into the body will reach the alveoli via the respiratory system. Mucociliary clearance in the upper airways, on the other hand, impedes the inhalation of bigger particles. Tiny particles can evade the restriction and remain for a long period in the deep lung (Vianello et al., 2013). As a result, employees in the synthetic textile, wool, and plastics industries, among others, are subjected to elevated concentrations of microplastics for lengthy periods each day and are extra prone to develop respiratory illnesses. Although the level of microplastics in the atmosphere is very low, continuous exposure to low quantities of microplastics might affect human health (Prata, 2018).

15.5.2 Soil

Soil ecosystems offer vital functions, like carbon sequestration, food production, gas exchange, soil structure preservation, soil-forming, pest control, degradation, and organic matter cycling. Microplastics present in soil significantly degrade its quality (Machado et al., 2018). Micro and nanoplastic relocation and trophic transmission in severely polluted soils, mainly in plastic-film-coated and wastewater-irrigated locations, represent a significant danger to the ecosystem. The vertical and horizontal movement of microplastics in the soil is influenced by variables, like soil features (e.g., soil accumulation, cracking, and macro-pores), soil biodiversity, and agricultural features (e.g., harvesting and plowing) (Guo et al., 2020). Micro and nanoplastics can move horizontally across great distances with the wind on the surface of the soil. Rillig et al. (2017) found that soil organisms (e.g., earthworms) are actively involved in the transportation of microplastic particles via ingestion and excretion. Once introduced into soil ecosystems, microplastics particles quickly integrate into soil matrices due to soil plowing, biodisturbance, or wet-dry periods (Zhou et al., 2020b), altering soil physical characteristics, soil makeup, bulk density, and water-holding capability. Machado et al. (2018) discovered that 0.4% polyether sulfone microfibre might considerably increase the water-holding capability and evaporation tendency of loamy soil while dramatically reducing water-stable soil aggregates and soil bulk density. Furthermore, the presence of microplastics may compromise soil physical integrity, leading to dryness cracking on the soil surface as well as substantial alterations in nutrient transportation and soil moisture (Wan et al., 2019). Micro and nanoplastics have an impact on both the chemical and physical characteristics of the soil. Conferring to Liu et al. (2017), polypropylene microplastics with high concentrations (>28%) enhanced the levels of nitrogen, organic carbon, and phosphorus in soil and stimulated the discharge of soil nutrients. According to Wang et al. (2016), plastic covering residues (67.5–337.5 kg ha^{-1}) decreased the amounts of carbon-based matter, accessible phosphorus, and potassium in soil, resulting in poor soil fertility. The pH value of soil is also affected by the kinds of microplastics. Polyethylene microplastics, for example, can significantly lower soil pH, but polylactic acid microplastics, on the other hand, can substantially raise soil pH (Wang et al., 2020). Because of their enormous surface areas and hydrophobicity, microplastics are thought to be a toxin vector. Ramos et al. (2015) demonstrated that pesticide concentrations in polyethylene film deposits 584–2284 μg pesticide g^{-1} plastic were substantially greater than in soil 13–32 μg pesticide g^{-1} soil. Microplastics may contaminate the soil by discharging hazardous compounds adsorbed on their surface while moving through soil ecosystems (Teuten et al., 2007). Some plastic additives will seep out during the breakdown of pieces, contaminating the surrounding soils. For example, when it comes to phthalate esters, the content of these compounds in greenhouse soil can reach 35.4 mg kg^{-1} (Zhou et al., 2020a).

Terrestrial plant growth is inextricably linked to the soil environment since it is a vital element of the soil system. Micro and nanoplastics in the soil can alter the cycle of plant nutrients, affecting seed germination and seedling development indirectly. Su et al. (2019) revealed that nano-sized microbeads might be endocytosed into tobacco cells, showing that small-size polymeric particles might penetrate the plant via absorbtion into the rhizosphere. Nanoplastics from the breakdown of microplastics in plants presents serious harm to the ecosystems and human health via migration and aggregation (Zarus et al., 2021). The assimilation of micro and nanoplastics into the soil may considerably reduce grain production, owing to changes in soil characteristics, such as its water-holding capacity, nutrition content, soil structure, and bulk density. Many microorganisms have a brief and concise life in soils, making the interaction between micro and nanoplastics and microorganisms in soil ecosystems a major concern. The mechanism of the micro and nanoplastics influence on microorganisms in the soil remains unknown at present. Previous research largely evaluated the possible effect of microplastics on soil microbes based on changes in the soil enzyme activity that are favorably connected to the soil microorganism movement. Microplastics' size, type, shape, and exposure doses can affect the microbial soil activity. For example, Machado et al. (2019) investigated the effects on soil microbial activity of various types of microplastic polyamide, polyethylene, polypropylene, polystyrene, and polyester, finding that polyamide and polyethylene have led to an apparent rise in the metabolic activity of microorganisms, while polyester and polystyrene have reduced metabolic activity. Adding polyacrylic and polyester microplastics at 0.05–0.4% w/w in dry soil had a detrimental impact on soil microbial activity as measured by fluorescein diacetate hydrolysis (Machado et al., 2019), but polypropylene microplastics at 7–28% of dry soil weight had a favorable impact (Liu et al., 2017). In these investigations, numerous characteristics, such as the form, concentration, type, and dimensions of the microplastics, were inconsistent. The harmful effects of microplastics on soil microbial activity based on individual variables are hard to generalize. Moreover, research into the co-existence of carbon-based pollutants or metals and soil micro and nanoplastics is hardly explored and must continue.

There is much less investigation of soil fauna, and it is challenging at the moment, due to the many sizes and lifestyles of soil animals, the complicated setup, and the unequal dispersion of soil fauna. Limited laboratory work on the ecotoxicological consequences of microplastics on soil animals has been conducted. Different soil fauna, such as collartail insects, earthworms, isopods, and nematodes, have been studied. The earthworms are the most important investigational animal among them. However, the real outcomes of various studies might differ significantly. For instance, low density <400 μm polyethylene microplastics applied to dry soil at a level of 0.2–1.2% (w/w) hindered the development and even caused mortality of earthworms (Huerta Lwanga et al., 2016). The findings collected from Cao et al. (2017) show that earthworm growth was significantly suppressed, and a fatal impact was also observed at 1–2% (w/w) microplastic concentrations. It is challenging to examine the influence of micro and nanoplastics on soil animals due to the restricted detection methods for these plastic particles. Moreover, whether the hazardous chemicals absorbed by micro and nanoplastics cause damage to soil animals has to be further explored.

15.5.3 Water

The ocean is the primary repository of plastic particles. The quantity of polymeric particles in the oceans is expected to rise as worldwide plastic production and demand increase. Micro and nanoplastics from cosmetic products like scrubs, toothpaste, and textile fiber particles can reach the marine environment via residential or industrial drainage systems. In addition, micro and nanoplastics are also discharged into the sea by atmospheric passage, seashores, and direct aquaculture, fishing, and shipping. Micro and nanoplastics can be carried to oceans all around the world owing to their durability and steadiness, plus the buoyancy of seawater. Micro and nanoplastics in freshwater are due mainly to wastewater release from sewage facilities, sewer runoff from rain water, surface runoff from agriculture, polymeric trash breakdown in water, and micro and nanoplastics accumulation in the air. Chemical consequences of adsorbed pollutants, the danger of polymeric additives, and possible damage to living organisms and people are examples of micro and nanoplastics environmental consequences. One of the significant concerns regarding micro and nanoplastics is the potential damage to aquatic creatures. Coral and bivalved organisms ingest microplastics which remain in them and shift between tissues (Cauwenberghe and

Janssen, 2014; Hall et al., 2015). In addition, microplastics were shown to induce hepatotoxicity, swelling (Lu et al., 2016), and neurotoxicity (Tang et al., 2020). Other detrimental consequences have also been documented, including olfactory mediated behavioral abnormalities (Shi et al., 2021), metabolic variations and sterility, intestinal tract blockage (Sharma and Chatterjee, 2017), and immunotoxicity (Shi et al., 2020; Tang et al., 2020).

Plastic additives are a concern because they may discharge potentially harmful chemicals into the environment and pose a serious threat to the biota, even in small amounts. Moreover, the bulky, precise surface area of micro and nanoplastics is thought to be helpful to the additives' escape. Aurisano et al. (2021) list more than 6000 compounds reported in plastics, and offer an outline of the problems and limitations in assessing their environmental and human health effect throughout the plastic products' life cycle. They discovered 1518 compounds from plastics that should be given priority for replacement with safer alternatives. Microplastics taken in by aquatic species may cause endocrine interference and potentially influence locomotion, growth, and reproduction, etc. (Barnes et al., 2009; Cole et al., 2011). Due to the wide area-to-volume ratios and surface hydrophobicity of micro and nanoplastic, these plastics can easily absorb harmful contaminants from the environment. Han et al. (2021) found microplastics may exacerbate the bioaccumulation of three veterinary antibiotics in *Mytilus Coruscus*, with synergistic immune-toxic consequences for mussels caused by microplastic antibiotic co-exposure. Zhou et al. (2020a) also observed that microplastics exacerbated bioaccumulation in an edible bivalve species of two waterborne veterinary antibiotics. Zhu et al. (2019) observed that triclosan-adsorbed microplastics had a higher inhibitory impact than single microplastics on microalgae development, largely because their excessive joint harmfulness might induce swelling and immunological injury, leading to mortality. Likewise, Zhou et al. (2021) revealed that the copresence of microplastics substantially exacerbated the toxic immunological effects. Sun et al. (2021) similarly observed substantial synergistic harmful blood effects due to the microplastics and polycyclic aromatic hydrocarbons. Bioaccumulation investigations utilizing other environmental pollutants demonstrated that trophic transfer plays a more significant role in the uptake than the waterborne exposure (Hasegawa and Nakaoka, 2021). Therefore, trophic transfer may also be the key contributor to the ingestion for micro and nanoplastics. The current investigations focus mainly on micro and nanoplastics spatial and chronological distribution, toxic consequences, analytical procedures, and additional features that cannot eliminate micro and nanoplastics from the environment, aiming instead at the continued rise in plastic manufacturing and the persistence of micro and nanoplastic over thousands of years. Consequently, finding effective and long-lasting techniques for removing plastic/microplastics is critical.

15.6 BIOLOGICAL EFFECTS

Engineered nanoparticles are characterized by a specific reactivity, functionality, or biological effect. As we have mentioned, micro and nanoplastics frequently arise from accidental anthropogenic processes rather than from designed ones. Even if they are deliberately created, they are not physiologically active, as such. However, certain polymer additions may, for example, inhibit biotic or abiotic breakdown. Therefore, it is necessary to examine the intended application and the characteristics of both engineered nanomaterials and micro and nanoplastics when assessing their possible ecological risks. Nanomaterials designed to have biocidal consequences are more hazardous than those intended to be inert to non-target species (Rist and Hartmann, 2018). Plastic particles containing biocidal chemicals, plasticizers, or flame retardants, on the other hand, are more likely to be dangerous to the environment since these compounds may leak out of the polymer matrix. Unique outcome mechanism (i.e., physical interactions between particles and organisms) is emphasized as significant in engineered nanomaterial, micro, and nanoplastics (Scherer et al., 2018). This includes swelling and disturbance of the energy equilibrium produced by the absorption of particles into the stomach, restricting food consumption. Various nanomaterials and nanoplastics were discovered to attach to microalgae surfaces, presumably generating a physical cellular shading effect (Hartmann et al., 2012). The physical effect of microplastics on organisms has been studied (Prata et al., 2020; Wright et al., 2013), and mechanisms that were identified as potentially significant include blocked digestive system, tissue abrasion, blocked feeding appendage in invertebrates, tissue embedding, blocked production of enzymes, reduced feeding stimulus, diluted nutrients, lower steroid hormone growth rates, and impaired reproduction. Microplastic experience can induce particle harmfulness in all living settings, including inflammatory lesions, oxidative anxiety, and enhanced uptake or translocation. The immune system's failure to eliminate synthetic particles can lead to persistent swelling and raise the danger of neoplasia. Table 15.1 shows a summary of impacts in response to physical particle properties that have been studied in several organisms. The capacity of microplastic to produce these impacts in living organisms depends on several things. Particles with a high ability to aggregate in living organisms and shift into tissues should have a greater physical influence (Wright et al., 2013). This is strongly associated with particle size, as discussed below. The shape is also essential, since irregular, sharp pieces inflict more harm than round, smooth particles. In the digestive system, fibers are more prone to collect. The ability of each species to absorb microplastics is also viewed as a significant element since it determines how long a living organism is subjected to polymeric particles (Wright et al., 2013).

It has been proven that the capacity of nano-sized particles to translocate in organisms is influenced by their size (Al-Sid-Cheikh et al., 2018). Engineered nanomaterials exhibit bio-uptake, biomagnification, and maternal transmission (Alimi et al., 2017; Karlsson et al., 2009; Rochman et al., 2019; Ruenraroengsak et al., 2012). Nanoplastics might be tiny enough to pass through biological layers by passive diffusion and retrieve specific endocytosis pathways as they approach the size of natural proteins (Jiang et al., 2008; Zhao and Stenzel, 2018). Engineered nanomaterials appear to be increasingly bioavailable to plants when particulate diameter drops below 20 nm (Schwab et al., 2016). Increased bioavailability reflects the greater toxicity in engineered nanomaterials than their bigger counterparts. Nanoscale particles, for example, caused more toxicity in plants exposed to cupric oxide than micrometre-sized particles (Karlsson et al., 2009). Size-dependent variations in consequences in nanomaterials are of special importance. Nanotechnology's entire

TABLE 15.1
Biological Consequences Reported in Organisms Following Exposure to Micro and Nanoplastics and Engineered Nanomaterials

Biological Consequences of Micro and Nanoplastics	Biological Consequences of Engineered Nanomaterials
DNA damage and variable gene expression are seen in echinoderms (Dell Torre et al., 2014), bivalves (Aurisano et al., 2021; Ogonowski et al., 2016; Su et al., 2019), and fish (Karami et al., 2016).	Algae oxidative stress (Von Moos and Slaveykova, 2014).
Cellular stress and reduced polychaetes metabolism (Wright et al., 2013).	Photosynthesis inhibition (shading) in algae (Sørensen et al., 2016).
Fish tissue damage (Karami et al. 2016; Lu et al. 2016; Pedà et al. 2016).	Fish histopathological alterations (Johari et al., 2015).
Transfer to crustacean tissues (Rosenkranz et al., 2009), mussels (Bianco and Passananti, 2020; Mazurais et al., 2015), and fish (Kashiwada, 2006).	Transfer into algal cells (Von Moos et al., 2014).
Polychaetes (Green et al., 2016), crustaceans (Watts et al., 2016), and bivalves (Rist et al., 2016) have impaired respiration.	Fish embryos with morphological malformations (Asharani et al., 2008).
Polychaetes (Besseling et al., 2014; Wegner et al., 2012), crustaceans (Chen et al., 2008; Nguyen et al., 2019), bivalves (Cole and Galloway, 2015; Rist et al., 2016; Wegner et al., 2012), and fish (Cedervall et al., 2012) have impaired feeding.	Reduced swimming speeds and mobility in crustaceans (Artells et al., 2013).
Crustaceans (Bergmann et al., 2019; Chen et al., 2020; Lacour, 2012), echinoderms (Nobre et al., 2015), bivalves (Sussarellu et al., 2016), and fish (Lönnstedt and Eklöv, 2016) have impaired development and reproduction.	Increased production of mucous in fish (Smith et al., 2007).
Crustaceans (Bergmann et al., 2019; Nguyen et al., 2019) and bivalves (Rist et al., 2016) have experienced decreased growth rates and biomass output.	Toxic consequences of discharged algae ions (Heinlaan et al., 2008).
Fish behavioral alterations (Cao et al., 2017; Liu et al., 2017; Rochman et al., 2019).	Reduced growth rates and algal biomass production (Hartmann et al., 2013).
Crustaceans (Nguyen et al., 2019; Prata et al., 2020), bivalves (Rist et al., 2016), and fish (Mazurais et al., 2015) all have higher death rates.	Inhibition of crustacean molting (Dabrunz et al., 2011).
Oxidative stress in human epithelial cells (Schirinzi et al., 2017).	Proinflammatory gene expression in epithelial cells, the neutrophil influx, and inflammation in rat's lungs (Brown et al., 2001).

goal is to make use of the new characteristics that come with a smaller scale. This includes the unique catalytic properties of various constituents on the nanoscale, such as titanium dioxide, gold, and cerium dioxide, in manufactured nanomaterials. These are largely inert as larger-sized (bulk) materials, but they become reactive when particle size and surface area decrease. Hence, there is a potential to be more toxic, even if the same substance is generally innocuous in bulk (micron) form, when particle size decreases (Nel et al., 2006). Furthermore, owing to their tiny size, engineered nanomaterials may be able to penetrate tissues and cells (Von Moos et al., 2014). Oxidative stress, tissue damage, photosynthesis suppression, poor development and growth, behavioral abnormalities, and higher death have all been observed as biological consequences of engineered nanomaterials in organisms (Table 15.1).

The question regarding micro and nanoplastics is whether a size drop is likely to make them more dangerous. To answer this, we shall look at the two primary concerns: new characteristics and digestion by organisms with the possible succeeding transmission into tissues. The new characteristics for tiny polymeric particles are related to their higher surface-to-volume ratio. With the decreasing particle size, there is a more significant percentage of molecules on the particle's surface. The surface is where interactions with the adjacent settings occur, including chemical reactions and biological interactions. Tiny particles, for example, might have a higher adsorption capacity on a mass basis than bigger particles (Wiesner et al., 2011), which has implications for the vector effects. A considerable portion of the molecules in nanoplastics are exposed to the surface, resulting in greatly enhanced surface reactivity compared with their micro- and macroscale counterparts, leading to a heightened importance of surface chemistries on interactions with biological systems. For example Miao et al. (2019) found that nanoscale polystyrene beads (100 nm) had distinct impacts on biofilm biological activity compared to larger polystyrene plastic particles, and positively charged nanoscale polystyrene had distinct effects when compared to carboxyl-functionalized counterparts. Engineered nanomaterials have been found to interact with proteins, causing variations in protein structure and the generation of reactive oxygen species and functioning as a vector for other pollutants. These actions might potentially increase the nanoplastics' toxicity. The second point of concern is the ability to overcome biological boundaries. Nano-sized polymeric particles, like nanoplastics, are possibly more dangerous owing to their easier shift into tissues and cells (Koelmans et al., 2015).

15.7 SUMMARY: LESSONS LEARNED AND THE WAY AHEAD

On the basis of the preceding explanation, microplastics were identified in a variety of environmental areas, and several investigations showed that microplastics had harmful consequences on species and were even a risk human health via food chains. Microplastics and their environmental impacts have become a growing global concern. The worldwide ecosystems today are contaminated by micro and nanoplastics as use of plastic goods continuously increases, prompting concern about their possible influence on the environment and ecosystems. Much greater attention should thus be paid to pollution by micro and nanoplastics. Micro and nanoplastics may access the environment via various paths and have been identified in many environmental areas, notably in the aquatic system as well as atmospheric and terrestrial settings. These micro and nanoplastics can enter the human body via several pathways and represent a serious hazard to human health. Additional methodical research and knowledge of the characteristics and distribution of micro and nanoplastics in the different ecosystems will help us manage global plastic pollution.

The previously prevailing subject of colloidal science was to some extent ignored when the environmental toxicology of nanomaterials arose as a technical topic during the previous decade. Over the years, it has been increasingly evident that numerous connections between the two areas may be made. As emphasized above, the connections among particle behavior, ecotoxicological consequences, and exposure demonstrate the extremely multidisciplinary character and complexity of this field of research. Cooperation amongst researchers with expertise in chemistry, ecology, and colloidal science is therefore necessary. In addition, it is essential to draw on environmental toxicology knowledge of nanomaterials, plus colloidal science in environmental behavior studies, and the impacts of micro and nanoplastics. This is the key to further understanding its possible environmental impacts. There must be general understanding and a continuing effort to establish acceptable test techniques for evaluating particulate contaminants rather than soluble substances. Incidental nanoparticles generated by plastic debris fragmentation are a key factor in the plastic waste life cycle. The lessons acquired from the 25 to 30 years of work on nano environmental health and safety should be used to comprehend the dimensions of plastic waste; one of these lessons is that multidisciplinary and multinational teams can draw on and share particular areas of knowledge from across the globe. Experts investigating nanoplastics in the environment can assist with this necessary transfer of knowledge by recruiting and visiting researchers, forming nano environmental health and safety research groups, and contacting research groups directly for collaboration with nano environmental health and safety. To facilitate partnerships, data sharing, data interpretation, and precise language and procedures harmonized among groups are essential. Size is the important part of the definition of nanoplastics; nevertheless, as far as engineered nanomaterials are concerned, we cannot restrict the cutoffs in the size of nanoplastics. It makes more sense to rely on the properties of the particles to define a nanoplastic.

We propose that the following factors be considered in work with micro and nanoplastics, based on experience in the field of engineered nanomaterials:

- Development of precise, standard terminology for classifying plastic particles.
- Comprehensive characterization of particles in exposure studies should be done.
- Addition of controls on biochemical leaching additives, monomers, etc.
- Expansion and utilization of the technical validation and comparison reference materials.
- Development and standardization of environmental toxicity test procedures, preparation of samples, and systematic approaches to reduce test artifacts.
- Research on the impact of micro and nanoplastic behavior and toxicity on environmental transformation processes (ageing).
- Advancement of analytical methods that introduce a nominal change in polymeric particulates in the preparation of samples to provide evidence on different physicochemical constraints and control complex mixed samples.

Although we must build on the current understanding of engineered nanomaterials, it is crucial to recognize where the parallels end and the differences begin. Micro and nanoplastics are likely to be environmentally, analytically, and methodologically different in certain ways from nanomaterials that are produced, which should be taken into account in preparing tests and in making knowledgeable judgments on the endpoints and methods of interest. Lastly, understanding the basic processes of effects linked with micro and nanoplastics is highly important: What characteristics make them dangerous? This is the way to go when it comes to replacing harmful plastic materials in consumer items and industrial applications with safer alternatives. When addressing future plastic production methods, such factors are critical for reducing environmental hazards and improving the possibility for plastic reuse and recycling.

REFERENCES

Abbasi, S., Keshavarzi, B., Moore, F., Turner, A., Kelly, F. J., Dominguez, A. O., and Jaafarzadeh, N. 2019. Distribution and potential health impacts of microplastics and microrubbers in air and Street dusts from asaluyeh county, Iran. *Environmental Pollution* 244: 153–64. https://doi.org/10.1016/j.envpol.2018.10.039.

Al-Sid-Cheikh, M., Rowland, S. J., Stevenson, K., Rouleau, C., Henry, T. B., and Thompson, R. C. 2018. Uptake, whole-body distribution, and depuration of nanoplastics by the scallop *Pecten maximus* at environmentally realistic concentrations. *Environmental Science and Technology* 52: 14480–86. https://doi.org/10.1021/acs.est.8b05266.

Alimi, O., Budarz, J. F., Hernandez, L. M., and Tufenkji, N. 2017. Microplastics and nanoplastics in aquatic environments: Aggregation, deposition, and enhanced contaminant transport. *Environmental Science and Technology* 52: 1704–24. https://doi.org/10.1021/acs.est.7b05559.

Allen, S., Allen, D., Phoenix, V. R., Le Roux, G., Durántez Jiménez, P., Simonneau, A., Binet, S., and Galop, D. 2019. Atmospheric transport and deposition of microplastics in a remote mountain catchment. *Nature Geoscience* 12 (5): 339–44. https://doi.org/10.1038/s41561-019-0335-5.

Anagnosti, L., Varvaresou, A., Pavlou, P., Protopapa, E., and Carayanni, V. 2021. Worldwide actions against plastic pollution from microbeads and microplastics in cosmetics focusing on European policies. Has the issue been handled effectively? *Marine Pollution Bulletin* 162: 111883. https://doi.org/10.1016/j.marpolbul.2020.111883.

Andrady, A. L. 2011. Microplastics in the marine environment. *Marine Pollution Bulletin* 62 (8): 1596–1605. https://doi.org/10.1016/j.marpolbul.2011.05.030.

Artells, E., Issartel, J., Auffan, M., Borschneck, D., Thill, A., Tella, M., Brousset, L., Rose, J., Bottero, J. Y., and Thiéry, A. 2013. Exposure to cerium dioxide nanoparticles differently affect swimming performance and survival in two daphnid species. *PLoS ONE* 8 (8): 1–11. https://doi.org/10.1371/journal.pone.0071260.

Arthur, C., Baker, J., and Bamford, H. 2009. Proceedings of the International Research Workshop on the Occurrence, Effects, and Fate of Microplastic Marine Debris.

Asharani, P. V., Wu, Y. L., Gong, Z., and Valiyaveettil, S. 2008. Toxicity of silver nanoparticles in zebrafish models. *Nanotechnology* 19 (25). https://doi.org/10.1088/0957-4484/19/25/255102.

Aurisano, N., Weber, R., and Fantke, P. 2021. Enabling a circular economy for chemicals in plastics. *Current Opinion in Green and Sustainable Chemistry* 31: 100513. https://doi.org/10.1016/j.cogsc.2021.100513.

Avio, C. G., Gorbi, S., Milan, M., Benedetti, M., Fattorini, D., D'Errico, G., Pauletto, M., Bargelloni, L., and Regoli, F. 2015. Pollutants bioavailability and toxicological risk from microplastics to marine mussels. *Environmental Pollution* 198: 211–22. https://doi.org/10.1016/j.envpol.2014.12.021.

Balakrishnan, G., Déniel, M., Nicolai, T., Chassenieux, C., and Lagarde, F. 2019. Towards more realistic reference microplastics and nanoplastics: Preparation of polyethylene micro/nanoparticles with a biosurfactant. *Environmental Science: Nano* 6: 315–24. https://doi.org/10.1039/C8EN01005F.

Barnes, D. K. A., Galgani, F., Thompson, R. C., Barlaz, M., Barnes, D. K. A., Galgani, F., Thompson, R. C., and Barlaz, M. 2009. Accumulation and fragmentation of plastic debris in global environments accumulation and fragmentation of plastic debris in global environments. *Philosophical Transactions of the Royal Society B* 364: 1985–98. https://doi.org/10.1098/rstb.2008.0205.

Benn, T. M., and Westerhoff, P. 2008. Nanoparticle silver released into water from commercially available sock fabrics. *Environmental Science and Technology* 42 (18): 7025–26. https://doi.org/10.1021/es801501j.

Bergmann, M., Mützel, S., Primpke, S., and Tekman, M. B. 2019. White and wonderful microplastics prevail in snow from the Alps to the Arctic. *Science Advances* 5: 1–10.

Besseling, E., Wang, B., Lürling, M., and Koelmans, A. A. 2014. Nanoplastic affects growth of S. Obliquus and reproduction of D. Magna. *Environmental Science and Technology* 48 (20): 12336–43. https://doi.org/10.1021/es503001d.

Bhat, M. A., Eraslan, F. N., Gedik, K., and Gaga, E. O. 2021. Impact of textile product emissions: Toxicological considerations in assessing indoor air quality and human health. In Malik, J. A., and Marathe, S. (Eds.), *Ecological and Health Effects of Building Materials*, 1st ed., 505–41. Springer, Nature Switzerland. https://doi.org/10.1007/978-3-030-76073-1_27.

Bhat, M. A., Gaga, E. O. 2022a. Air pollutant emissions in the pristine Kashmir valley from the Brick Kilns. In Öztürk, M., Khan, S. M., Altay, V., Efe, R., Egamberdieva, D., and Khassanov, F. O. (Eds.), *Biodiversity, Conservation and Sustainability in Asia*, 2nd ed., 959–979. Springer, Cham. https://doi.org/10.1007/978-3-030-73943-0_53.

Bhat, M. A., Gaga, E. O. 2022b. Compendium of a road transport emission inventory for Srinagar city of Kashmir. In Öztürk, M., Khan, S. M., Altay, V., Efe, R., Egamberdieva, D., and Khassanov, F. O. (Eds.), *Biodiversity, Conservation and Sustainability in Asia*, 2nd ed., 997–1011. Springer, Cham. https://doi.org/10.1007/978-3-030-73943-0_55.

Bianco, A., and Passananti, M. 2020. Atmospheric micro and nanoplastics: An enormous microscopic problem. *Sustainability* 12 (18): 7327. https://doi.org/10.3390/su12187327.

Brown, D. M., Wilson, M. R., MacNee, W., Stone, V., and Donaldson, K. 2001. Size-dependent proinflammatory effects of ultrafine polystyrene particles: A role for surface area and oxidative stress in the enhanced activity of ultrafines. *Toxicology and Applied Pharmacology* 175: 191–99. https://doi.org/10.1006/taap.2001.9240.

Browne, M. A., Galloway, T., and Thompson, R. 2007. Microplastic—an emerging contaminant of potential concern? *Integrated Environmental Assessment and Management* 3 (4): 559–66. https://doi.org/10.1002/ieam.5630030315.

Cai, L., Wang, J., Peng, J., Tan, Z., Zhan, Z., Tan, X., and Chen, Q. 2017. Characteristic of microplastics in the atmospheric fallout from Dongguan City, China : Preliminary research and first evidence. *Environmental Science and Pollution Research* 24: 24928–35. https://doi.org/10.1007/s11356-017-0116-x.

Cao, D., Wang, X., Luo, X., Liu, G., and Zheng, H. 2017. Effects of Polystyrene Microplastics on the Fitness of Earthworms in an Agricultural Soil. In *IOP Conference Series: Earth and Environmental Science*, 61:012148. https://doi.org/10.1088/1755-1315/61/1/012148.

Cauwenberghe, L. V., and Janssen, C. R. 2014. Microplastics in bivalves cultured for human consumption. *Environmental Pollution* 193: 65–70. https://doi.org/10.1016/j.envpol.2014.06.010.

Cedervall, T., Hansson, L. A., Lard, M., Frohm, B., and Linse, S. 2012. Food chain transport of nanoparticles affects behavior and fat metabolism in fish. *PLoS ONE* 7 (2): 1–6. https://doi.org/10.1371/journal.pone.0032254.

Chen, G., Feng, Q., and Wang, J. 2020. Mini-review of microplastics in the atmosphere and their risks to humans. *Science of the Total Environment* 703: 135504. https://doi.org/10.1016/j.scitotenv.2019.135504.

Chen, Z., Westerhoff, P., and Herckes, P. 2008. Quantification of C60 fullerene concentrations in water. *Environmental Toxicology and Chemistry* 27 (9): 1852–59. https://doi.org/10.1897/07-560.1.

Cole, M., and Galloway, T. S. 2015. Ingestion of nanoplastics and microplastics by pacific oyster larvae. *Environmental Science and Technology* 49 (24): 14625–32. https://doi.org/10.1021/acs.est.5b04099.

Cole, M., Lindeque, P., Halsband, C., and Galloway, T. S. 2011. Microplastics as contaminants in the marine environment: A review. *Marine Pollution Bulletin* 62: 2588–97. https://doi.org/10.1016/j.marpolbul.2011.09.025.

Dabrunz, A., Duester, L., Prasse, C., Seitz, F., Rosenfeldt, R., Schilde, C., Schaumann, G. E., and Schulz, R. 2011. Biological surface coating and molting inhibition as mechanisms of TiO2 nanoparticle toxicity in *Daphnia magna*. *PLoS ONE* 6 (5): 1–7. https://doi.org/10.1371/journal.pone.0020112.

Dawson, A. L., Kawaguchi, S., King, C. K., Townsend, K. A., King, R., Huston, W. M., and Bengtson Nash, S. M. 2018. Turning microplastics into nanoplastics through digestive fragmentation by Antarctic krill. *Nature Communications* 9: 1–8. https://doi.org/10.1038/s41467-018-03465-9.

Della Torre, C., Bergami, E., Salvati, A., Faleri, C., Cirino, P., Dawson, K. A., and Corsi, I. 2014. Accumulation and embryotoxicity of polystyrene nanoparticles at early stage of development of sea urchin embryos *Paracentrotus lividus*. *Environmental Science and Technology* 48 (20): 12302–11. https://doi.org/10.1021/es502569w.

Dris, R., Gasperi, J., Mirande, C., Mandin, C., Guerrouache, M., Langlois, V., and Tassin, B. 2017. A first overview of textile fibers, including microplastics, in indoor and outdoor environments. *Environmental Pollution* 221: 453–58. https://doi.org/10.1016/j.envpol.2016.12.013.

Dris, R., Gasperi, J., Rocher, V., Saad, M., Renault, N., and Tassin, B. 2015. Microplastic contamination in an urban area: A case study in Greater Paris. *Environmental Chemistry* 12 (5): 592–99. https://doi.org/10.1071/EN14167.

Duis, K., and Coors, A. 2016. Microplastics in the aquatic and terrestrial environment: Sources (with a specific focus on personal care products), fate and effects. *Environmental Sciences Europe* 28 (1): 1–25. https://doi.org/10.1186/s12302-015-0069-y.

Eraslan, F. N., Bhat, M. A., Gaga, E. O., and Gedik, K. 2021. Comprehensive analysis of research trends in volatile organic compounds emitted from building materials : A bibliometric analysis. In Malik, J. A., and Marathe, S. (Eds.), *Ecological and Health Effects of Building Materials*, 1st ed., 87–109. Springer Nature, Switzerland. https://doi.org/10.1007/978-3-030-76073-1_6.

Eriksen, M., Lebreton, L. C. M., Carson, H. S., Thiel, M., Moore, C. J., Borerro, J. C., Galgani, F., and Ryan, P. G. 2014. Plastic pollution in the world's oceans : More than 5 trillion plastic pieces weighing over 250, 000 tons afloat at sea. *PLoS One* 9 (12): 1–15. https://doi.org/10.1371/journal.pone.0111913.

European Chemicals Agency (ECHA). 2012. Guidance for Monomers and Polymers - Guidance for the Implementation of REACH. https://echa.europa.eu/guidance-documents/guidance-on-reach?panel=registration#registration%3EGuidanceonRegistration%3C/a%3E%3C/li%3E%3Cli%3E%3Ca data-cke-saved-href=.

European Commission. 2011. Commission recommendation of 18 October 2011 on the definition of nanomaterial. *Official Journal of the European Union*. https://doi.org/10.7748/ns.24.26.6.s4.

Free, C. M., Jensen, O. P., Mason, S. A., Eriksen, M., Williamson, N. J., and Boldgiv, B. 2014. High-levels of microplastic pollution in a large, remote, mountain lake. *Marine Pollution Bulletin* 85 (1): 156–63. https://doi.org/10.1016/j.marpolbul.2014.06.001.

Frias, J. P. G. L., and Nash, R. 2019. Microplastics: Finding a consensus on the definition. *Marine Pollution Bulletin* 138: 145–47. https://doi.org/10.1016/j.marpolbul.2018.11.022.

Fytianos, G., Rahdar, A., and Kyzas, G. Z. 2020. Nanomaterials in cosmetics: Recent updates. *Nanomaterials* 10 (5): 1–16. https://doi.org/10.3390/nano10050979.

Gangadoo, S., Owen, S., Rajapaksha, P., Plaisted, K., Cheeseman, S., Haddara, H., and Truong, V. K., et al. 2020. Nano-plastics and their analytical characterisation and fate in the marine environment: From source to sea. *Science of the Total Environment* 732: 138792. https://doi.org/10.1016/j.scitotenv.2020.138792.

Gasperi, J., Dris, R., Mirande, C., Mandin, C., Langlois, V., and Tassin, B. 2015. First Overview of Microplastics in Indoor and Outdoor Air. In *15th EuCheMS International Conference on Chemistry and the Environment Leipzig, Germany*. https://www.researchgate.net/publication/281657363.

Geyer, R., Jambeck, J. R., and Law, K. L. 2017. Production, use, and fate of all plastics ever made. *Science Advances* 3 (7): 25–29. https://doi.org/10.1126/sciadv.1700782.

Gigault, J., El Hadri, H., Nguyen, B., Grassl, B., Rowenczyk, L., Tufenkji, N., Feng, S., and Wiesner, M. 2021. Nanoplastics are neither microplastics nor engineered nanoparticles. *Nature Nanotechnology* 16 (5): 501–7. https://doi.org/10.1038/s41565-021-00886-4.

Gigault, J., ter Halle, A., Baudrimont, M., Pascal, P. Y., Gauffre, F., Phi, T. L., El Hadri, H., Grassl, B., and Reynaud, S. 2018. Current opinion: What is a nanoplastic? *Environmental Pollution* 235: 1030–34. https://doi.org/10.1016/j.envpol.2018.01.024.

Gouin, T., Avalos, J., Brunning, I., Brzuska, K., de Graaf, J., Kaumann, J., and Koning, T., et al. 2015. Use of micro-plastic beads in cosmetic products in Europe and their estimated emissions to the North Sea environment. *SOFW-Journa* 3 (141): 40–46.

Green, D. S., Boots, B., Sigwart, J., Jiang, S., and Rocha, C. 2016. Effects of conventional and biodegradable microplastics on a marine ecosystem engineer (*Arenicola marina*) and sediment nutrient cycling. *Environmental Pollution* 208: 426–34. https://doi.org/10.1016/j.envpol.2015.10.010.

Guo, J. J., Huang, X. P., Xiang, L., Wang, Y. Z., Li, Y. W., Li, H., Cai, Q. Y., Mo, C. H., and Wong, M. H. 2020. Source, migration and toxicology of microplastics in soil. *Environment International* 137: 105263. https://doi.org/10.1016/j.envint.2019.105263.

Hall, N. M., Berry, K. L. E., Rintoul, L., and Hoogenboom, M. O. 2015. Microplastic ingestion by scleractinian corals. *Mar Biol* 162: 725–32. https://doi.org/10.1007/s00227-015-2619-7.

Han, Y., Zhou, W., Tang, Y., Shi, W., Shao, Y., Ren, P., and Zhang, J. 2021. Science of the total environment microplastics aggravate the bioaccumulation of three veterinary antibiotics in the thick shell mussel *Mytilus coruscus* and induce synergistic immunotoxic effects. *Science of the Total Environment* 770: 145273. https://doi.org/10.1016/j.scitotenv.2021.145273.

Hartmann, N. B., Engelbrekt, C., Zhang, J., Ulstrup, J., Kusk, K. O., and Baun, A. 2013. The challenges of testing metal and metal oxide nanoparticles in algal bioassays: Titanium dioxide and gold nanoparticles as case studies. *Nanotoxicology* 7 (6): 1082–94. https://doi.org/10.3109/17435390.2012.710657.

Hartmann, N. B., Hüffer, T., Thompson, R. C., Hassellöv, M., Verschoor, A., Daugaard, A. E., and Rist, S., et al. 2019. Are we speaking the same language? Recommendations for a definition and categorization framework for plastic debris. *Environmental Science and Technology* 53 (3): 1039–47. https://doi.org/10.1021/acs.est.8b05297.

Hartmann, N. B., Legros, S., Von der Kammer, F., Hofmann, T., and Baun, A. 2012. The potential of TiO 2 nanoparticles as carriers for cadmium uptake in lumbriculus variegatus and *Daphnia magna*. *Aquatic Toxicology* 118–119: 1–8. https://doi.org/10.1016/j.aquatox.2012.03.008.

Hasegawa, T., and Nakaoka, M. 2021. Trophic transfer of microplastics from mysids to fish greatly exceeds direct ingestion from the water column. *Environmental Pollution* 273: 116468. https://doi.org/10.1016/j.envpol.2021.116468.

Heinlaan, M., Ivask, A., Blinova, I., Dubourguier, H. C., and Kahru, A. 2008. Toxicity of nanosized and bulk ZnO, CuO and TiO2 to bacteria *Vibrio fischeri* and crustaceans *Daphnia magna* and *Thamnocephalus platyurus*. *Chemosphere* 71 (7): 1308–16. https://doi.org/10.1016/j.chemosphere.2007.11.047.

Hendren, C. O., Mesnard, X., Dröge, J., and Wiesner, M. R. 2011. Estimating production data for five engineered nanomaterials as a basis for exposure assessment. *Environmental Science and Technology* 45: 2562–69. https://doi.org/10.1021/es103300g.

Holmes, L. A., Turner, A., and Thompson, R. C. 2014. Interactions between trace metals and plastic production pellets under estuarine conditions. *Marine Chemistry* 167: 25–32. https://doi.org/10.1016/j.marchem.2014.06.001.

Horton, A. A., Walton, A., Spurgeon, D. J., Lahive, E., and Svendsen, C. 2017. Microplastics in freshwater and terrestrial environments: Evaluating the current understanding to identify the knowledge gaps and future research priorities. *Science of the Total Environment* 586: 127–41. https://doi.org/10.1016/j.scitotenv.2017.01.190.

Huerta Lwanga, E., Gertsen, H., Gooren, H., Peters, P., Salánki, T., Van Der Ploeg, M., Besseling, E., Koelmans, A. A., and Geissen, V. 2016. Microplastics in the terrestrial ecosystem: Implications for *Lumbricus terrestris* (Oligochaeta, lumbricidae). *Environmental Science and Technology* 50 (5): 2685–91. https://doi.org/10.1021/acs.est.5b05478.

Jambeck, J. R., Geyer, R., Wilcox, C., Siegler, T. R., Perryman, M., Andrady, A., Narayan, R., and Law, K. L. 2015. Plastic waste inputs from land into the ocean. *Science* 347(6223): 768–71.

Jiang, W., Kim, B. Y. S., Rutka, J. T., and Chan, W. C. W. 2008. Nanoparticle-mediated cellular response is size-dependent. *Nature Nanotechnology* 3: 145–50. https://doi.org/10.1038/nnano.2008.30.

Johari, S. A., Kalbassi, M. R., Yu, I. J., and Lee, J. H. 2015. Chronic effect of waterborne silver nanoparticles on rainbow trout (Oncorhynchus mykiss): Histopathology and bioaccumulation. *Comparative Clinical Pathology* 24: 995–1007. https://doi.org/10.1007/s00580-014-2019-2.

Karami, A., Romano, N., Galloway, T., and Hamzah, H. 2016. Virgin microplastics cause toxicity and modulate the impacts of phenanthrene on biomarker responses in African catfish (Clarias gariepinus). *Environmental Research* 151: 58–70. https://doi.org/10.1016/j.envres.2016.07.024.

Karlsson, H. L., Gustafsson, J., Cronholm, P., and Möller, L. 2009. Size-dependent toxicity of metal oxide particles-a comparison between nano- and micrometer size." *Toxicology Letters* 188: 112–18. https://doi.org/10.1016/j.toxlet.2009.03.014.

Kashiwada, S. 2006. Distribution of nanoparticles in the see-through medaka (Oryzias latipes). *Environmental Health Perspectives* 114 (11): 1697–1702. https://doi.org/10.1289/ehp.9209.

Klein, M., and Fischer, E. K. 2019. Microplastic abundance in atmospheric deposition within the metropolitan area of Hamburg, Germany. *Science of the Total Environment* 685: 96–103. https://doi.org/10.1016/j.scitotenv.2019.05.405.

Koelmans, A. A., Besseling, E., and Shim, W. J. 2015. Nanoplastics in the aquatic environment. Critical review. In Bergmann, M., Gutow, L., and Klages, M. (Eds.), *Marine Anthropogenic Litter*, 325–40. https://doi.org/10.1007/978-3-319-16510-3_12.

Koelmans, A. A. 2019. Proxies for nanoplastic. *Nature Nanotechnology* 14 (4): 307–8. https://doi.org/10.1038/s41565-019-0416-z.

Lacour, S. 2012. Emerging questions for emerging technologies: Is there a law for the nano? In Brayner, R., Fiévet, F. and Coradin, T. (Eds.), *Nanomaterials: A Danger or a Promise?: A Chemical and Biological Perspective*, 357–78. Springer, London. https://doi.org/10.1007/978-1-4471-4213-3_14.

Lambert, S., and Wagner, M. 2016a. Chemosphere characterisation of nanoplastics during the degradation of polystyrene. *Chemosphere* 145: 265–68. https://doi.org/10.1016/j.chemosphere.2015.11.078.

Lambert, S., and Wagner, M. 2016b. Formation of microscopic particles during the degradation of different polymers. *Chemosphere* 161: 510–17. https://doi.org/10.1016/j.chemosphere.2016.07.042.

Lassen, C., Hansen, S. F., Magnusson, K., Hartmann, N. B., Jensen, P. R., Nielsen, T. G., and Brinch, A. 2015. Microplastics occurrence, effects and sources of releases to the environment in Denmark. *Danish Environmental Protection Agency.* http://mst.dk/service/publikationer/publikationsarkiv/2015/nov/rapport-om-mikroplast%0AGeneral.

Liu, H., Yang, X., Liu, G., Liang, C., Xue, S., Chen, H., Ritsema, C. J., and Geissen, V. 2017. Response of soil dissolved organic matter to microplastic addition in Chinese loess soil. *Chemosphere* 185: 907–17. https://doi.org/10.1016/j.chemosphere.2017.07.064.

Liu, K., Wang, X., Fang, T., Xu, P., Zhu, L., and Li, D. 2019. Source and potential risk assessment of suspended atmospheric microplastics in Shanghai. *Science of the Total Environment* 675: 462–71. https://doi.org/10.1016/j.scitotenv.2019.04.110.

Liu, P., Zhan, X., Wu, X., Li, J., Wang, H., and Gao, S. 2020. Effect of weathering on environmental behavior of microplastics: Properties, sorption and potential risks. *Chemosphere* 242: 125193. https://doi.org/10.1016/j.chemosphere.2019.125193.

Lönnstedt, O. M., and Eklöv, P. 2016. Environmentally relevant concentrations of microplastic particles influence larval fish ecology. *Science* 352: 1213–1216. https://doi.org/10.1126/science.aad8828.

Lu, Y., Zhang, Y., Deng, Y., Jiang, W., Zhao, Y., Geng, J., Ding, L., and Ren, H. 2016. Uptake and accumulation of polystyrene microplastics in zebrafish (Danio rerio) and toxic effects in liver. *Environmental Science and Technology* 50 (7): 4054–60. https://doi.org/10.1021/acs.est.6b00183.

Machado, A. A. S., Lau, C. W., Kloas, W., Bergmann, J., Bachelier, J. B., Faltin, E., Becker, R., Görlich, A. S., and Rillig, M. C. 2019. Microplastics can change soil properties and affect plant performance. *Environmental Science and Technology* 53: 6044–52. https://doi.org/10.1021/acs.est.9b01339.

Machado, A. A. S., Lau, C. W., Till, J., Kloas, W., Lehmann, A., Becker, R., and Rillig, M. C. 2018. Impacts of microplastics on the soil biophysical environment. *Environmental Science and Technology* 52 (17): 9656–65. https://doi.org/10.1021/acs.est.8b02212.

Mazurais, D., Ernande, B., Quazuguel, P., Severe, A., Huelvan, C., Madec, L., and Mouchel, O., et al. 2015. Evaluation of the impact of polyethylene microbeads ingestion in European sea bass (Dicentrarchus labrax) larvae. *Marine Environmental Research* 112: 78–85. https://doi.org/10.1016/j.marenvres.2015.09.009.

Miao, L., Hou, J., You, G., Liu, Z., Liu, S., Li, T., Mo, Y., Guo, S., and Qu, H. 2019. Acute effects of nanoplastics and microplastics on periphytic biofilms depending on particle size, concentration and surface modification. *Environmental Pollution* 255: 113300. https://doi.org/10.1016/j.envpol.2019.113300.

Mitrano, D. M., Beltzung, A., Frehland, S., Schmiedgruber, M., Cingolani, A., and Schmidt, F. 2019. Synthesis of metal-doped nanoplastics and their utility to investigate fate and behavior in complex environmental systems. *Nature Nanotechnology* 14: 362–68. https://doi.org/10.1038/s41565-018-0360-3.

Von Moos, N., Bowen, P., and Slaveykova, V. I. 2014. Bioavailability of inorganic nanoparticles to planktonic bacteria and aquatic microalgae in freshwater. *Environmental Science: Nano* 1: 214–32. https://doi.org/10.1039/c3en00054k.

Von Moos, N, and Slaveykova, V. I. 2014. Oxidative stress induced by inorganic nanoparticles in bacteria and aquatic microalgae - state of the art and knowledge gaps. *Nanotoxicology* 8: 605–30. https://doi.org/10.3109/17435390.2013.809810.

Nel, A., Xia, T., Mädler, L., and Li, N. 2006. Toxic potential of materials at the nanolevel. *Science* 311: 622–27. https://doi.org/10.1126/science.1114397.

Nguyen, B., Claveau-Mallet, D., Hernandez, L. M., Xu, E. G., Farner, J. M., and Tufenkji, N. 2019. Separation and analysis of microplastics and nanoplastics in complex environmental samples. *Accounts of Chemical Research* 52 (4): 858–66. https://doi.org/10.1021/acs.accounts.8b00602.

Nobre, C. R., Santana, M. F. M., Maluf, A., Cortez, F. S., Cesar, A., Pereira, C. D. S., and Turra, A. 2015. Assessment of microplastic toxicity to embryonic development of the sea urchin Lytechinus variegatus (Echinodermata: Echinoidea). *Marine Pollution Bulletin* 92: 99–104. https://doi.org/10.1016/j.marpolbul.2014.12.050.

Oehlmann, J., Schulte-Oehlmann, U., Kloas, W., Jagnytsch, O., Lutz, I., Kusk, K. O., and Wollenberger, L., et al. 2009. A critical analysis of the biological impacts of plasticizers on wildlife. *Philosophical Transactions of the Royal Society B: Biological Sciences* 364: 2047–62. https://doi.org/10.1098/rstb.2008.0242.

Ogonowski, M., Schür, C., Jarsén, Åsa, and Gorokhova, E. 2016. The effects of natural and anthropogenic microparticles on individual fitness in *Daphnia magna*. *PLoS ONE* 11 (5): 1–20. https://doi.org/10.1371/journal.pone.0155063.

Organisation for Economic Cooperation and Development (OECD). 2016. Environment Directorate Joint Meeting of the Chemicals Committee and the Working Party on Chemicals, Pesticides and Biotechnology. http://www.oecd.org/officialdocuments/publicdisplaydocumentpdf/?cote=env/jm/mono(2016)2&doclanguage=en.

Pastrana, H., Avila, A., and Tsai, C. S. J. 2018. Nanomaterials in cosmetic products: The challenges with regard to current legal frameworks and consumer exposure. *NanoEthics* 12 (2): 123–37. https://doi.org/10.1007/s11569-018-0317-x.

Pauly, J. L., Stegmeier, J., Cheney, T., Zhang, P. J., Allaart, A., and Mayer, G. 1998. Inhaled cellulosic and plastic fibers found in human lung tissue. *Cancer Epidemiology, Biomarkers & Prevention* 7: 419–28.

Pedà, C., Caccamo, L., Fossi, M. C., Gai, F., Andaloro, F., Genovese, L., Perdichizzi, A., Romeo, T., and Maricchiolo, G. 2016. Intestinal alterations in European sea bass Dicentrarchus labrax (Linnaeus, 1758) exposed to microplastics: Preliminary results. *Environmental Pollution* 212: 251–56. https://doi.org/10.1016/j.envpol.2016.01.083.

Pessoni, L., Veclin, C., El Hadri, H., Cugnet, C., Davranche, M., Pierson-Wickmann, A. C., Gigault, J., Grassl, B., and Reynaud, S. 2019. Soap- and metal-free polystyrene latex particles as a nanoplastic model. *Environmental Science: Nano* 6: 2253–58. https://doi.org/10.1039/c9en00384c.

Petersen, F., and Hubbart, J. A. 2021. The occurrence and transport of microplastics: The state of the science. *Science of the Total Environment* 758: 143936. https://doi.org/10.1016/j.scitotenv.2020.143936.

Petosa, A. R., Jaisi, D. P., Quevedo, I. R., Elimelech, M., and Tufenkji, N. 2010. Aggregation and deposition of engineered nanomaterials in aquatic environments: Role of physicochemical interactions. *Environmental Science and Technology* 44 (17): 6532–49. https://doi.org/10.1021/es100598h.

Prata, J. C. 2018. Airborne microplastics: Consequences to human health? *Environmental Pollution* 234: 115–26. https://doi.org/10.1016/j.envpol.2017.11.043.

Prata, J. C., da Costa, J. P., Lopes, I., Duarte, A. C., and Rocha-Santos, T. 2020. Environmental exposure to microplastics: An overview on possible human health effects. *Science of the Total Environment* 702: 134455. https://doi.org/10.1016/j.scitotenv.2019.134455.

Raj, S., Jose, S., Sumod, U. S., and Sabitha, M. 2012. Nanotechnology in cosmetics: Opportunities and challenges. *Journal of Pharmacy and Bioallied Sciences* 4 (3): 186–93. https://doi.org/10.4103/0975-7406.99016.

Ramos, L., Berenstein, G., Hughes, E. A., Zalts, A., and Montserrat, J. M. 2015. Polyethylene film incorporation into the horticultural soil of small periurban production units in Argentina. *Science of the Total Environment* 523: 74–81. https://doi.org/10.1016/j.scitotenv.2015.03.142.

Rillig, M. C., Ziersch, L., and Hempel, S. 2017. Microplastic transport in soil by earthworms. *Scientific Reports* 7: 1–6. https://doi.org/10.1038/s41598-017-01594-7.

Rist, S. E., Assidqi, K., Zamani, N. P., Appel, D., Perschke, M., Huhn, M., and Lenz, M. 2016. Suspended micro-sized PVC particles impair the performance and decrease survival in the Asian green mussel Perna viridis. *Marine Pollution Bulletin* 111: 213–20. https://doi.org/10.1016/j.marpolbul.2016.07.006.

Rist, S., and Hartmann, N. B. 2018. Aquatic ecotoxicity of microplastics and nanoplastics: Lessons learned from engineered nanomaterials. In Wagner, M. and Lambert, S. (Eds.), *Freshwater Microplastics*, 1st ed., 25–49. Springer, Cham. https://doi.org/10.1007/978-3-319-61615-5.

Rochman, C. M., Brookson, C., Bikker, J., Djuric, N., Earn, A., Bucci, K., and Athey, S., et al. 2019. Rethinking microplastics as a diverse contaminant suite. *Environmental Toxicology and Chemistry* 38 (4): 703–11. https://doi.org/10.1002/etc.4371.

Rosenkranz, P., Chaudhry, Q., Stone, V., and Fernandes, T. F. 2009. A comparison of nanoparticle and fine particle uptake by *Daphnia magna. Environmental Toxicology and Chemistry* 28 (10): 2142–49. https://doi.org/10.1897/08-559.1.

Ruenraroengsak, P., Novak, P., Berhanu, D., Thorley, A. J., Valsami-Jones, E., Gorelik, J., Korchev, Y. E., and Tetley, T. D. 2012. Respiratory epithelial cytotoxicity and membrane damage (Holes) caused by amine-modified nanoparticles. *Nanotoxicology* 6: 94–108. https://doi.org/10.3109/17435390.2011.558643.

Sander, M., Kohler, H., and McNeil, K. 2019. Assessing the environmental transformation of nanoplastic through 13 c-labelled polymers. *Nature Nanotechnology* 14: 301–3.

Scherer, C., Weber, A., Lambert, S., and Wagner, M. 2018. Interactions of microplastics with freshwater Biota. In Wagner, M. and Lambert, S. (Eds.), *Freshwater Microplastics*, 1st ed., 153–80. Springer, Heidelberg. https://doi.org/10.1007/978-3-319-61615-5.

Schirinzi, G. F., Pérez-Pomeda, I., Sanchís, J., Rossini, C., Farré, M., and Barceló, D. 2017. Cytotoxic effects of commonly used nanomaterials and microplastics on cerebral and epithelial human cells. *Environmental Research* 159: 579–87. https://doi.org/10.1016/j.envres.2017.08.043.

Schwab, F., Zhai, G., Kern, M., Turner, A., Schnoor, J. L., and Wiesner, M. R. 2016. Barriers, pathways and processes for uptake, translocation and accumulation of nanomaterials in plants - critical review. *Nanotoxicology* 10 (3): 257–78. https://doi.org/10.3109/17435390.2015.1048326.

Schwabl, P., Koppel, S., Konigshofer, P., Bucsics, T., Trauner, M., Reiberger, T., and Liebmann, B. 2019. Detection of various microplastics in human stool: A prospective case series. *Annals of Internal Medicine* 171 (7): 453–57. https://doi.org/10.7326/M19-0618.

Sharma, S., and Chatterjee, S. 2017. Microplastic pollution, a threat to marine ecosystem and human health : A short review. *Environ Sci Pollut Res (2017)* 24: 21530–47. https://doi.org/10.1007/s11356-017-9910-8.

Shi, W., Han, Y., Sun, S., Tang, Y., Zhou, W., Du, X., and Liu, G. 2020. Immunotoxicities of microplastics and sertraline, alone and in combination, to a bivalve species : Size-dependent interaction and potential toxication mechanism. *Journal of Hazardous Materials* 396: 122603. https://doi.org/10.1016/j.jhazmat.2020.122603.

Shi, W., Sun, S., Han, Y., Tang, Y., Zhou, W., Du, X., and Liu, G. 2021. Microplastics impair olfactory-mediated behaviors of goldfish carassius auratus. *Journal of Hazardous Materials* 409: 125016. https://doi.org/10.1016/j.jhazmat.2020.125016.

Smith, C. J., Shaw, B. J., and Handy, R. D. 2007. toxicity of single walled carbon nanotubes to rainbow trout, (Oncorhynchus mykiss): Respiratory toxicity, organ pathologies, and other physiological effects. *Aquatic Toxicology* 82 (2): 94–109. https://doi.org/10.1016/j.aquatox.2007.02.003.

Sørensen, S. N., Engelbrekt, C., Lützhøft, H. C. H., Jiménez-Lamana, J., Noori, J. S., Alatraktchi, F. A., Delgado, C. G., Slaveykova, V. I., and Baun, A. 2016. A multimethod approach for investigating algal toxicity of platinum nanoparticles. *Environmental Science and Technology* 50: 10635–43. https://doi.org/10.1021/acs.est.6b01072.

Su, Y., Ashworth, V., Kim, C., Adeleye, A. S., Rolshausen, P., Roper, C., White, J., and Jassby, D. 2019. Delivery, uptake, fate, and transport of engineered nanoparticles in plants: A critical review and data analysis. *Environmental Science: Nano* 6: 2311–31. https://doi.org/10.1039/c9en00461k.

Sun, Q., Ren, S. Y., and Ni, H. G. 2020. Incidence of microplastics in personal care products: An appreciable part of plastic pollution. *Science of the Total Environment* 742: 140218. https://doi.org/10.1016/j.scitotenv.2020.140218.

Sun, S., Shi, W., Tang, Y., Han, Y., Du, X., Zhou, W., Zhang, W., Sun, C., and Liu, G. 2021. The toxic impacts of microplastics (MPs) and polycyclic aromatic hydrocarbons (PAHs) on haematic parameters in a marine bivalve species and their potential mechanisms of action. *Science of the Total Environment* 783: 147003. https://doi.org/10.1016/j.scitotenv.2021.147003.

Sussarellu, R., Suquet, M., Thomas, Y., Lambert, C., Fabioux, C., Pernet, M. E. J., and Le Goïc, N., et al. 2016. Oyster reproduction is affected by exposure to polystyrene microplastics. *Proceedings of the National Academy of Sciences of the United States of America* 113 (9): 2430–35. https://doi.org/10.1073/pnas.1519019113.

Syberg, K., Khan, F. R., Selck, H., Palmqvist, A., Banta, G. T., Daley, J., Sano, L., and Duhaime, M. B. 2015. Microplastics: Addressing ecological risk through lessons learned. *Environmental Toxicology and Chemistry* 34 (5): 945–53. https://doi.org/10.1002/etc.2914.

Tang, Y., Rong, J., Guan, X., Zha, S., Shi, W., Han, Y., and Du, X. 2020. Immunotoxicity of microplastics and two persistent organic pollutants alone or in combination to a bivalve Species*. *Environmental Pollution* 258: 113845. https://doi.org/10.1016/j.envpol.2019.113845.

Tang, Y., Zhou, W., Sun, S., Du, X., Han, Y., Shi, W., and Liu, G. 2020. Immunotoxicity and neurotoxicity of bisphenol A and microplastics alone or in combination to a Bivalve species, Tegillarca Granosa*. *Environmental Pollution* 265: 115115. https://doi.org/10.1016/j.envpol.2020.115115.

Teuten, E. L., Rowland, S. J., Galloway, T. S., and Galloway, T. S. 2007. Potential for plastics to transport hydrophobic contaminants. *Environ. Sci. Technol* 41: 7759–64. https://pubs.acs.org/doi/abs/10.1021/es071737s.

Toussaint, B., Raffael, B., Angers-Loustau, A., Gilliland, D., Kestens, V., Petrillo, M., Rio-Echevarria, I. M., and Van den Eede, G. 2019. Review of micro- and nanoplastic contamination in the food chain. *Food Additives and Contaminants - Part A Chemistry, Analysis, Control, Exposure and Risk Assessment* 36 (5): 639–73. https://doi.org/10.1080/19440049.2019.1583381.

UK Department for Environment, Food & Rural Affairs. 2016. Microbead Ban Announced to Protect Sealife Government. 2016. https://www.gov.uk/government/news/microbead-ban-announced-to-protect-sealife.

US Congress. 2015. Public Law 114–Prohibition Against Sale or Distribution of Rinse-off Cosmetics Containing Plastic Microbeads. https://www.congress.gov/114/plaws/publ114/PLAW-114publ114.pdf.

Vianello, A., Boldrin, A., Guerriero, P., Moschino, V., Rella, R., Sturaro, A., and Da Ros, L. 2013. Microplastic particles in sediments of lagoon of Venice, Italy: First observations on occurrence, spatial patterns and identification. *Estuarine, Coastal and Shelf Science* 130: 54–61. https://doi.org/10.1016/j.ecss.2013.03.022.

Vianello, A., Jensen, R. L., Liu, L., and Vollertsen, J. 2019. Simulating human exposure to indoor airborne microplastics using a breathing thermal manikin. *Scientific Reports* 9 (1): 1–11. https://doi.org/10.1038/s41598-019-45054-w.

Wagner, M., and Lambert, S. 2018a. Microplastics are contaminants of emerging concern in freshwater environments: An overview. In Wagner, M. and Lambert, S. (Eds.), *Freshwater Microplastics*, 1–23. https://doi.org/10.1007/978-3-319-61615-5.

Wagner, M., and Lambert, S. 2018b. Understanding the risks of microplastics: A social-ecological risk perspective. In *Freshwater Microplastics*, edited by M. Wagner and S. Lambert, 223–37. https://doi.org/10.1007/978-3-319-61615-5.

Wagner, S., and Reemtsma, T. 2019. Things we know and don't know about nanoplastic in the environment. *Nature Nanotechnology* 14 (4): 300–301. https://doi.org/10.1038/s41565-019-0424-z.

Wan, Y., Wu, C., Xue, Q., and Hui, X. 2019. Effects of plastic contamination on water evaporation and desiccation cracking in soil. *Science of the Total Environment* 654: 576–82. https://doi.org/10.1016/j.scitotenv.2018.11.123.

Wang, J., Lv, S., Zhang, M., Chen, G., Zhu, T., Zhang, S., Teng, Y., Christie, P., and Luo, Y. 2016. Effects of plastic film residues on occurrence of phthalates and microbial activity in soils. *Chemosphere* 151: 171–77. https://doi.org/10.1016/j.chemosphere.2016.02.076.

Wang, D., Su, L., Ruan, H. D., Chen, J., Lu, J., Lee, C. H., and Jiang, S. Y. 2021. Quantitative and qualitative determination of microplastics in oyster, seawater and sediment from the coastal areas in Zhuhai, China. *Marine Pollution Bulletin* 164: 112000. https://doi.org/10.1016/j.marpolbul.2021.112000.

Wang, F., Zhang, X., Zhang, S., Zhang, S., and Sun, Y. 2020. Interactions of microplastics and cadmium on plant growth and arbuscular mycorrhizal fungal communities in an agricultural soil. *Chemosphere* 254: 126791. https://doi.org/10.1016/j.chemosphere.2020.126791.

Wang, Y., Westerhoff, P., and Hristovski, K. D. 2012. Fate and biological effects of silver, titanium dioxide, and C 60 (Fullerene) nanomaterials during simulated wastewater treatment processes. *Journal of Hazardous Materials* 201–202: 16–22. https://doi.org/10.1016/j.jhazmat.2011.10.086.

Watts, A. J. R., Urbina, M. A., Goodhead, R., Moger, J., Lewis, C., and Galloway, T. S. 2016. Effect of microplastic on the gills of the shore crab carcinus maenas. *Environmental Science and Technology* 50 (10): 5364–69. https://doi.org/10.1021/acs.est.6b01187.

Wegner, A., Besseling, E., Foekema, E. M., Kamermans, P., and Koelmans, A. A. 2012. Effects of nanopolystyrene on the feeding behavior of the blue mussel (Mytilus edulis L.). *Environmental Toxicology and Chemistry* 31 (11): 2490–97. https://doi.org/10.1002/etc.1984.

Wiesner, M. R., Lowry, G. V., Casman, E., Bertsch, P. M., Matson, C. W., Di Giulio, R. T., Liu, J., and Hochella, M. F. 2011. Meditations on the ubiquity and mutability of nano-sized materials in the environment. *ACS Nano* 5 (11): 8466–70. https://doi.org/10.1021/nn204118p.

Wright, S. L., Rowe, D., Thompson, R. C., and Galloway, T. S. 2013. Microplastic ingestion decreases energy reserves in marine Worms. *Current Biology* 23: R1031–33. https://doi.org/10.1016/j.cub.2013.10.068.

Wright, S. L., Thompson, R. C., and Galloway, T. S. 2013. The physical impacts of microplastics on marine organisms: A review. *Environmental Pollution (Barking, Essex : 1987)* 178: 483–92. https://doi.org/10.1016/j.envpol.2013.02.031.

Wright, S. L., Ulke, J., Font, A., Chan, K. L. A., and Kelly, F. J. 2019. Atmospheric microplastic deposition in an urban environment and an evaluation of transport. *Environment International* 136: 105411. https://doi.org/10.1016/j.envint.2019.105411.

Xiong, X., Zhang, K., Chen, X., Shi, H., Luo, Z., and Wu, C. 2018. Sources and distribution of microplastics in China's largest inland Lake. *Environmental Pollution* 235: 899–906. https://doi.org/10.1016/j.envpol.2017.12.081.

Zarus, G. M., Muianga, C., Hunter, C. M., and Pappas, R. S. 2021. A review of data for quantifying human exposures to micro and nanoplastics and potential health risks. *Science of the Total Environment* 756: 144010. https://doi.org/10.1016/j.scitotenv.2020.144010.

Zhang, Y., Kang, S., Allen, S., Allen, D., Gao, T., and Sillanpää, M. 2020a. Atmospheric microplastics: A review on current status and perspectives. *Earth-Science Reviews* 203 (February): 103118. https://doi.org/10.1016/j.earscirev.2020.103118.

Zhang, J., Wang, L., and Kannan, K. 2020b. Microplastics in house dust from 12 countries and associated human exposure. *Environment International* 134 (November 2019): 105314. https://doi.org/10.1016/j.envint.2019.105314.

Zhang, S., Wang, J., Yan, P., Hao, X., Xu, B., Wang, W., and Aurangzeib, M. 2021. Non-biodegradable microplastics in soils: A brief review and challenge. *Journal of Hazardous Materials* 409: 124525. https://doi.org/10.1016/j.jhazmat.2020.124525.

Zhao, J., and Stenzel, M. H. 2018. Entry of nanoparticles into cells: The importance of nanoparticle properties. *Polymer Chemistry* 9: 259–72. https://doi.org/10.1039/c7py01603d.

Zhou, W., Han, Y., Tang, Y., Shi, W., Du, X., Sun, S., and Liu, G. 2020a. Contaminants in aquatic and terrestrial environments microplastics aggravate the bioaccumulation of two waterborne veterinary antibiotics in an edible bivalve species : Potential mechanisms and implications for human health. *Environmental Science & Technology* 54: 8115–8122. https://doi.org/10.1021/acs.est.0c01575.

Zhou, Y., Wang, J., Zou, M., Jia, Z., Zhou, S., and Li, Y. 2020b. Microplastics in soils: A review of methods, occurrence, fate, transport, ecological and environmental risks. *Science of the Total Environment* 748: 141368. https://doi.org/10.1016/j.scitotenv.2020.141368.

Zhou, W., Tang, Y., Du, X., Han, Y., Shi, W., Sun, S., Zhang, W., Zheng, H., and Liu, G. 2021. Fine polystyrene microplastics render immune responses more vulnerable to two veterinary antibiotics in a bivalve species. *Marine Pollution Bulletin* 164: 111995. https://doi.org/10.1016/j.marpolbul.2021.111995.

Zhu, Z., Wang, S., Zhao, F., Wang, S., Liu, F., and Liu, G. 2019. Joint toxicity of microplastics with triclosan to marine microalgae skeletonema costatum. *Environmental Pollution* 246: 509–17. https://doi.org/10.1016/j.envpol.2018.12.044.

Index

O

P